Regression Models for Categorical Dependent Variables Using Stata

Second Edition

Regression Models for Categorical Dependent Variables Using Stata

Second Edition

J. SCOTT LONG
Department of Sociology
Indiana University
Bloomington, Indiana

JEREMY FREESE
Department of Sociology
University of Wisconsin-Madison
Madison, Wisconsin

A Stata Press Publication
StataCorp LP
College Station, Texas

Stata Press, 4905 Lakeway Drive, College Station, Texas 77845

To our parents

Contents

List of Figures

Preface

Our goal in writing this book was to make it routine to carry out the complex calculations necessary to fully interpret regression models for categorical outcomes. Interpreting these models is complex because the models are nonlinear. Most software packages that fit these models do not provide options that make it simple to compute the quantities that are useful for interpretation. In this book, we briefly describe the statistical issues involved in interpretation, and then we show how you can use Stata to make these computations. As you read this book, we strongly encourage you to be at your computer so that you can experiment with the commands as you read. To facilitate this, we include two appendices. Appendix A summarizes each of the commands that we have written for interpreting regression models. Appendix B provides information on the datasets that we use as examples. We have also written a command called spex, standing for Stata postestimation examples, which makes it simple to load the datasets and run the sample programs from our book. We find this incredibly useful for teaching and for our own exploration of the methods.

Many of the commands that we discuss are not part of official Stata, but instead they are commands (in the form of ado-files) that we have written. To follow the examples in this book, you must install these commands. Details on how to do this are given in chapter 2. Although the book assumes that you are using Stata 9 or later, most commands will work in Stata 8 or Stata 7, but some of the output will appear differently. For details, see *http://www.indiana.edu/~jslsoc/spost.htm*. The screenshots that we present are from Stata 9 for Windows. If you are using a different operating system, your screen might not look the same. See the StataCorp publication *Getting Started with Stata* for your operating system for further details. All the examples, however, should work on all computing platforms that support Stata.

We use several conventions throughout the book. Stata commands, variable names, filenames, and output are presented in a typewriter-style font; for example, logit, lfp, age, wc, hc, and k5. Italics are used to indicate that something should be substituted for the word in italics. For example, logit *variablelist* indicates that the command logit is to be followed by a list of variables. When output from Stata is shown, the command is preceded by a period (which is the Stata prompt). For example,

```
. logit lfp age wc hc k5, nolog
Logistic regression                              Number of obs   =        753
   (output omitted )
```

If you want to reproduce the output, you do not type the period before the command. And, as just illustrated, when we have deleted part of the output, we indicate this with (*output omitted*). Keystrokes are set in this font. For example, Alt-f means that you are to hold down the Alt key and press f. The headings for sections that discuss advanced topics are tagged with an asterisk (*). These sections can be skipped with no loss of continuity with the rest of the book.

As we wrote this book and developed the accompanying software, many people provided suggestions and commented on early drafts. In particular, we thank Jun Xu, Ben Jann, Tait Medina, Simon Cheng, Claudia Geist, Lowell Hargens, and Patricia McManus. Pravin Trivedi was always willing to discuss problems and to provide sage advice. Many people at StataCorp provided their expertise in many ways. More than this, though, we are grateful for their engaging and encouraging our project in ways that are exemplary for academic publishing and that have made working with them enjoyable throughout. We particularly thank Richard Gates, Lisa Gilmore, Jeff Pitblado, and Gabe Waggoner. Last, we are indebted to David Drukker at StataCorp for his encouragement, valuable advice, and patience throughout the many stages of this project, not to mention his gentle way of imposing deadlines.

What's new in the second edition

The second edition has many changes, both large and small. Many of the changes reflect enhancements made in Stata 9. We discuss new regression models, including the zero-truncated Poisson and the zero-truncated negative binomial models, the hurdle model for counts, the stereotype logistic regression model, the rank-ordered logit model, and the multinomial probit model. We also discuss new Stata commands, such as `estat`, which provides a uniform way to access statistics useful for postestimation interpretation.

Also we have extended our suite of programs, known as SPost. The biggest change, and the most requested, is the inclusion of confidence intervals for predictions computed by `prvalue` and `prgen`. This work, completed with Jun Xu of Indiana University, has many applications that are illustrated throughout the book. We also added some new commands. `misschk` is a tool for examining the patterns of missing data. `leastlikely` provides a simple way to find observations that your model does not fit well. `asprvalue` computes predicted probabilities for the conditional logit model and other models for alternative-specific data that are discussed in the new chapter 7. And, `countfit` provides many measures of fit for comparing count models. Other commands have been expanded to work with more models, and we have added options that make the commands easier to use. (Users who complained about getting the tick marks right when using `prgen` will particularly appreciate the `gap` option!) Unfortunately, and despite encouragement from many users, SPost does not work with all the regression models fitted by Stata.

Getting help from the authors

We are gratified that many people have bought our book, but as a consequence, we have received many emails with questions and suggestions. We would like to respond to everyone who contacts us, but our time is limited. Here are things that you can do to make it easier for us to answer your questions, which will also increase the odds that you will get a prompt answer.

1. Make sure that you have the latest version of the Stata executable and ado-files. See [U] **28 Using the Internet to keep up to date** and [R] **update**.

2. Make sure that you have the last version of SPost. See page 9.

3. Check *http://www.indiana.edu/~jslsoc/spost.htm* for any new information.

4. Make sure that you do not have anything but letters, numbers, and underscores in your value labels. Some commands get hung up when value labels include special characters.

5. Look at the sample files in the `spost9_do` package. It is sometimes easiest to figure out how to use a command by seeing how others use it.

If none of these suggestions solves your problem, send us an email at *spostsup@indiana.edu*.

It is hard to figure out some problems by seeing just the log file. So, we suggest that you send a do-file, the resulting log file, and a small dataset (extract cases and variables from your full data).

1. Do not refer to specific directories. For example, do not include something like `use c:\data\project3\sample.dta` because it will not run on our machines unless we either edit your file or create your directory structure. The same is true if you use something like `log using c:\data\project3\problem, text`.

2. Send output in text rather than SMCL format. To do this, add the `text` option to your `log` command.

3. Include `prwhich` at the start of your do-file. This tells us which versions of the commands you are using. Then include `about` to tell us which version of Stata you are using.

4. Do not send a `.zip` file since the Indiana University email server will reject them. If you compress the data, change the suffix from `.zip` to something else.

Here is an example of what a do-file might look like:

```
capture log close
log using yourname.log, text replace
* prchange generates a variable not found error.
* scott long - jslong@indiana.edu - 4July2005
about
which prchange
use sample.dta
logit y x1 x2
* the following command causes the error
prchange, x(x1=1 z2=3)
log close
```

Wisconsin and Indiana
November 2005

Jeremy Freese
Scott Long

Part I

General Information

Our book is about using Stata for fitting and interpreting regression models with categorical outcomes. The book is divided into two parts. Part I contains general information that applies to all the regression models that are considered in detail in part II.

- **Chapter 1** is a brief orienting discussion that also includes *critical information* about installing a collection of Stata commands that we have written to facilitate the interpretation of regression models. Without these commands, you cannot do many of the things we suggest in the later chapters.

- **Chapter 2** includes both an introduction to Stata for those who have not used the program and more advanced suggestions for using Stata effectively for data analysis.

- **Chapter 3** considers issues of estimation, testing, assessing fit, and interpretation that are common to all the models considered in later chapters. We discuss both the statistical issues involved and the Stata commands that carry out these operations.

Chapters 4–8 of part II are organized by the type of outcome being modeled. Chapter 9 deals primarily with complications on the right-hand side of the model, such as including nominal variables and allowing interactions. The material in the book is supplemented on our web site at *http://www.indiana.edu/~jslsoc/spost.htm*, which includes data files, examples, and a list of frequently asked questions (FAQs). Although the book assumes that you are running Stata 9, most of the information also applies to Stata 8 and Stata 7; our web site includes special instructions for users of these releases.

1 Introduction

1.1 What is this book about?

Our book shows you efficient and effective ways to use regression models for categorical and count outcomes. It is a book about data analysis and is not a formal treatment of statistical models. To be effective in analyzing data, you want to spend your time thinking about substantive issues and not laboring to get your software to generate the results of interest. Accordingly, good data analysis requires good software and good technique.

Although we believe that these points apply to all data analysis, they are particularly important for the regression models that we examine. The reason is that these models are *nonlinear*, and consequently the simple interpretations that are possible in linear models are no longer appropriate. In nonlinear models, the effect of each variable on the outcome depends on the level of *all* variables in the model. Because of this nonlinearity, which we discuss in more detail in chapter 3, no method of interpretation can fully describe the relationships among the independent variables and the outcome. Rather, a series of *postestimation* explorations are needed to uncover the most important aspects of these relationships. In general, if you limit your interpretations to the standard output, that output constrains and can even distort how you understand your results.

In the linear regression model (LRM), most of the work of interpretation is complete once the estimates are obtained. You simply read off the coefficients, which can be interpreted as for a unit increase in x_k, y is expected to increase by β_k units, holding all other variables constant. In nonlinear models, such as logit or negative binomial regression, much more computation is necessary after the estimates are obtained. With few exceptions, the software that fits regression models does not provide much help with these analyses. Consequently, the computations are tedious, time consuming, and error prone. All in all, it is not fun work. In this book, we show how postestimation analysis can be accomplished easily using Stata and the set of new commands that we have written. These commands make sophisticated postestimation analysis routine and even enjoyable. With the tedium removed, the data analyst can focus on the substantive issues.

1.2   Which models are considered?

Regression models analyze the relationship between an explanatory variable and an outcome variable while controlling for the effects of other variables. The linear regression model (LRM) is probably the most commonly used statistical method in the social sciences. As mentioned, a key advantage of the LRM is the ease of interpreting results. Unfortunately, this model applies only to cases in which the dependent variable is continuous.[1] Using the LRM when it is not appropriate produces coefficients that are biased and inconsistent, and there is nothing advantageous about the simple interpretation of results that are incorrect.

Fortunately, many appropriate models exists for categorical outcomes, and these models are the focus of our book. We cover cross-sectional models for four kinds of dependent variables.

Binary outcomes (dichotomous or dummy variables) have two values, such as whether a citizen voted in the last election, whether a patient was cured after receiving some medical treatment, or whether a respondent attended college.

Ordinal or *ordered* outcomes have more than two categories, and these categories are assumed to be ordered. For example, a survey might ask if you would be "very likely", "somewhat likely", or "not at all likely" to take a new subway to work, or if you agree with the president on "all issues", "most issues", "some issues", or "almost no issues".

Nominal outcomes also have more than two categories but are not ordered. Examples include the mode of transportation a person takes to work (e.g., bus, car, train) or an individual's employment status (e.g., employed, unemployed, out of the labor force).

Finally, *count* variables count the number of times something has happened, such as the number of articles written by a student after receiving the Ph.D., or the number of patents a biotechnology company has obtained.

The specific cross-sectional models that we consider, along with the corresponding Stata commands, are

Binary outcomes: binary logit (`logit`), binary probit (`probit`), and the complementary log-log regression model (`cloglog`)

Ordinal and nominal outcomes: ordered logit (`ologit`) and ordered probit (`oprobit`) for ordinal outcomes; the stereotype logistic regression (`slogit`) for ordinal and nominal outcomes; multinomial logit (`mlogit`), multinomial probit with uncorrelated errors (`mprobit`), conditional logit (`clogit`), and alternative-specific multinomial probit with correlated errors (`asmprobit`) for nominal outcomes; and rank-ordered logit (`rologit`) for ranked outcomes

---

1. Using the LRM with binary dependent variables leads to the linear probability model (LPM). We do not consider the LPM further, given the advantages of models such as logit and probit. See Long (1997, 35–40) for details.

Count outcomes: Poisson regression (`poisson`), negative binomial regression (`nbreg`), zero-truncated Poisson (`ztp`), zero-truncated negative binomial (`ztnb`), hurdle regression, zero-inflated Poisson regression (`zip`), and zero-inflated negative binomial regression (`zinb`)

Although this book covers models for many different types of outcomes, they are all models for cross-sectional data. We do not consider models for survival or event-history data, even though Stata has a powerful set of commands for dealing with these data (see the entry for `st` in the *Survival Analysis Reference Manual*). Likewise, we do not consider any models for panel data, even though Stata contains commands for fitting these models (see the entry for `xt` in the *Longitudinal/Panel Data Reference Manual*).

1.3 Whom is this book for?

We expect that readers of this book will vary considerably in their knowledge of both statistics and Stata. With this in mind, we have tried to structure the book to accommodate the diversity of our audience. Minimally, however, we assume that readers have a solid familiarity with ordinary least-squares regression for continuous dependent variables and that they are comfortable using the basic features of the operating system of their computer. Although we have provided sufficient information about each model so that you can read each chapter without prior exposure to the models discussed, we strongly recommend that you do *not* use this book as your sole source of information on the models (section 1.6 recommends more readings). Our book will be most useful if you have already studied the models considered or are studying these models in conjunction with this book.

We assume that you have access to a computer that is running Stata 9 or later and that you have access to the Internet to download commands, datasets, and sample programs that we have written (see section 1.5 for details on obtaining these). For information about obtaining Stata, see the StataCorp web site at *http://www.stata.com*. Although most of the commands in later chapters also work in Stata 8 and Stata 7, there are some differences. For details, see our web site at *http://www.indiana.edu/~jslsoc/spost.htm*.

1.4 How is the book organized?

Chapters 2 and 3 introduce materials that are necessary for working with the models we present in the later chapters:

Chapter 2: Introduction to Stata reviews the basic features of Stata that are necessary to get new or inexperienced users up and running with the program. This introduction is by no means comprehensive, so we include information on how to get more help. New users should work through the brief tutorial that we provide in section 2.17. Users already skilled with Stata can skip this chapter, although even these readers might benefit from quickly reading it.

Chapter 3: Estimation, testing, fit, and interpretation provides a review of
using Stata for regression models. It includes details on how to fit models, test
hypotheses, compute measures of model fit, and interpret results. We focus on
those issues that apply to all the models considered in part II. We also provide
detailed descriptions of the add-on commands that we have written to make these
tasks easier. Even if you are an advanced user, we recommend that you look over
this chapter before jumping ahead to the chapters on specific models.

Chapters 4–8 cover models for a different type of outcome:

Chapter 4: Models for binary outcomes begins with an overview of how the
binary logit and probit models are derived and how they can be fitted. After the
model has been fitted, we show how Stata can be used to test hypotheses, compute
residuals and influence statistics, and calculate scalar measures of model fit. Then
we describe postestimation commands that assist in interpretation using predicted
probabilities, discrete and marginal change in the predicted probabilities, and,
for the logit model, odds ratios. Because binary models provide a foundation on
which some models for other kinds of outcomes are derived, and because chapter 4
provides more detailed explanations of common tasks than later chapters do, we
recommend reading this chapter even if you are interested mainly in another type
of outcome.

Chapter 5: Models for ordinal outcomes introduces the ordered logit and ordered
probit models. We show how these models are fitted and how to test hypotheses
about coefficients. We also consider two tests of the parallel regression assumption.
In interpreting results, we discuss similar methods as those described in chapter 4,
as well as interpretation in terms of a latent dependent variable.

Chapter 6: Models for nominal outcomes with case-specific data focuses
on the multinomial logit model. We show how to test a variety of hypotheses
that involve multiple coefficients and discuss two tests of the assumption of the
independence of irrelevant alternatives. Although the methods of interpretation
are again similar to those presented in chapter 4, the model's many parameters
often complicate interpretation. To deal with this complexity, we present two
graphical methods of representing results. The multinomial probit model without
correlated errors is discussed briefly, and then the multinomial logit model is used
to explain the stereotype logit model. This model, which is often used with ordinal
outcomes, also has applications with nominal outcomes.

Chapter 7: Models for nominal outcomes with alternative-specific data introduces models for situations in which you have at least some variables that
vary over the alternatives for each individual, such as an individual's similarity
to each candidate in an election. We first show you how to rearrange data into
the format required for these models. Then we describe the conditional logit
model, which is equivalent to the multinomial logit model when only case-specific
regressors are used, but which also allows alternative-specific regressors. Then

we discuss the alternative-specific multinomial probit model, which is more than just a probit version of the conditional logit model because it allows correlations between alternative-specific error terms, thus relaxing the assumption of the independence of irrelevant alternatives. Last, we present the rank-ordered logistic regression model, which can be used when you have information about the ranking of outcomes as opposed to only information about the selected or most preferred outcome.

Chapter 8: Models for count outcomes begins with the Poisson and negative binomial regression models, including a test to determine which model is appropriate for your data. We also show how to incorporate differences in exposure time into the estimation. Next we consider interpretation for changes in the predicted rate and changes in the predicted probability of observing a given count. The rest of the chapter deals with models that specifically address problems associated with having too many zeros or none at all. We start with zero-truncated models for which zeros are missing from the outcome variable, perhaps due to the way the data were collected. We then merge a binary model and a zero-truncated model to create the hurdle model. The rest of the chapter considers fitting and interpreting zero-inflated count models, which are designed to account for the many zero counts often found in count outcomes.

Chapter 9 returns to issues that affect all models.

Chapter 9: More topics deals with several topics, but the primary concern is with complications among independent variables. We consider the use of ordinal and nominal independent variables, nonlinearities among the independent variables, and interactions. The proper interpretation of the effects of these types of variables requires special adjustments to the commands considered in earlier chapters. Many of these examples involve writing small programs with macros and loops. We then comment briefly on how to modify our commands to work with other estimation commands. Finally, we discuss several features in Stata that we think make data analysis easier and more enjoyable.

1.5 What software do you need?

To get the most out of this book, you should read it while you are at a computer where you can experiment with the commands as they are introduced. We assume that you are using Stata 9 or later. If you are running Stata 8 or Stata 7, most of the commands work in the same way, but a few options will not work and the output might look different. The SPost commands that ran in Stata 8 (and usually with Stata 7) will continue to be available if you install the package `spostado`, with examples contained in `spostst8`.[2] The version 9 commands require Stata 9 or later.

---

2. `spostado` and `spostst8` are the names used in the prior edition of our book. We are keeping the names and contents of those packages unchanged so that they will continue to work as described in that edition. For Stata 9, the packages are called `spost9_ado` and `spost9_do`, following a naming scheme we plan to keep as new versions of Stata are released.

Advice to new Stata users If you have never used Stata, you might find the instructions in this section to be confusing. It might be easier if you only skim the
material now and return to it after you have read the introductory sections of
chapter 2.

1.5.1   Updating Stata 9

Before you work through our examples, we strongly recommend that you have the
latest version of `wstata.exe` (or `wsestata.exe` if you are using Stata/SE) and the
official Stata ado-files. *You should do this even if you have just installed Stata* because
the CD that you received might not have the latest changes to the program. If you are
connected to the Internet and are in Stata, you can update Stata by selecting Official
Updates from the Help menu. Stata responds with the following screen:

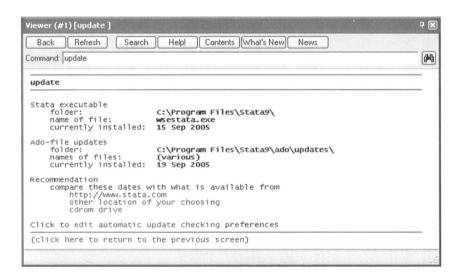

This screen tells you the current dates of your files. If you do not have Stata set up
to automatically update, you can update your files to the latest versions by clicking
on `http://www.stata.com`. We suggest that you do this every few months. Or, if
you encounter something that you think is a bug in Stata or in our commands, update
your copy of Stata to see if the problem has been resolved. After you update, read the
resulting screen carefully. You might need to click on `update swap` to complete the
update.

1.5.2 Installing SPost

From our point of view, one of the best things about Stata is how easy it is to add your own commands. This means that if Stata does not have a command you need or some command does not work the way you like, you can program the command yourself, and it will work as if it were part of official Stata. Indeed, we have created a suite of programs, referred to collectively as SPost (for Stata Postestimation commands), for the postestimation interpretation of regression models. These commands must be installed before you can try the examples in later chapters.

What is an ado-file? Programs that add commands to Stata are contained in files that end in the extension .ado (hence the name). For example, prvalue.ado is the program for the command prvalue. Hundreds of ado-files are included with the official Stata package, but experienced users can write their own ado-files to add new commands. However, for Stata to use a command implemented as an ado-file, *the ado-file must be located in one of the directories where Stata looks for ado-files.* If you type the command sysdir, Stata lists the directories that Stata searches for ado-files in the order that it searches them. However, if you follow our instructions below, you should not have to worry about managing these directories.

Installing SPost using search

Installation should be simple if you are connected to the Internet. If you have installed a prior version of SPost, we suggest that you begin by uninstalling it. To do this, enter the command ado. This will list all packages (a package is a collection of related files) that have been installed, with each package marked with a number in square brackets. Scan the list and record the number of the package containing the SPost ado-files and any other packages related to SPost. Although the ado-files are now contained in the package spost9_ado, other names were used in the past. Uninstall these packages with the command ado uninstall [#], where # is the package number (you must include the brackets; for example, ado uninstall [3]). You must run ado uninstall once for each package to be uninstalled.

The search *word*, net command searches an online database that StataCorp uses to keep track of user-written additions to Stata. Typing search spost, net brings up the names and descriptions of several packages in the Results window. One of these packages should be labeled spost9_ado from http://www.indiana.edu/~jslsoc/stata. The label is in blue, which means that it is a clickable link.[3] After you click on the link, a new Viewer window opens with information about our commands and another link saying "click here to install". If you click on this link, Stata attempts to install the

---

3. If you click on a link and immediately get a beep with an error message saying that Stata is busy, Stata is probably waiting for you to press a key. Most often, this occurs when Stata is displaying output that does not fit on one screen.

package. After a delay during which files are downloaded, Stata responds with one of
the following messages:

`installation complete` means that the package has been successfully installed and
that you can now use the commands. Just above the "installation complete"
message, Stata tells you the directory where the files were installed.

`all files already exist and are up-to-date` means that your system already has
the latest version of the package. You can stop.

`the following files exist and are different` indicates that your system already
has files with the same names as those in the package being installed and that
the existing files differ from those in the package. The names of those files are
listed and you are given several options. Assuming that the files listed are earlier
versions of our programs, you should select the option "Force installation replacing
already-installed files". This might sound ominous, but it is not. Because the files
on our web site are the latest versions, you want to replace your current files with
these new files. After you accept this option, Stata updates your files to newer
versions.

`cannot write in directory` *directory-name* means that you do not have write privileges to the directory where Stata wants to install the files. Usually, this occurs
only when you are using Stata on a network. Then we recommend that you contact your network administrator and ask if our commands can be installed using
the instructions given above. If you cannot wait for a network administrator to
install the commands or to give you the needed write access, you can install the
programs to any directory where you have write permission, including a zip disk
or your directory on a network. For example, suppose that you want to install
SPost to your directory called `d:\username` (which can be any directory where
you have write access). You should use the following commands:

```
. cd d:\username
d:\username
. mkdir ado
. sysdir set PERSONAL "d:\username\ado"
. net set ado PERSONAL
. net search spost
(contacting http://www.stata.com)
```

Then follow the installation instructions that we provided earlier for installing
SPost. If you get the error "could not create directory" after typing `mkdir ado`,
you probably do not have write privileges to the directory.

If you install ado-files to your own directory, each time you begin a new session
you must tell Stata where these files are located. You do this by typing `sysdir
set PERSONAL` *directory*, where *directory* is the location of the ado-files you have
installed. For example,

```
. sysdir set PERSONAL d:\username\ado
```

Installing SPost using net install

You can also install the commands entirely from the Command window. (If you have already installed SPost, you do not need to read this section.) While you are online, enter

> `. net from http://www.indiana.edu/~jslsoc/stata/`

The available packages will be listed. To install `spost9_ado`, type

> `. net install spost9_ado`

`net get` can be used to download supplementary files (e.g., datasets, sample do-files) from our web site. For example, to download the package `spost9_do`, type

> `. net get spost9_do`

These files are placed in the current working directory (see chapter 2 for a full discussion of the working directory).

1.5.3   What if commands do not work?

This section assumes that you have installed SPost but that some of the commands do not work. Here are some things to consider:

1. If you get the error message `unrecognized command`, there are several possibilities.

   a. If you discover that commands that used to work do not work anymore, you could be working on a different computer or on a different station in a computer lab. Because user-written ado-files work seamlessly in Stata, you might not realize that these programs must be installed on each machine that you use.

   b. If you sent a do-file that contains SPost commands to another person, and they cannot get the commands to work, let them know that they need to install SPost.

   c. If you get the error message `unrecognized command:` *strangename* after typing one of our commands, where *strangename* is not the name of the command that you typed, it means that Stata cannot find an ancillary ado-file that the command needs. We recommend that you install the SPost files again.

2. If you are getting an error message that you do not understand, click on the blue return code beneath the error message for more information about the error.

3. Make sure that Stata is properly installed and up to date. Typing `verinst` will verify that Stata has been properly installed. Typing `update query` will tell you if the version you are running is up to date and what you need to type to update it. If you are running Stata over a network, your network administrator may need to do this for you.

4. Often what appears to be a problem with one of our commands is actually a mistake you have made (we know because we make them, too). For example, make sure that you are not using = when you should be using ==.

5. Because our commands work after you have fitted a model, make sure that there were no problems with the last model fitted. If Stata was not successful in fitting your model, our commands will not have the information needed to operate properly.

6. Irregular value labels can cause Stata programs to fail. We recommend using labels that have fewer than eight characters and contain no spaces or special characters other than underscores (_). If your variables (especially your dependent variable) do not meet this standard, try changing your value labels with the `label` command (details are given in section 2.15).

7. Unusual values of the outcome categories can also cause problems. For ordinal or nominal outcomes, some of our commands require that all the outcome values be integers between 0 and 99. For these types of outcomes, we recommend using consecutive integers starting with 1.

Before attempting to contact us about a problem, please check our web site *http://www.indiana.edu/~jslsoc/spost.htm* for new information about SPost. You should also read the information at *http://www.indiana.edu/~jslsoc/spost_help.htm* and check page xxxi in the Preface.

1.5.4   Uninstalling SPost

Stata keeps track of the packages that it has installed, which makes it easy for you to uninstall them in the future. If you want to uninstall our commands, simply type `ado uninstall spost9_ado`.

1.5.5   Using spex to load data and run examples

Experimenting with the postestimation commands that we discuss requires that you have fitted the appropriate model. In our examples, we show you how to open a dataset and fit models as you would when you were working with your own data. Accordingly, we begin with a `use` command to load the data and then use an estimation command, such as `logit`, to fit the model. To make it simpler to experiment with the methods in later chapters, we have written the command `spex` (Stata postestimation examples). If you type `spex logit`, for example, it will automatically load the data and fit the model that serves as our main logit example. Typing `spex` *commandname* will produce our primary example for that estimation command. Or, you can specify the name of any dataset that we use, `spex` *datasetname*, and `spex` will load those data but not fit any model. By default, `spex` looks for the dataset on our web site. If it does not find it there, it will look in the current working directory and all the directories where Stata searches for ado-files. Specifying the option `user` tells `spex` to look in your current working directory first. For more information, type `help spex`.

1.5.6   More files available on the web site

In addition to the SPost commands, we have provided other packages that you might find useful. For example, the package called `spost9_do` contains the do-files and datasets needed to reproduce the examples from this book. The package `spostst8` contains the do-files and datasets to reproduce the results from Long (1997). To obtain these packages, type `net search spost` and follow the instructions you will be given. **Important**: if a package does not contain ado-files, Stata will download the files to the current working directory. Consequently, you need to change your working directory to wherever you want the files to go *before* you select "click here to get". More information about working directories and changing your working directory is provided in section 2.5.

1.6   Where can I learn more about the models?

There are many valuable sources for learning more about the regression models that are covered in this book. Not surprisingly, we recommend

Long, J. Scott. 1997. *Regression Models for Categorical and Limited Dependent Variables*. Thousand Oaks, CA: Sage. This book provides more details about the models discussed in our book.

We also recommend the following:

Cameron, A. C. and P. K. Trivedi. 2005. *Microeconometrics: Methods and Applications*. New York: Cambridge University Press. This graduate textbook provides an excellent introduction to the methods and models discussed in this book.

Cameron, A. C. and P. K. Trivedi. 1998. *Regression Analysis of Count Data*. Cambridge: Cambridge University Press. This is the definitive reference for count models.

Greene, W. H. 2003. *Econometric Analysis*. 5th ed. Upper Saddle River, NJ: Prentice Hall. Although this book focuses on models for continuous outcomes, several later chapters deal with models for categorical outcomes.

Hardin, J. and J. Hilbe. 2001. *Generalized Linear Models and Extensions*. College Station, TX: Stata Press. This is a thorough review of the generalized linear model or GLM approach to modeling and includes detailed information on the use of these models with Stata.

Hosmer, Jr., D. W., and S. Lemeshow. 2000. *Applied Logistic Regression*. 2nd ed. New York: Wiley. This book, written primarily for biostatisticians and medical researchers, provides much useful information about logit models for binary, ordinal, and nominal outcomes. Often the authors discuss how their recommendations can be executed using Stata.

Powers, D. A. and Y. Xie. 2000. *Statistical Methods for Categorical Data Analysis.* San Diego: Academic Press. This book considers all the models discussed in our book, with the exception of count models, and includes loglinear models and models for event history analysis.

Train, K. 2003. *Discrete Choice Methods with Simulation.* Cambridge: Cambridge University Press. This is an outstanding review of models for a wide range of models for discrete choice and includes details on new methods of estimation using simulation.

2 Introduction to Stata

This book is about fitting and interpreting regression models using Stata, and to earn our pay we must get to these tasks quickly. With that in mind, this chapter is a relatively concise introduction to Stata 9 for those with little or no familiarity with the package. Experienced Stata users can skip this chapter, although a quick reading might be useful. We focus on teaching the reader what is necessary to work through the examples later in the book and to develop good working techniques for using Stata for data analysis. By no means are the discussions exhaustive; often we show you either our favorite approach or the approach that we think is simplest. One of the great things about Stata is that there are usually several ways to accomplish the same thing. If you find a better way than what we have shown you, use it!

You cannot learn how to use Stata simply by reading. Accordingly, we strongly encourage you to try the commands as we introduce them. We have also included a tutorial in section 2.17 that covers many of the basics of using Stata. Indeed, you might want to try the tutorial first and then read our detailed discussions of the commands.

Although people who are new to Stata should find this chapter sufficient for understanding the rest of the book, if you want further instruction, look at the resources listed in section 2.3. We also assume that you know how to load Stata on the computer you are using and that you are familiar with your computer's operating system. By this, we mean that you should be comfortable copying and renaming files, working with subdirectories, closing and resizing windows, selecting options with menus and dialog boxes, and so on.

(Continued on next page)

2.1   The Stata interface

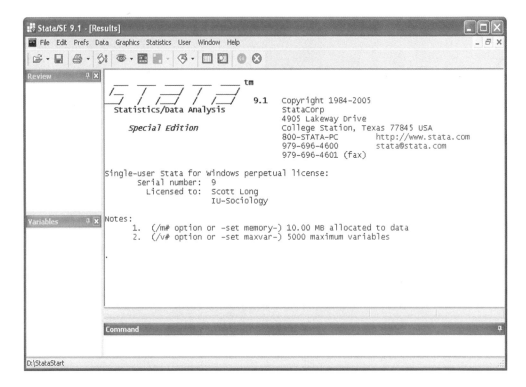

Figure 2.1: Opening screen in Stata for Windows.

When you launch Stata, you will see a screen with several smaller windows located within the larger Stata window, as shown in figure 2.1. This screen shot is for Windows using the default windowing preferences. If the defaults have been changed or you are running Stata under Unix or MacOS, your screen will look slightly different.[1] Figure 2.2 shows what Stata looks like after several commands have been entered and data have been loaded into memory. In both figures, four windows are shown.

---

1. Our screen shots and descriptions are based on Stata for Windows. Please refer to the books *Getting Started with Stata for Macintosh* or *Getting Started with Stata for Unix* for examples of the screens for those operating systems.

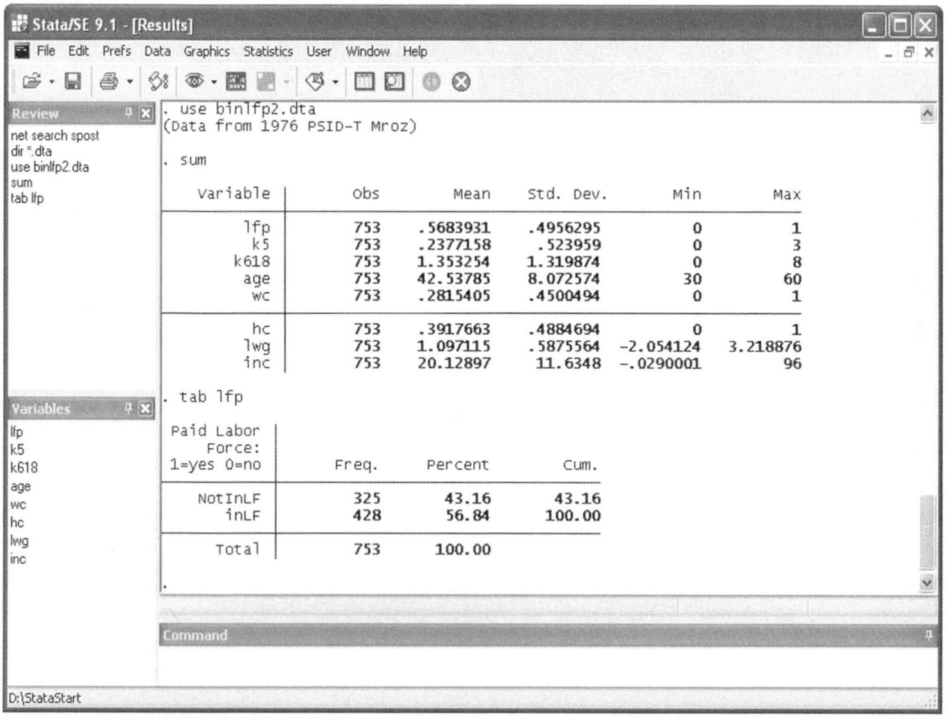

Figure 2.2: Example of Stata windows after several commands have been entered and data have been loaded.

The Command window is where you type commands that are executed when you press Enter. As you type commands, you can edit them at any time *before* pressing Enter. Pressing PageUp brings the most recently used command into the Command window; pressing PageUp again retrieves the command before that; and so on. Once a command has been retrieved to the Command window, you can edit it and press Enter to run the modified command.

The Results window contains output from the commands entered in the Command window. The Results window also echoes the command that generated the output, where the commands are preceded by a ".". as shown in figure 2.2. The scroll bar on the right lets you scroll back through output that is no longer on the screen. You can also use the scroll wheel on the mouse to scroll back and forth. Within the window, you can highlight text and right-click the mouse to see options for copying the highlighted text. In Stata for Windows and Stata for Macintosh, the Copy Table option copies the selected lines to the clipboard, and Copy Table as HTML allows you to copy the selected text as an HTML table (see page 94 for more information). Right-clicking also gives you the option to print the contents of the window. Only the most recent output is available this way; earlier lines are lost unless you have saved them to a log file (discussed below). Details on setting the size of the scrollback buffer are given below.

The Review window lists the commands that have been entered from the Command window. If you click on a command in this window, it is pasted into the Command window, where you can edit it before execution of the command. If you double-click on a command in the Review window, it is pasted into the Command window and immediately executed.

The Variables window lists the names of variables that are in memory, including those loaded from disk files and those created with Stata commands. If you click on a name, it is pasted into the Command window.

The Command and Results windows illustrate the important point that Stata has its origins in a command-based system. This means that you tell Stata what to do by typing commands that consist of one line of text and then pressing Enter.[2] Beginning with Stata 8, there is a complete graphical user interface (GUI) for accessing all nonprogramming commands. At the risk of seeming old-fashioned, however, we still prefer the command-based interface. Although it can take longer to learn, once you learn it, you should find it much faster to use. If you currently prefer using pulldown menus, stick with us, and you will likely change your mind.

Because Stata 8 and later have a complete GUI, you can do almost anything in Stata by pointing and clicking. Some of the most important tasks can be performed by clicking on icons on the toolbar at the top of the screen. Although we on occasion mention the use of these icons, for the most part we stick with text commands. Indeed, even if you do click on an icon or issue a command from a dialog box, Stata shows you how this could be done with a text command. For example, if you click on the **Data Browser** button, , Stata opens a spreadsheet for examining your data. Meanwhile, ". browse" is written to the Results window. This means that instead of clicking the icon, you could have typed browse. Overall, not only is the range of things you can do with menus limited, but almost everything you can do with the mouse can also be done with commands, often more efficiently. Therefore, and because it makes things much easier to automate later, we describe things mainly in terms of commands. Even so, we encourage you to explore the tasks available through menus and the toolbar and to use them when preferred.

Changing the scrollback buffer size

How far back you can scroll in the Results window is controlled by the command

```
set scrollbufsize #
```

---

2. For now, we consider entering only one command at a time, but in section 2.9 we show you how to run a series of commands at once using "do-files".

where $10,000 \leq \# \leq 500,000$. By default, the buffer size is 32,000 bytes. When you change the size of the scroll buffer using `set scrollbufsize`, the change will take effect the next time you launch Stata. Unless memory is a problem, we set the buffer to its maximum.

Changing the display of variable names in the Variables window

The Variables window displays both the names of variables in memory and their variable labels. By default, 32 columns are reserved for the name of the variable. The maximum number of characters to display for variable names is controlled by the command

    set varlabelpos #

where $8 \leq \# \leq 32$. By default, the size is 32. In figure 2.2, none of the variable labels are shown since the 32 columns take up the width of the window. If you use short variable names, it is useful to `set varlabelpos` to a smaller number so that you can see the variable labels.

Tip: Changing defaults We prefer a larger scroll buffer and less space for variable names. We could enter the command `set varlabelpos 14` at the start of each Stata session, but it is easier to add the commands to `profile.do`, a file that is automatically run each time Stata begins. We show you how to do this in chapter 9.

2.2   Abbreviations

Commands and variable names can often be abbreviated. For variable names, the rule is easy: *any variable name can be abbreviated to the shortest string that uniquely identifies it.* For example, if there are no other variables in memory that begin with a, the variable `age` can be abbreviated as `a` or `ag`. If you have the variables `income` and `income2` in your data, neither of these variable names can be abbreviated.

There is no general rule for abbreviating commands, but as one would expect, typically the most common and general command names can be abbreviated. For example, four of the most often used commands are `summarize`, `tabulate`, `generate`, and `regress`, and these can be abbreviated as `su`, `ta`, `g`, and `reg`, respectively. From now on, when we introduce a Stata command that can be abbreviated, we underline the shortest abbreviation (e.g., `generate`). But, although very short abbreviations are easy to type, they can be confusing when you are getting started. Accordingly, when we use abbreviations, we stick with at least three-letter abbreviations.

2.3   How to get help

2.3.1   Online help

If you find our description of a command incomplete, or if we use a command that is not explained, you can use Stata to find more information. Use the `help` command when you know the name of the command and want to find out more about it. For example, `help regress` tells you about the `regress` command. `search` is used when you do not know the name of the command or where something is documented. `help` and `search` are typed in the Command window, with results for `help` displayed in a Viewer window and results for `search` returned to the Results window. You can also open the Viewer by clicking on . At the top of the Viewer, there is a line labeled Command, where you can type commands, such as `help`. The Viewer is particularly useful for reading long help files.

`help` lists a shortened version of the documentation in the manual for any command. You can even type `help help` for help on using `help`. The output from `help` often makes reference to other commands, which are shown in blue. (Anything in the Results window that is in blue type is a clickable link.) Here clicking on a command name in blue type is the same as typing `help` for that command.

`search` is so useful for tracking down information that we encourage you to type `help search` and read more about this powerful command, because here we provide only a few details. By default, `search` *word* ... searches Stata's index on your local machine and lists the entries that match your query. For example, `search gen` lists information on `generate` but also links to many related commands. Or, if you want to run a truncated regression model but cannot remember the name of the command, you could try `search truncated` to get information on commands related to truncation. The resulting commands are listed in blue, so you can click on the name; details appear in the Viewer. If you want to extend your search to the Internet, add the option `all`; for example, `search truncated, all`. When you search the Internet, you get information from the Stata web site, including FAQs (frequently asked questions), articles that have appeared in the *Stata Journal* (often abbreviated SJ), and information about user-written commands that are not part of official Stata. For example, when you installed the SPost commands, you might have used `search spost, all` to find the links for installation. To get a better idea of how the `all` option works, try `search truncated, all` and compare the results with those from `search truncated`. If you always want to include a search of the Internet, you can `set searchdefault all, permanently` and `search` will automatically search the Internet each time you use the command. Or, you can leave the `search` default to search your local machine and use the `findit` command, which is equivalent to `search` *word* ..., `all`.

Tip: Help with error messages Error messages in Stata can be terse and sometimes confusing. Whereas the error message is printed in red, errors also have a *return code* (e.g., `r(199)`) listed in blue. Clicking on the return code provides a more detailed description of the error.

2.3.2 Manuals

The Stata manuals are extensive, and it is worth taking an hour to browse them to get an idea of the many features in Stata. In general, we find that learning how to read the manuals (and use the help system) is more efficient than asking someone else, and it allows you to save your questions for the really hard stuff. For those new to Stata, we recommend the *Getting Started* manual (which is specific to your platform) and the first part of the *User's Guide*. As you become more acquainted with Stata, the reference manuals will become increasingly valuable for detailed information about commands, including a discussion of the statistical theory related to the commands and references for further reading.

2.3.3 Other resources

The *User's Guide* also discusses more sources of information about Stata. Most importantly, the Stata web site (*http://www.stata.com*) contains many useful resources, including links to tutorials and an extensive FAQ section that discusses both introductory and advanced topics. You can also get information on the NetCourses offered by Stata, which are 4- to 7-week courses offered over the Internet. Another excellent set of online resources is provided by UCLA's Academic Technology Services at *http://www.ats.ucla.edu/stat/stata/*.

There is also a Statalist listserver that is not part of StataCorp, although many programmers and statisticians from StataCorp participate. This list is a wonderful resource for information on Stata and statistics. You can submit questions and usually receive answers very quickly. Monitoring the listserver is also a quick way to pick up insights from Stata veterans. For details on joining the list, visit *http://www.stata.com/statalist/*.

2.4 The working directory

The *working directory* is the default directory for any file operations such as using data, saving data, or logging output. If you type `cd` or `pwd` in the Command window, Stata displays the name of the current working directory. To load a data file stored in the working directory, you just type `use` *filename* (e.g., `use binlfp2`). If a file is not in the working directory, you must specify the full path (e.g., `use d:\spostdata\examples\binlfp2`).

At the beginning of each Stata session, we like to change our working directory to the directory where we plan to work, since this is easier than repeatedly entering the path name for the directory. For example, typing cd d:\spostdata changes the working directory to d:\spostdata. If the directory name includes spaces, you must put the path in quotation marks (e.g., cd "d:\my work\").

You can list the files in your working directory by typing dir or ls, which are two names for the same command. With this command, you can use the * wildcard. For example, dir *.dta lists all files with the extension .dta.

2.5   Stata file types

Stata uses and creates many types of files, which are distinguished by extensions at the end of the filename. Some of the extensions used by Stata are

.ado	Programs that add commands to Stata, such as the SPost commands.
.class	Files that define classes in the Stata class system.
.dlg	Programs that define the appearance and functionality of dialog boxes.
.do	Batch files that execute a set of Stata commands.
.dta	Data files in Stata's format.
.emf	Graphs saved as Windows Enhanced Metafiles.
.gph	Graphs saved in Stata's proprietary format.
.hlp	The text displayed when you use the help command. For example, fitstat.hlp has help for fitstat.
.log	Output saved as plain text by the log using command.
.mata	Original source code for programs written in Mata.
.mlib	Libraries of programs written in Mata.
.smcl	Output saved in the SMCL format by the log using command.

The most important of these for a new user are the .smcl, .log, .dta, and .do files, which we will now discuss.

2.6   Saving output to log files

Stata does not automatically save the output from your commands. To save your output to print or examine later, you must open a *log file*. Once a log file is opened, both the commands and the output they generate are saved. Because the commands are recorded, you can tell exactly how the results were obtained. The syntax for the log command is

log using *filename* [, append replace [smcl | text] name(*logname*)]

By default, the log file is saved to your working directory. You can save it to a different directory by typing the full path (e.g., log using d:\project\mylog, replace).

Options

append means that if the file exists, new output should be added to the end of the
existing file.

replace indicates that you want to replace the log file if it already exists. For example, log using mylog creates the file mylog.smcl. If this file already exists, Stata
generates an error message. So, you could use log using mylog, replace, and the
existing file would be overwritten by the new output.

smcl and text specify the format in which the log is to be recorded.

smcl is the default option that requests that the log be written using the Stata
Markup and Control Language (SMCL) with the file suffix .smcl. SMCL files
contain special codes that add solid horizontal and vertical lines, bold and italic
typefaces, and hyperlinks to the Results window. The disadvantage of SMCL is
that the special features can be viewed only within Stata. If you open a SMCL
file in a text editor, your results will appear amidst a jumble of special codes.

text specifies that the log should be saved as plain text (ASCII), which is the preferred format for loading the log into a text editor for printing. Instead of adding
the text option, such as log using mywork, text, you can specify plain text by
including the .log extension (for example, log using mywork.log).

If you open multiple log files, you may choose a different format for each file.

name(*logname*) specifies an optional name for the log file to be used while it is open.
This lets you have multiple log files open, each with a different name. You can then
close, temporarily suspend, or resume them individually.

Tip: Plain text logs by default We prefer plain text for output rather than SMCL.
Typing set logtype text at the beginning of a Stata session makes plain text
the default for log files for the current session. Typing set logtype text,
permanently makes plain text the default for future sessions.

If you have a question and would like to send us a log file, make sure that it
is in text format rather than in SMCL. Before you send anything, please check
http://www.indiana.edu/~jslsoc/spost_help.htm for information on what you can
do to increase your odds of getting an answer!

2.6.1 Closing a log file

To close a log file, type

```
. log close
```

Also when you exit Stata, the log file closes automatically.

2.6.2   Viewing a log file

Regardless of whether a log file is open or closed, a log file can be viewed by selecting File→Log→View from the menu, and the log file will be displayed in the Viewer. When in the Viewer, you can print the log by selecting File→Print Viewer.... You can also view the log file by clicking on ⬛, which opens the log in the Viewer. If the Viewer window gets lost behind other windows, you can click on ⬛ to bring the Viewer to the front.

2.6.3   Converting from SMCL to plain text or PostScript

If you want to convert a log file in SMCL format to plain text, you can use the `translate` command. For example,

```
. translate mylog.smcl mylog.log, replace
(file mylog.log written in .log format)
```

tells Stata to convert the SMCL file `mylog.smcl` to a plain-text file called `mylog.log`. Or, you can convert a SMCL file to a PostScript file, which is useful if you are using TeX or LaTeX or if you want to convert your output into Adobe's Portable Document Format. For example,

```
. translate mylog.smcl mylog.ps, replace
(file mylog.ps written in .ps format)
```

Converting can also be done through the menus by selecting File→Log→Translate.

2.7   Using and saving datasets

2.7.1   Data in Stata format

Stata uses its own data format with the extension `.dta`. The `use` command loads such data into memory. Pretend that we are working with the file `nomocc2.dta` in directory `d:\spostdata`. We can load the data by typing

```
. use d:\spostdata\nomocc2, clear
```

where the `.dta` extension is assumed by Stata. The `clear` option erases all data currently in memory and proceeds with loading the new data. Stata does not give an error if you include `clear` when there are no data in memory. If `d:\spostdata` was our working directory, we could use the simpler command

```
. use nomocc2, clear
```

If you have changed the data by deleting cases, merging in another file, or creating new variables, you can save the file with the `save` command. For example,

```
. save d:\spostdata\nomocc3, replace
```

where again we did not need to include the `.dta` extension. Also we saved the file with a different name so that we can use the original data later. The `replace` option indicates that if the file `nomocc3.dta` already exists, Stata should overwrite it. If the file does not already exist, `replace` is ignored. If `d:\spostdata` was our working directory, we could save the file with

```
. save nomocc3, replace
```

`save` stores the data in a format that can be read only by Stata 8 or later. (Stata 8 and Stata 9 share the same dataset format.) If you use the command `saveold` instead of `save`, the dataset is written so that it can be read by Stata 7, but if your data contain multiple missing-value codes, a feature that became available in Stata 8, all the missing-value codes will be mapped to the smallest missing value (`.`).

Tip: compress before saving Before saving a file, run `compress`, which checks each variable to determine if it can be saved in a more compact form. For instance, binary variables fit into the `byte` type, which takes up only one-fourth of the space of the `float` type. If you run `compress`, it might make your data file much more compact, and at worst it will do no harm.

2.7.2   Data in other formats

To load data from another statistical package, such as SAS or SPSS, you need to convert it into Stata's format. The easiest way to do this is with a conversion program such as Stat/Transfer (*http://www.stattransfer.com*). We recommend obtaining one of these programs if you are using more than one statistical package or if you often share data with others who use different packages.

If you are moving data between Stata and SAS, you can use `fdasave`, `fdause`, and `fdadescribe` to convert datasets to and from the SAS XPORT Transport format. The commands begin with `fda` since the U.S. Food and Drug Administration has adopted the SAS XPORT format for new drug and other applications. Although a program like Stat/Transfer might be easier to use than these commands, if you move back and forth between SAS and Stata frequently, it is worth spending some time learning the `fda` commands. Type `help fdasave` or `search fda` for details.

Alternatively, but less conveniently, most statistical packages allow you to save and load data in ASCII format. You can load an ASCII file with the `infile` or `infix` commands and export it with the `outfile` command. The reference manual entry for `infile` contains an extensive discussion that is particularly helpful for reading in ASCII data, or you can type `help infile`.

2.7.3   Entering data by hand

Data can also be entered by hand using a spreadsheet-style editor. Although we do not recommend using the editor to change existing data (because it is too easy to make a mistake), we find that it is useful for entering small datasets. To enter the editor, click on ▦ or type `edit` on the command line. The *Getting Started* manual has a tutorial for the editor, but most people who have used a spreadsheet before will be immediately comfortable with the editor.

As you use the editor, *every* change that you make to the data is reported in the Results window and is captured by the log file, if it is open. For example, if you change `age` for the fifth observation to 32, Stata reports `replace age = 32 in 5`. This tells you that instead of using the editor, you could have changed the data with a `replace` command. When you close the editor, Stata asks if you really want to keep the changes or revert to the unaltered data.

2.8   Size limitations on datasets[*]

If you get the error message `r(900): no room to add more observations` when trying to load a dataset or the message `r(901): no room to add more variables` when trying to add a new variable, you may need to allocate more memory. Typing `memory` shows how much memory is allocated to Stata and how much it is using. You can increase the amount of memory by typing `set memory #k` (for KB) or `#m` (for MB). For example, `set memory 32000k` or `set memory 32m` sets the memory to 32 MB.[3] If you have variables in memory, you must type `clear` before you can set the memory.

If you get the error `r(1000): system limit exceeded--see manual` when you try to load a dataset or add a variable, your dataset might have too many variables or the dataset might be too wide. Intercooled Stata is limited to a maximum of 2,047 variables, and the dataset can be up to 24,564 units wide (a binary variable has width 1, a double-precision variable has width 8, and a string variable has width equal to its length). Stata/SE allows 32,767 variables, and the dataset can be up to 393,192 units wide. String variables can be up to 244 characters. File transfer programs such as Stat/Transfer can drop specified variables and optimize variable storage. You can use these programs to create multiple datasets that each contain only the variables necessary for specific analyses.

2.9   Do-files

You can execute commands in Stata by typing one command at a time into the Command window and pressing Enter, as we have been doing. This interactive mode is

---

3. Stata can use virtual memory if you need to allocate memory beyond that which is physically available on a system, but we find that virtual memory makes Stata unbearably slow.

useful when you are learning Stata, exploring your data, or experimenting with alternative specifications of your regression model. You can also create a text file that contains a series of commands and then tell Stata to execute all the commands in that file, one after the other. These files, which are known as *do-files* because they use the extension .do, have the same function as "syntax files" in SPSS or "batch files" in other statistics packages. For more serious or complex work, we *always* use do-files because they make it easier to redo the analysis with small modifications later and because they provide an exact record of what has been done.

To get an idea of how do-files work, consider the file `example.do` saved in the working directory:

```
log using example, replace text
use binlfp2, clear
tabulate hc wc, row nolabel
log close
```

To execute a do-file, you type the command

```
do dofilename
```

from the Command window. For example, `do example` tells Stata to run each of the commands in `example.do`. (If the do-file is not in the working directory, you need to specify the directory, such as `do d:\spostdata\example`.) Executing `example.do` begins by opening the log `example.log`, and then loads `binlfp2.dta`, and finally constructs a table with `hc` and `wc`. Here is what the output looks like:

(Continued on next page)

```
-------------------------------------------------------------------------------
        log:  f:\spostdata\example.log
   log type:  text
  opened on:  26 September 2005, 15:44:45

. use http://www.stata-press.com/data/lf2/binlfp2, clear
(Data from 1976 PSID-T Mroz)

. tabulate hc wc, row nolabel

+-----------------+
| Key             |
|-----------------|
|    frequency    |
| row percentage  |
+-----------------+

   Husband |    Wife College: 1=yes
   College: |          0=no
1=yes 0=no  |          0            1 |     Total
-----------+----------------------+----------
        0  |        417           41 |       458
           |      91.05         8.95 |    100.00
-----------+----------------------+----------
        1  |        124          171 |       295
           |      42.03        57.97 |    100.00
-----------+----------------------+----------
    Total  |        541          212 |       753
           |      71.85        28.15 |    100.00

. log close
        log:  f:\spostdata\example.log
   log type:  text
  closed on:  26 September 2005, 15:44:45
-------------------------------------------------------------------------------
```

2.9.1 Adding comments

Stata has several different methods for denoting comments. We will make extensive use
of two methods. First, on any given line, Stata treats everything that comes after // or
after * as comments that are simply echoed to the output. Second, on any given line,
Stata ignores whatever comes after /// and treats the next line as a continuation of the
current line. For example, the following do-file executes the same commands as the one
above but includes comments:

```
//
// ==> short simple do-file
// ==> for didactic purposes
//
log using example, replace  // this comment is ignored
// next we load the data
use binlfp2, clear
// tabulate husband´s and wife´s education
tabulate hc wc,    /// the next line is treated as a continuation of this one
      row nolabel
// close up
log close
// make sure there is a cr at the end!
```

If you look at the do-files on our web site that reproduce the examples in this book, you will see that we use many comments. They are extremely helpful if others will be using your do-files or log files, or if there is a chance that you will use them again later.

2.9.2 Long lines

Sometimes you need to execute a command that is longer than the text that can fit onto a screen. If you are entering the command interactively, the Command window simply pushes the left part of the command off the screen as space is needed. Before entering a long command line in a do-file, however, you can use `#delimit ;` to tell Stata to interpret ";" as the end of a command. After the long command is entered, you can enter `#delimit cr` to return to using the carriage return as the end-of-line delimiter. For example,

```
#delimit ;
recode income91 1=500 2=1500 3=3500 4=4500 5=5500 6=6500 7=7500 8=9000
9=11250 10=13750 11=16250 12=18750 13=21250 14=23750 15=27500 16=32500
17=37500 18=45000 19=55000 20=67500 21=75000 *=. ;
#delimit cr
```

Instead of the `#delimit` command, we could have used `///`. For example,

```
recode income91 1=500 2=1500 3=3500 4=4500 5=5500 6=6500 7=7500 8=9000    ///
    9=11250 10=13750 11=16250 12=18750 13=21250 14=23750 15=27500 16=32500 ///
    17=37500 18=45000 19=55000 20=67500 21=75000 *=.
```

2.9.3 Stopping a do-file while it is running

If you are running a command or a do-file that you want to stop before it completes execution, click on ⊗ or press Ctrl-Break.

2.9.4 Creating do-files

Using Stata's Do-file Editor

Do-files can be created with Stata's built-in Do-file Editor. To use the editor, enter the command `doedit` to create a file to be named later or `doedit` *filename* to create or edit a file named *filename*.do. You can also click on ◈. The Do-file Editor is easy to use and works like most text editors (see *Getting Started* for further details). After you finish your do-file, select Tools→Do to execute the file or click on ▤↓.

Using other editors to create do-files

Because do-files are plain text files, you can create do-files with any program that
creates text files. Specialized text editors work much better than word processors such
as WordPerfect or Microsoft Word. Among other things, with word processors it is easy
to forget to save the file as plain text. Our own preference for creating do-files is TextPad
(*http://www.textpad.com*), which runs in Windows. This program has many features
that make it faster to create do-files. For example, you can create a "clip library" that
contains frequently entered material, and you can obtain a syntax file from our web site
that provides color coding of reserved words for Stata.

If you use an editor other than Stata's built-in editor, you cannot run the do-file by
clicking on an icon or selecting from a menu. Instead, you must switch from your editor
and then enter the command do *filename*.

Two nice features of the Do-file Editor are that you can highlight a section of a file
and Stata will execute only the commands that you have highlighted, and you can select
Tools→Do to Bottom and Stata will execute commands starting wherever the cursor is
located to the end of the file.

Warning Stata executes commands when it encounters a carriage return (i.e., the
Enter key). If you do not include a carriage return after the last line in a do-file,
that last line will not be executed. TextPad has a feature to enter that final,
pesky carriage return automatically. To set this option in TextPad, select the
option "Automatically terminate the last line of the file" in the preferences for the
editor.

2.9.5   Recommended structure for do-files

This is the basic structure that we recommend for do-files:

```
// including version number ensures compatibility with later Stata releases
version 9
// if a log file is open, close it
capture log close
// don´t pause when output scrolls off the page
set more off
// log results to file myfile.log
log using myfile, replace text
// * myfile.do - written 19 oct 2005 to illustrate do-files
//
// * your commands go here
//
// close the log file.
log close
```

Although the comments (which you can remove) should explain most of the file, there
are a few points that we need to explain.

- The `version 9` command indicates that the program was written for use in Stata 9. This command tells any future version of Stata that you want the commands that follow to work just as they did when you ran them in Stata 9. This prevents the problem of old do-files not running correctly in newer releases of the program.

- The command `capture log close` is very useful. Suppose that you have a do-file that starts with `log using mylog, replace`. You run the file and it "crashes" before reaching `log close`, which means that the log file remains open. If you revise the do-file and run it again, an error is generated when it tries to open the log file because the file is already open. The prefix `capture` tells Stata not to stop the do-file if the command that follows produces an error. Accordingly, `capture log close` closes the log file if it is open. If it is not open, the error generated by trying to close an already-closed file is ignored.

Tip: The command `cmdlog` is much like the `log` command, except that it creates a text file with extension `.txt` that saves all subsequent commands that are entered in the Command window (it does not save commands that are executed within a do-file). This is handy because it allows you to use Stata interactively and then make a do-file based on what you have done. You simply load the cmdlog that you saved, rename it to *newname*`.do`, delete commands you no longer want, and execute the new do-file. Your interactive session is now documented as a do-file. The syntax for opening and closing cmdlog files is the same as that for `log` (i.e., `cmdlog using` to open and `cmdlog close` to close), and you can have log and cmdlog files open simultaneously.

2.10 Using Stata for serious data analysis

Voltaire is said to have written *Candide* in three days. Creative work often rewards such inspired, seat-of-the-pants, get-the-details-later activity. *Data management does not.* Instead, effective data management rewards forethought, carefulness, double- and triple-checking of details, and meticulous, albeit tedious, documentation. Errors in data management are astonishingly (and painfully) easy to make. Moreover, tiny errors can have disastrous implications that can cost hours and even weeks of work. The extra time it takes to conduct data management carefully is rewarded many times over by the reduced risk of errors. That is, it helps prevent you from getting incorrect results that you do not know are incorrect. With this in mind, we begin with some broad, perhaps irritatingly practical, suggestions for doing data analysis efficiently and effectively.

1. *Ensure replicability by using do-files and log files for everything.* For data analysis to be credible, you must be able to reproduce *entirely and exactly* the trail from the original data to the tables in your paper. Thus any permanent changes you make to the data should be made by running do-files rather than by using the interactive mode. If you work interactively, be sure that the first thing you do is to open a `log` or `cmdlog` file. Then when you are done, you can use these files to create a do-file to reproduce your interactive results.

2. *Document your do-files.* Reasoning that is obvious today can be baffling in 6 months. We use comments extensively in our do-files, which are invaluable for remembering what we did and why we did it.

3. *Keep a research log.* For serious work, you should keep a diary that includes a description of *every* program you run, the research decisions that are being made (e.g., the reasons for recoding a variable in a particular way), and the files that are created. A good research log allows you to reproduce everything you have done starting with the original data. We cannot overemphasize how helpful such notes are when you return to a project that was put on hold, when you are responding to reviewers, or when you are moving on to the next stage of your research.

4. *Develop a system for naming files.* Usually it makes the most sense to have each do-file generate one log file with the same prefix (e.g., `clean_data.do`, `clean_data.log`). Names are easiest to organize when brief, but they should be long enough and logically related enough to make sense of the task the file does. Scott prefers to keep the names short and organized by major task (e.g., `recode01.do`), whereas Jeremy likes longer names (e.g., `make_income_vars.do`). Either is fine as long as it works for you.

5. *Use new names for new variables and files.* Never change a dataset and save it with the original name. If you drop three variables from `pcoms1.dta` and create two new variables, call the new file `pcoms2.dta`. When you transform a variable, give it a new name rather than simply replacing or recoding the old variable. For example, if you have a variable `workmom` with a five-point attitude scale, and you want to create a binary variable indicating positive and negative attitudes, create a new variable called `workmom2`.

6. *Use labels and notes.* When you create a new variable, give it a variable label. If it is a categorical variable, assign value labels. You can add a note about the new variable using the `notes` command (described below). When you create a new dataset, you can also use `notes` to document what it is.

7. *Double-check every new variable.* Cross-tabulating or graphing the old variable and the new variable are often effective for verifying new variables. As we describe below, using `list` with a subset of cases is similarly effective for checking transformations. Be sure to at least look carefully at the frequency distributions and summary statistics of variables in your analysis. You would not believe how many times puzzling regression results turn out to involve miscodings of variables that would have been immediately apparent by looking at the descriptive statistics.

8. *Practice good archiving.* If you want to retain hard copies of all your analyses, develop a system of binders for doing so rather than a set of intermingling piles on your desk. Back up everything. Make off-site backups or keep any on-site backups in a fireproof box. Should cataclysm strike, you will have enough other things to worry about without also having lost months or years of work.

2.11 Syntax of Stata commands

Think about the syntax of commands in everyday, spoken English. They usually begin with a verb telling the other person what they are supposed to do. Sometimes the verb is the entire command: "Help!" or "Stop!" Sometimes the verb needs to be followed by an object that indicates who or what the verb is to be performed on: "Help Dave!" or "Stop the car!" Sometimes the verb is followed by a qualifier that gives specific conditions under which the command should or should not be performed: "Give me a piece of pizza *if it doesn't have mushrooms*" or "Call me *if you get home before nine*". Verbs can also be followed by adverbs that specify that the action should be performed in some way that is different from how it might normally be, such as when a teacher commands her students to "Talk *clearly*" or "Walk *single file*".

Stata follows an analogous logic, albeit with some other wrinkles that we will introduce later. The basic syntax of a command has four parts:

1. *Command*: What action do you want performed?

2. *Names of variables, files, or other objects*: On what things is the command to be performed?

3. *Qualifier on observations*: On which observations should the command be performed?

4. *Options*: What special things should be done in executing the command?

All commands in Stata require the first of these parts, just as it is hard in English to issue spoken commands without a verb. Each of the other three parts can be required, optional, or not allowed, depending on the particular command and circumstances. Here is an example of a command that features all four parts and uses `binlfp2.dta`, which we loaded earlier:

(Continued on next page)

```
. tabulate hc wc if age>40, row
```

```
┌─────────────────┐
│ Key             │
├─────────────────┤
│    frequency    │
│  row percentage │
└─────────────────┘
```

Husband College: 1=yes 0=no	Wife College: 1=yes 0=no NoCol	College	Total
NoCol	263	23	286
	91.96	8.04	100.00
College	58	91	149
	38.93	61.07	100.00
Total	321	114	435
	73.79	26.21	100.00

If you want to suppress the key, you can add the option `nokey`. For example, `tabulate hc wc, row nokey`.

`tabulate` is a command for making one- or two-way tables of frequencies. Here we want a two-way table of the frequencies of variables `hc` by `wc`. By putting `hc` first, we make this the row variable and `wc` the column variable. By specifying `if age>40`, we specify that the frequencies should include observations only for those older than 40. The option `row` indicates that row percentages should be printed as well as frequencies. These allow us to see that in 61% of the cases in which the husband had attended college, the wife had also done so, whereas wives had attended college only in 8% of cases in which the husbands had not. Notice the comma preceding `row`: *whenever options are specified, they are at the end of the command with a single comma to indicate where the list of options begins.* The precise ordering of multiple options after the comma is never important.

Next we provide more information on each of the four components.

2.11.1 Commands

Commands define the tasks that Stata is to perform. A great thing about Stata is that the set of commands is deliciously open ended. It expands not just with new releases of Stata but also when users add their own commands, such as our SPost commands. Each new command is stored in its own file, ending with the extension `.ado`. Whenever Stata encounters a command that is not in its built-in library, it searches various directories for the appropriate ado-file. The list of the directories it searches (and the order that it searches them) can be obtained by typing `adopath`.

2.11.2   Variable lists

Variable names are case sensitive. For example, you could have three different variables named income, Income, and inCome. Of course, this is not a good idea because it leads to confusion. To keep life simple, we stick exclusively to lowercase names. Starting with Stata 7, Stata allows variable names up to 32 characters long, compared with the eight-character maximum imposed by earlier versions of Stata and many other statistics packages. In practice, we try not to give variables names more than eight characters, as this makes it easier to share data with people who use other packages. Also we recommend using short names because longer variable names become unwieldy to type. (Although variable names can be abbreviated to whatever initial set of characters identifies the variable uniquely, we worry that too much reliance on this feature might cause one to make mistakes.)

If you list no variables, many commands assume that you want to perform the operation on every variable in the dataset. For example, the summarize command provides summary statistics on the listed variables:

```
. summarize age inc k5
    Variable |       Obs        Mean    Std. Dev.        Min        Max
-------------+--------------------------------------------------------
         age |       753    42.53785    8.072574         30         60
         inc |       753    20.12897     11.6348   -.0290001         96
          k5 |       753    .2377158     .523959          0          3
```

We could also get summary statistics on every variable in our dataset by just typing

```
. summarize
    Variable |       Obs        Mean    Std. Dev.        Min        Max
-------------+--------------------------------------------------------
         lfp |       753    .5683931    .4956295          0          1
          k5 |       753    .2377158     .523959          0          3
        k618 |       753    1.353254    1.319874          0          8
         age |       753    42.53785    8.072574         30         60
          wc |       753    .2815405    .4500494          0          1
-------------+--------------------------------------------------------
          hc |       753    .3917663    .4884694          0          1
         lwg |       753    1.097115    .5875564   -2.054124   3.218876
         inc |       753    20.12897     11.6348   -.0290001         96
```

You can also select all variables that begin or end with the same letters by using the wildcard operator *. For example,

```
. summarize k*
    Variable |       Obs        Mean    Std. Dev.        Min        Max
-------------+--------------------------------------------------------
          k5 |       753    .2377158     .523959          0          3
        k618 |       753    1.353254    1.319874          0          8
```

2.11.3   if and in qualifiers

Stata has two qualifiers that restrict the sample that is analyzed: if and in. in performs operations on a range of consecutive observations. Typing summarize in 20/100 gives summary statistics based only on the 20th through 100th observations. in restrictions depend on the current sort order of the data, meaning that if you re-sort your data, the 81 observations selected by the restriction summarize in 20/100 might be different.[4]

In practice, if conditions are used much more often than in conditions. if restricts the observations to those that fulfill a specified condition. For example, summarize if age<50 provides summary statistics for only those observations where age is less than 50. Here is a list of the elements that can be used to construct logical statements for selecting observations with if:

Operator	Definition	Example
==	Equal to	if female==1
!=	Not equal to	if female!=1
>	Greater than	if age>20
>=	Greater than or equal to	if age>=21
<	Less than	if age<66
<=	Less than or equal to	if age<=65
&	And	if age==21 & female==1
|	Or	if age==21|educ>16

There are two important things about the if qualifier:

1. Use a double equal sign (e.g., summarize if female==1) to specify a condition to test. When assigning a value to something, such as when creating a new variable, use a single equal sign (e.g., gen newvar=1). Putting these examples together results in gen newvar=1 if female==1.

2. The missing-value codes are the largest positive numbers. This implies that Stata treats missing cases as positive infinity when evaluating if expressions. In other words, if you type summarize ed if age>50, the summary statistics for ed are calculated on all observations where age is greater than 50, including cases where the value of age is missing. You must be careful of this when using if with > or >= expressions. If you type summarize ed if age<., Stata gives summary statistics for cases where age is not missing. (Note that . is the smallest of the 27 missing-value codes. See section 2.12.3 for more details on missing values.) Entering summarize ed if age>50 & age<. provides summary statistics for those cases where age is greater than 50 and is not missing.

---

4. In Stata 6 and earlier, some official Stata commands changed the sort order of the data, but fortunately this quirk was removed in Stata 7. As of Stata 7, no properly written Stata command should change the sort order of the data, although readers should beware that user-written programs may not always follow proper Stata programming practice.

Examples of if qualifier

If we wanted summary statistics on income for only those respondents who were between the ages of 25 and 65, we would type

```
. summarize income if age>=25 & age<=65
```

If we wanted summary statistics on income for only female respondents who were between the ages of 25 and 65, we would type

```
. summarize income if age>=25 & age<=65 & female==1
```

If we wanted summary statistics on income for the remaining female respondents—that is, those who are younger than 25 or older than 65—we would type

```
. summarize income if (age<25 | age>65) & age<. & female==1
```

We need to include & age<. because Stata treats missing codes as positive infinity. The condition (age<25 | age>65) would otherwise include those cases for which age is missing.

Tip: Removing the separator If you do not like the horizontal separator that appears after every five variables in the output for summarize, you can remove the lines with the option sep(0).

2.11.4 Options

Options are set off from the rest of the command by a comma. Options can often be abbreviated, although whether and how they can be abbreviated varies across commands. In this book, we rarely cover all the available options available for any given command, but you can check the manual or use help for more options that might be useful for your analyses.

2.12 Managing data

2.12.1 Looking at your data

There are two easy ways to look at your data.

browse opens a spreadsheet in which you can scroll to look at the data, but you cannot change the data. You can look and change data with the edit command, but this is risky. We much prefer making changes to our data using do-files, even when we are changing the value of only one variable for one observation. The browser is also available by clicking on , whereas the data editor is available by clicking on .

`list` creates a list of values of specified variables and observations. `if` and `in` qualifiers can be used to look at just a portion of the data, which is sometimes useful for checking that transformations of variables are correct. For example, if you want to confirm that the variable `lninc` has been correctly constructed as the natural log of `inc`, typing `list inc lninc in 1/20` lets you see the values of `inc` and `lninc` for the first 20 observations.

2.12.2 Getting information about variables

There are several methods for obtaining basic information about your variables. Here are five commands that we find useful. Which one you use depends mostly on the kind and level of detail you need.

`describe` provides information on the size of the dataset and the names, labels, and types of variables. For example,

```
. use http://www.stata-press.com/data/lf2/binlfp2, clear
(Data from 1976 PSID-T Mroz)

. describe

Contains data from binlfp2.dta
  obs:            753                          Data from 1976 PSID-T Mroz
  vars:             8                          30 Apr 2001 16:17
  size:        13,554 (98.7% of memory free)   (_dta has notes)

                storage  display    value
variable name    type    format     label     variable label

lfp             byte     %9.0g      lfplbl     Paid Labor Force: 1=yes 0=no
k5              byte     %9.0g                 # kids < 6
k618            byte     %9.0g                 # kids 6-18
age             byte     %9.0g                 Wife's age in years
wc              byte     %9.0g      collbl     Wife College: 1=yes 0=no
hc              byte     %9.0g      collbl     Husband College: 1=yes 0=no
lwg             float    %9.0g                 Log of wife's estimated wages
inc             float    %9.0g                 Family income excluding wife's

Sorted by:  lfp
```

`summarize` provides summary statistics. By default, `summarize` presents the number of nonmissing observations, the mean, the standard deviation, the minimum values, and the maximum. Adding the `detail` option includes more information. For example,

```
. summarize age, detail
                        Wife's age in years

              Percentiles       Smallest
    1%             30                30
    5%             30                30
   10%             32                30      Obs                  753
   25%             36                30      Sum of Wgt.          753

   50%             43                        Mean            42.53785
                                Largest      Std. Dev.       8.072574
   75%             49                60
   90%             54                60      Variance        65.16645
   95%             56                60      Skewness         .150879
   99%             59                60      Kurtosis        1.981077
```

tabulate creates the frequency distribution for a variable. For example,

```
. tabulate hc
     Husband |
    College: |
  1=yes 0=no |     Freq.      Percent        Cum.

       NoCol |       458        60.82       60.82
     College |       295        39.18      100.00

       Total |       753       100.00
```

If you do not want the value labels included, type

```
. tabulate hc, nolabel
     Husband |
    College: |
  1=yes 0=no |     Freq.      Percent        Cum.

           0 |       458        60.82       60.82
           1 |       295        39.18      100.00

       Total |       753       100.00
```

If you want a two-way table, type

```
. tabulate hc wc
     Husband | Wife College: 1=yes
    College: |          0=no
  1=yes 0=no |    NoCol    College |     Total

       NoCol |      417         41 |       458
     College |      124        171 |       295

       Total |      541        212 |       753
```

By default, tabulate does not tell you the number of missing values for either variable. Specifying the missing option includes missing values. We recommend this option whenever you are generating a frequency distribution to check that some transformation was done correctly. The options `row`, `col`, and `cell` request row, column, and cell percentages along with the frequency counts. The option `chi2` reports the χ^2 for a test that the rows and columns are independent.

`tab1` presents univariate frequency distributions for each variable listed. For example,

```
. tab1 hc wc

-> tabulation of hc

  Husband
  College:
 1=yes 0=no |     Freq.       Percent        Cum.

      NoCol |       458         60.82       60.82
    College |       295         39.18      100.00

      Total |       753        100.00
-> tabulation of wc

     Wife
  College:
 1=yes 0=no |     Freq.       Percent        Cum.

      NoCol |       541         71.85       71.85
    College |       212         28.15      100.00

      Total |       753        100.00
```

`dotplot` generates a quick graphical summary of a variable, which is useful for quickly checking your data. For example, the command `dotplot age` leads to the following graph:

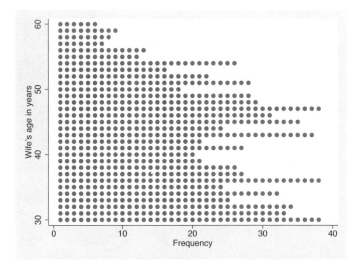

This graph will appear in a new window called the Graph window. Details on saving, printing, and enhancing graphs are given in section 2.16.

codebook summarizes a variable in a format designed for printing a codebook. For example, codebook age produces

```
. codebook age
```

age					Wife's age in years

type:	numeric (byte)				
range:	[30,60]		units:	1	
unique values:	31		missing .:	0/753	
mean:	42.5378				
std. dev:	8.07257				
percentiles:	10%	25%	50%	75%	90%
	32	36	43	49	54

2.12.3 Missing values

Although numeric missing values are automatically excluded when Stata fits models, they are stored as the largest positive values. Twenty-seven missing values are available, with the ordering

$$\text{all numbers} < . < .a < .b < \cdots < .z$$

This way of handling missing values can have unexpected consequences when determining samples. For instance, the expression if age>65 is true when age has a value greater than 65 and when age is missing. Similarly, the expression occupation!=1 is true if occupation is not equal to 1 or occupation is missing. When expressions such as these are required, be sure to explicitly exclude any unwanted missing values. For instance, age>65 & age<. would be true only for those people whose age is not missing and who are over 65. Similarly, occupation!=1 & occupation <. would be true only when the occupation is not missing and not equal to 1.

The different missing values can be used to record the distinct reasons why a variable is missing. For instance, consider a survey that asked people about their driving records. The variable that records whether someone received a ticket after being involved in an accident could be missing because the respondent had not been involved in any accidents or because the person refused to answer the question.

2.12.4 Selecting observations

As previously mentioned, you can select cases with the if and in qualifiers. For example, summarize age if wc==1 provides summary statistics on age for only those observations where wc equals 1. Sometimes it is simpler to remove the cases with either the drop or keep commands. drop removes observations from memory (not from the .dta file) based on an if or in specification. The syntax is

drop [*in*] [*if*]

Only observations that do *not* meet those conditions are left in memory. For example, `drop if wc==1` keeps only those cases where `wc` is not equal to 1, including observations with missing values on `wc`.

`keep` has the same syntax as `drop` and deletes all cases *except* those that meet the condition. For example, `keep if wc==1` keeps only those cases where `wc` is 1; all other observations, including those with missing values for `wc`, are dropped from memory. After selecting the observations that you want, you can save the remaining variables to a new dataset with the `save` command.

2.12.5 Selecting variables

You can also select which variables you want to keep. The syntax is

`drop` *variable_list*

`keep` *variable_list*

With `drop`, all variables are kept except those that are explicitly listed. With `keep`, only those variables that are explicitly listed are kept. After selecting the variables that you want, you can save the remaining variables to a new dataset with the `save` command.

2.13 Creating new variables

The variables that you analyze are often constructed differently from the variables in the original dataset. Here we consider basic methods for creating new variables. Our examples always create a new variable from an old variable rather than transforming an existing variable. Even though you can simply transform an existing variable, we find that this leads to mistakes.

2.13.1 generate command

`generate` creates new variables. For example, to create `age2` as an exact copy of `age`, type

```
. generate age2 = age
. summarize age2 age
```

Variable	Obs	Mean	Std. Dev.	Min	Max
age2	753	42.53785	8.072574	30	60
age	753	42.53785	8.072574	30	60

The results of `summarize` show that the two variables are identical. We used a single equal sign because we are making a variable equal to some value.

Observations excluded by `if` or `in` qualifiers in the `generate` command are coded as missing. For example, to generate `age3` that equals `age` for those over 40 but is otherwise missing, type

```
. gen age3 = age if age>40
(318 missing values generated)

. summarize age3 age
    Variable |        Obs        Mean    Std. Dev.        Min        Max
-------------+--------------------------------------------------------
        age3 |        435     48.3977    4.936509         41         60
         age |        753    42.53785    8.072574         30         60
```

Whenever **generate** (or **gen**, as it can be abbreviated) produces missing values, it tells you how many cases are missing.

generate can also create variables that are mathematical functions of existing variables. For example, we can create **agesq** that is the square of **age** and **lnage** that is the natural log of **age**:

```
. gen agesq = age^2
. gen lnage = ln(age)
```

For quick reference, here is a list of the standard mathematical operators

Operator	Definition	Example
+	Addition	gen y = a+b
−	Subtraction	gen y = a-b
/	Division	gen density = pop/area
*	Multiplication	gen y = a*b
^	Take to a power	gen y = a^3

and some particularly useful functions:

Function	Definition	Example
ln	Natural log	gen lnwage = ln(wage)
exp	Exponential	gen y = exp(a)
sqrt	Square root	gen agesqrt = sqrt(age)

For a complete list of functions in Stata, type **help functions**.

Tip: Although **gen** *newvar=oldvar* is the most intuitive way of creating a copy of the values of *oldvar* as *newvar*, sometimes **clonevar** *newvar=oldvar* is a better alternative. **clonevar** copies not only the values of *oldvar* but also the variable label, value label, and other attributes.

2.13.2   replace command

replace has the same syntax as generate but is used to change values of a variable that already exists. For example, say we want to make a new variable, age4, that equals age if age is over 40 but equals 40 for all persons aged 40 and under. First, we create age4 equal to age. Then we replace those values we want to change:

```
. gen age4 = age
. replace age4 = 40 if age<40
(298 real changes made)
. summarize age4 age
    Variable |     Obs        Mean    Std. Dev.       Min        Max
-------------+--------------------------------------------------------
        age4 |     753    44.85126    5.593896         40         60
         age |     753    42.53785    8.072574         30         60
```

replace reports how many values were changed. This is useful in verifying that the command did what you intended. Also summarize confirms that the minimum value of age is 30 and that age4 now has a minimum of 40 as intended.

Warning Of course, we could have simply changed the original variable: replace age = 40 if age<40. But, if we did this and saved the data, there would be no way to return to the original values for age if we later needed them.

2.13.3   recode command

The values of *existing* variables can also be changed using the recode command. With recode you specify a set of correspondences between old values and new ones. For example, you might want old values of 1 and 2 to correspond to new values of 1, old values of 3 and 4 to correspond to new values of 2, and so on. This is particularly useful for combining categories. To use this command, we recommend that you start by making a copy of an existing variable. Then recode the copy. Or, to be more efficient, you can use the generate(*newvariablename*) option with recode. With this option, Stata creates a new variable instead of overwriting the old one. recode is best explained by example, several of which we include below (for more, type help recode).

To change 1 to 2 and 3 to 4 but leave all other values unchanged, type

```
. recode origvar (1=2) (3=4), generate(myvar1)
(23 differences between origvar and mvar1)
```

To change 2 to 1 and change all other values (including missing) to 0:

```
. recode origvar (2=1) (*=0), gen(myvar2)
(100 differences between origvar and myvar2)
```

where the asterisk indicates all values, including missing values, that have not been explicitly recoded.

To change 2 to 1 and change all other values *except missing* to 0:

```
. recode origvar (2=1) (nonmissing=0), gen(myvar3)
(89 differences between origvar and myvar3)
```

To change values from 1 to 4 inclusive to 2 and keep other values unchanged:

```
. recode origvar (1/4=2), gen(myvar4)
(40 differences between origvar and myvar4)
```

To change values 1, 3, 4, and 5 to 7 and keep other values unchanged:

```
. recode origvar (1 3 4 5=7), gen(myvar5)
(55 differences between origvar and myvar5)
```

To change all values from the minimum through 5 to the minimum:

```
. recode origvar (min/5=min), gen(myvar6)
(56 differences between origvar and myvar6)
```

To change missing values to 9:

```
. recode origvar (missing=9), gen(myvar7)
(11 differences between origvar and myvar7)
```

To change values of -999 to missing:

```
. recode origvar (-999=.), gen(myvar8)
(56 differences between origvar and myvar8)
```

recode can be used to recode several variables at once if they are all to be recoded the same way. Just include all the variable names before the instructions on how they are to be recoded, and include all the names for new variables (if you do not want the old variables to be overwritten) within the parentheses of the generate() option.

2.13.4 Common transformations for RHS variables

For the models we discuss in later chapters, you can use many of the tricks you learned for coding right-hand-side (i.e., independent) variables in the linear regression model. Here are some useful examples. Details on how to interpret such variables in regression models are given in chapter 9.

Breaking a categorical variable into a set of binary variables

To use a *j*-category nominal variable as an independent variable in a regression model, you need to create a set of $j - 1$ binary variables, also known as dummy variables or indicator variables. To show how to do this, we use educational attainment (degree), which is coded as 0 = no diploma, 1 = high school diploma, 2 = associate's degree, 3 = bachelor's degree, and 4 = postgraduate degree, with some missing data. We want to make four binary variables with the "no diploma" category serving as our reference category. We also want observations that have missing values for `degree` to have missing values in each of the dummy variables that we create. The simplest way to do this is to use the `generate` option with `tabulate`:

```
. use http://www.stata-press.com/data/lf2/gsskidvalue2, clear
(1993 and 1994 General Social Survey)
. tabulate degree, generate(edlevel)
```

rs highest degree	Freq.	Percent	Cum.
lt high school	801	17.47	17.47
high school	2,426	52.92	70.40
junior college	273	5.96	76.35
bachelor	750	16.36	92.71
graduate	334	7.29	100.00
Total	4,584	100.00	

The `generate(name)` option creates a new binary variable for each category of the specified variable. Here `degree` has five categories, so five new variables are created. These variables all begin with `edlevel`, the root that we specified with the `generate(edlevel)` option. We can check the five new variables by typing `summarize edlevel*`:

```
. summarize edlevel*
```

Variable	Obs	Mean	Std. Dev.	Min	Max
edlevel1	4584	.1747382	.3797845	0	1
edlevel2	4584	.5292321	.4991992	0	1
edlevel3	4584	.059555	.2366863	0	1
edlevel4	4584	.1636126	.369964	0	1
edlevel5	4584	.0728621	.2599384	0	1

By cross-tabulating the new `edlevel1` by the original `degree`, we can see that `edlevel1` equals 1 for individuals with no high school diploma and equals 0 for everyone else except the 14 observations with missing values for `degree`:

```
. tabulate degree edlevel1, missing
```

rs highest degree	degree==lt high school			Total
	0	1	.	
lt high school	0	801	0	801
high school	2,426	0	0	2,426
junior college	273	0	0	273
bachelor	750	0	0	750
graduate	334	0	0	334
.	0	0	14	14
Total	3,783	801	14	4,598

One limitation of using the generate(*name*) option of tabulate is that it works only when there is a one-to-one correspondence between the original categories and the dummy variables that we wish to create. So, let's suppose that we want to combine high school graduates and those with associate's degrees when creating our new binary variables. Say also that we want to treat those without high school diplomas as the omitted category. The following is one way to create the three binary variables that we need:

```
. gen hsdeg = (degree==1 | degree==2) if degree<.
(14 missing values generated)
. gen coldeg = (degree==3) if degree<.
(14 missing values generated)
. gen graddeg = (degree==4) if degree<.
(14 missing values generated)
. tabulate degree coldeg, missing
```

rs highest degree	coldeg 0	1	.	Total
lt high school	801	0	0	801
high school	2,426	0	0	2,426
junior college	273	0	0	273
bachelor	0	750	0	750
graduate	334	0	0	334
.	0	0	14	14
Total	3,834	750	14	4,598

To understand how this works, you need to know that when Stata is presented with an expression (e.g., degree==3) where it expects a value, it evaluates the expression and assigns it a value of 1 if true and 0 if false. Consequently, gen coldeg = (degree==3) creates the variable coldeg that equals 1 whenever degree equals 3 and 0 otherwise. By adding if degree<. to the end of the command, we assign these values *only* to observations in which the value of degree is not missing. If an observation has a missing value for degree, these cases are given a missing value.

More examples of creating binary variables

Binary variables are used so often in regression models that it is worth providing more examples of generating them. In the dataset that we use in chapter 5 (ordwarm2.dta), the independent variable for respondent's education (ed) is measured in years. We can create a dummy variable that equals 1 if the respondent has at least 12 years of education and 0 otherwise:

```
. gen ed12plus = (ed>=12) if ed<.
```

We might also want to create a set of variables that indicates whether an individual has less than 12, between 13 and 16, or 17 or more years of education. This is done as follows:

```
. gen edlt13 = (ed<=12) if ed<.
. gen ed1316 = (ed>=13 & ed<=16) if ed<.
. gen ed17plus = (ed>17) if ed<.
```

Tip: Naming dummy variables Whenever possible, we name dummy variables so that 1 corresponds to "yes" and 0 to "no". With this convention, a dummy variable called female is coded 1 for women (i.e., yes, the person is female) and 0 for men. If the dummy variable were named sex, there would be no immediate way to know what 0 and 1 mean.

The recode command can also be used to create binary variables. The variable warm contains responses to the question of whether working women can have as warm a relationship with their children as women who do not work: $1 =$ strongly disagree, $2 =$ disagree, $3 =$ agree, and $4 =$ strongly agree. To create a dummy indicating agreement as opposed to disagreement, type

```
. gen wrmagree = warm
. recode wrmagree 1=0 2=0 3=1 4=1
(wrmagree: 2293 changes made)
. tabulate wrmagree warm
```

| | Mom can have warm relations with child | | | | |
wrmagree	SD	D	A	SA	Total
0	297	723	0	0	1,020
1	0	0	856	417	1,273
Total	297	723	856	417	2,293

Nonlinear transformations

Nonlinear transformations of the independent variables are commonly used in regression models. For example, researchers often include both age and age^2 as explanatory variables to allow the effect of a 1-year increase in age to change as one gets older. We can create a squared term as

```
. gen agesq = age*age
```

Likewise, income is often logged so that the impact of each additional dollar decreases as income increases. The new variable can be created as

```
. gen lnincome = ln(income)
(495 missing values generated)
```

We can use the minimum and maximum values reported by summarize as a check on our transformations:

```
. summarize age agesq income lnincome
    Variable │      Obs        Mean    Std. Dev.         Min          Max
─────────────┼──────────────────────────────────────────────────────────
         age │     4598    46.12375     17.33162          18           99
       agesq │     4598     2427.72     1798.477         324         9801
      income │     4103     34790.7     22387.45        1000        75000
    lnincome │     4103    10.16331     .8852605    6.907755     11.22524
```

Interaction terms

In regression models, you can include interactions by taking the product of two independent variables. For example, we might think that the effect of family income differs for men and women. If sex is measured as the dummy variable `female`, we can construct an interaction term as follows:

```
. gen feminc = female * income
(495 missing values generated)
```

2.14 Labeling variables and values

Variable labels provide descriptive information about what a variable measures. For example, the variable `agesq` might be given the label "age squared", or `warm` could have the label "Mother has a warm relationship". *Value* labels provide descriptive information about the different values of a categorical variable. For example, value labels might indicate that the values 1–4 correspond to survey responses of strongly agree, agree, disagree, and strongly disagree. Adding labels to variables and values is not much fun, but in the long run, it can save much time and prevent misunderstandings. Also many of the commands in SPost produce output that is more easily understood if the dependent variable has value labels.

2.14.1 Variable labels

The `label variable` command attaches a label of up to 80 characters to a variable. For example,

(Continued on next page)

```
. label variable agesq "Age squared"
. describe agesq

              storage  display     value
variable name  type    format      label      variable label

agesq          float   %9.0g                   Age squared
```

If no label is specified, any existing variable label is removed. For example,

```
. label variable agesq
. describe agesq

              storage  display     value
variable name  type    format      label      variable label

agesq          float   %9.0g
```

Tip: Use short labels Although variable labels of up to 80 characters are allowed, we recommend that you use short labels whenever possible. Output often does not show all 80 characters. For the same reason, we also find it useful to put the most important information at the beginning of the label. That way, if the label is truncated, you will still see the critical information.

Tip: Searching variable labels Typing lookfor *string* will search the data file and list all variables in which *string* appears in either the variable name or label.

2.14.2   Value labels

Beginners often find value labels in Stata confusing. Remember that Stata splits the process of labeling values into two steps: creating labels and then attaching the labels to variables.

Step 1 defines a set of labels *without* reference to a variable. Here are some examples of value labels:

```
. label define yesno 1 yes 0 no
. label define posneg4 1 veryN 2 negative 3 positive 4 veryP
. label define agree4 1 StrongA 2 Agree 3 Disagree 4 StrongD
. label define agree5 1 StrongA 2 Agree 3 Neutral 4 Disagree 5 StrongD
```

First, each *set* of labels is given a unique name (e.g., yesno, agree4). Second, individual labels are associated with a specific value. Third, none of our labels has spaces in them (e.g., we use StrongA not Strong A). Although you can have spaces if you place the label within quotes, some commands crash when they encounter blanks in value labels. So, it is easier not to do it. We have also found that the period, colon, and left curly bracket ({) in value labels can cause similar problems. Fourth, our labels are eight letters or shorter in length because some programs have trouble with value labels longer than eight letters.

Step 2 assigns the value labels to variables. Let's say that variables `female`, `black`, and `anykids` all imply yes/no categories with 1 as yes and 0 as no. To assign labels to the values, we would use the following commands:

```
. label values female yesno
. label values black yesno
. label values anykids yesno
. describe female black anykids

              storage  display   value
variable name  type    format    label     variable label

female         byte    %9.0g     yesno     Female
black          byte    %9.0g     yesno     Black
anykids        byte    %9.0g     yesno     R have any children?
```

The output for `describe` shows which value labels were assigned to which variables. The new value labels are reflected in the output from `tabulate`:

```
. tabulate anykids

R have any
 children?      Freq.       Percent        Cum.

       no       1,267         27.64       27.64
      yes       3,317         72.36      100.00

    Total       4,584        100.00
```

For the `degree` variable that we looked at earlier, we assign labels with

```
. label define degree 0 "no_hs" 1 "hs" 2 "jun_col" 3 "bachelor" 4 "graduate"
. label values degree degree
. tabulate degree

rs highest
   degree       Freq.       Percent        Cum.

    no_hs         801         17.47       17.47
       hs       2,426         52.92       70.40
  jun_col         273          5.96       76.35
 bachelor         750         16.36       92.71
 graduate         334          7.29      100.00

    Total       4,584        100.00
```

We used underscores (_) instead of spaces.

If you want a list of the value labels being used in your current dataset, use the command `labelbook`, which provides a detailed list of all value labels, including which labels are assigned to which variables. This can be useful both in setting up a complex dataset and for documenting your data.

2.14.3 notes command

The `notes` command allows you to add notes to the dataset as a whole or to specific variables. Because the notes are saved in the dataset, the information is always available when you use the data. Here we add one note describing the dataset and two that describe the `income` variable:

```
. notes: General Social Survey extract for Stata book
. notes income: self-reported family income, measured in dollars
. notes income: refusals coded as missing
```

We can review the notes by typing `notes`:

```
. notes
_dta:
    1.  General Social Survey extract for Stata book
income:
    1.  self-reported family income, measured in dollars
    2.  refusals coded as missing
```

If we save the dataset after adding notes, the notes become a permanent part of the dataset.

2.15 Global and local macros

Although macros are most often used when writing ado-files, they are also very useful in do-files. Later in the book, and especially in chapter 9, we use macros extensively. Accordingly, we will discuss them briefly here. Readers with less familiarity with Stata might want to skip this section for now and read it later when macros are used in our examples.

In Stata, you can assign values or strings to *macros*. Whenever Stata encounters the macro name, it automatically substitutes the contents of the macro. For example, pretend that you want to generate a series of two-by-two tables where you want cell percentages, requiring the `cell` option; missing values, requiring the `missing` option; values printed instead of value labels, requiring the `nolabel` option; and the chi-squared test statistic, requiring the `chi2` option. Even if you use the shortest abbreviations, this would require typing ", ce m nol ch" at the end of each `tab` command. Instead, you could use the following command to define a global macro called `myopt`:

```
. global myopt = ", cell miss nolabel chi2 nokey"
```

Then whenever you type `$myopt` (the $ tells Stata that `myopt` is a global macro), Stata substitutes , cell miss nolabel chi2 nokey. If you type

```
. tab lfp wc $myopt
```

Stata interprets this as if you had typed

```
. tab lfp wc, cell miss nolabel chi2 nokey
```

Global macros are "global" because, once they are set, they can be accessed by any do (or ado) program in the current session. The flip side is that the global macros that you are using can be reset by any of the do- or ado-files that you use along the way. By contrast, "local" macros can be accessed only within the do- or ado-file in which they are defined. When the do- or ado-file program terminates, the local macro disappears. We prefer using local macros whenever possible because you do not have to worry about conflicts with other programs or do-files that try to use the same macro name for a different purpose. Local macros are defined using the `local` command, and they are referenced by placing the name of the local macro in single quotes; for example, `myopt`. The two single quote marks use different symbols (on many keyboards, the left single quote is in the upper left-hand corner, whereas the right single quote is next to the Enter key). If the operations we just performed were in a do-file, we could have produced the same output with the following lines:

```
. local opt = ", cell miss nolabel chi2 nokey"
. tab lfp wc `opt`
  (output omitted)
```

Local and global macros can also be used as a shorthand way to refer to lists of variables. For example, you could use these commands to create lists of variables:

```
. local demogvars "age white female"
. local edvars "highsch college graddeg"
```

Then when you run regression models, you could use the command

```
. regress y `demogvars` `edvars`
```

which Stata would translate into

```
. regress y age white female highsch college graddeg
```

Or, you could use the command

```
. regress y `demogvars` `edvars` x1 x2 x3
```

which Stata would translate into

```
. regress y age white female highsch college graddeg x1 x2 x3
```

This technique has several advantages. First, it is easier to write the commands since you do not have to keep retyping a long list of variables. Second, if you change the set of demographic variables that you want to use, you have to do it only in one place, which reduces the chance of errors.

Often when you use a local macro name for a list of variables, the list becomes longer than one line. As with other Stata commands that extend over one line, you can use ///, as in

```
local vars age age squared income education female occupation dadeduc dadocc ///
    momeduc momocc
```

You can also define macros to equal the result of computations. After entering `global four = 2+2`, the value 4 will be substituted for `$four`. Also Stata contains many *macro functions* in which items retrieved from memory are assigned to macros. For example, to display the variable label that you have assigned to the variable `wc`, you can type

```
. global wclabel : variable label wc
. display "$wclabel"
Wife College: 1=yes 0=no
```

We have only scratched the surface of the potential of macros. Macros are immensely flexible and are indispensable for a variety of advanced tasks that Stata can perform. Perhaps most importantly, macros are essential for doing any meaningful Stata programming. If you look at the ado-files for the commands we have written for this book, you will see many instances of macros, and even of macros within macros. For users interested in advanced applications, the `macro` entry in the *Programming Reference Manual* should be read closely.

2.16 Graphics

Stata has a very extensive and powerful graphics system. Not only can you create many different kinds of graphs, but you have a great deal of control over almost all aspects of a graph's appearance. The cost of this is that the syntax for making a graph can get complicated. Here we provide a brief introduction to graphics in Stata, focusing on the types of graphs that we use in later chapters. Our hope is to provide a basic understanding of how the graphics system works so that you can start using it. For more information, we suggest the following. Stata has a manual dedicated to the syntax for graphics. Although we find the *Stata Graphics Reference Manual* to be an invaluable reference when you already have a good understanding of what you want to do, we find it less helpful when you want to be reminded of the way to do something, what an option is called, or to get ideas about what kinds of graph to use. In this regard, we find Mitchell's 2004 *A Visual Guide to Stata Graphics* to be extremely useful. This book shows hundreds of graphs drawn in Stata, along with the commands used to generate them. The book is organized in a way that makes it easy to scan the pictures until you see a graph that does what you want. Then you can look at the text to find out which options to use.

The way we use Stata to make graphs differs from how we use Stata to fit models or do virtually anything else. Namely, when making graphs, we make extensive use of dialog boxes. If you pull down the **Graphics** menu (or type **Alt-g**), you will see a list of the plot types and families of plot types available in Stata:

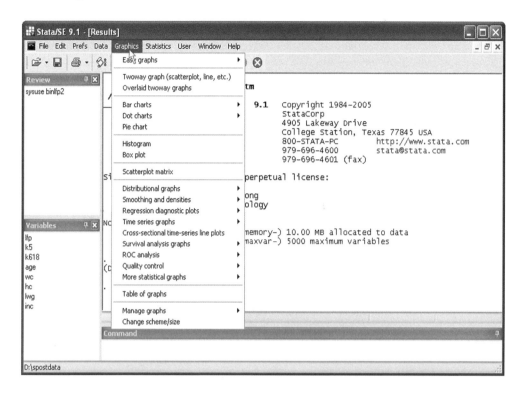

Selecting any of these will call up a dialog box. (The first time you open a dialog box there can be a considerable delay, but it is shorter the next time.) Selecting Twoway graph (scatterplot, line, etc.) from the Graphics menu displays the following:

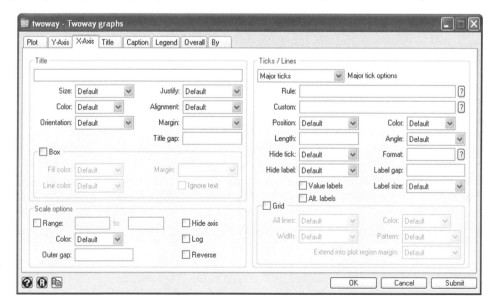

You can make selections from each tab and then click **Submit** or **OK**. (**Submit** leaves the dialog box open, whereas **OK** closes it before generating the graph.) The dialog box will translate your options into the commands Stata uses to draw the graph. These commands are echoed to the Results window, while the graph appears in a Graph window. Next we tweak the options until we have the graph the way we want it. Then we copy the command from the Results window and paste it into a do-file, so that we can reproduce the graph later. We can also edit the do-file to modify the graph.

In the rest of this section, we describe the basic syntax for Stata graphics, because it is helpful to understand how this syntax works even if you ultimately use dialog boxes to do the bulk of the work. We focus on plots of one or more outcomes against a single explanatory variable. For this, we use the commands `graph twoway scatter` and `graph twoway connected`. (In later chapters, we will introduce a few other types of graphs as they are needed.) Namely, we consider plots of one or more outcomes against an independent variable using the command `graph twoway`. The syntax of this command has the form:

`graph twoway` *plottype* ...

The *Stata Graphics Reference Manual* lists 38 different *plottypes* for the `graph twoway` command. Since we discuss only two (`scatter` and `connected`) *plottypes* here, interested readers are encouraged to consult the *Graphics Reference Manual* or to type `help graph` for more information.[5]

Graphs that you create in Stata are drawn in their own window, which should appear on top of the four windows we discussed above. If the Graph window is hidden, you can bring it to the front by clicking on ▓. You can make the Graph window larger or smaller by clicking and dragging the borders.

2.16.1   graph command

The type of graph that we use most often shows how the predicted probability of observing a given outcome changes as a continuous variable changes over a specified range. For example, in chapter 4 we show you how to compute the predicted probability of a woman being in the labor force according to the number of children she has and the family's income. In later chapters, we show you how to compute these predictions, but for now you can simply load them into memory with the command `use lfpgraph2, clear`. The variable `income` is family income measured in thousands of dollars, excluding any contribution made by the woman of the household, whereas the next three variables show the predicted probabilities of working for a woman who has no children under six (`kid0p1`), one child under six (`kid1p1`), or two children under six (`kid2p1`). Because there are only 11 values, we can easily list them:

---

5. We also use a third plottype called called `rarea`, but we will postpone describing that until later.

```
. use http://www.stata-press.com/data/lf2/lfpgraph2, clear
(Sample predictions to plot.)
. list income kid0p1 kid1p1
```

	income	kid0p1	kid1p1
1.	10	.7330963	.3887608
2.	18	.6758616	.3256128
3.	26	.6128353	.2682211
4.	34	.54579	.2176799
5.	42	.477042	.1743927
6.	50	.409153	.1381929
7.	58	.3445598	.1085196
8.	66	.285241	.0845925
9.	74	.2325117	.065553
10.	82	.18698	.0505621
11.	90	.1486378	.0388569

We see that as annual income increases the predicted probability of being in the labor force decreases. Also by looking across any row, we see that for a given level of income the probability of being in the labor force decreases as the number of young children increases. We want to display these patterns graphically.

graph twoway scatter can be used to draw a scatterplot in which the values of one or more y-variables are plotted against values of an x-variable. Here income is the x-variable, and the predicted probabilities kid0p1, kid1p1, and kid2p1 are the y-variables. Thus for each value of x, we have three values of y. In making scatterplots with graph twoway scatter, the y-variables are listed first, and the x-variable is listed last. If we type

```
. graph twoway scatter kid0p1 kid1p1 kid2p1 income, ytitle(Probability)
```

we obtain the following graph:

(Continued on next page)

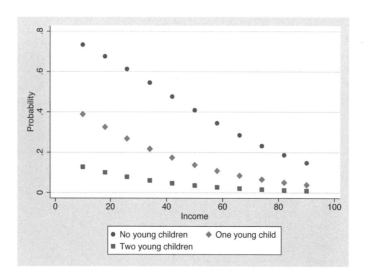

Our simple scatterplot shows the pattern of decreasing probabilities as income or number of children increases.

This simple command produces a reasonable first graph. Indeed, Stata's default settings are usually a good place to begin. Even so, we can make a more effective graph by including more options. Below we focus only on adding titles and changing the labels for the axes. Remember that if you want to change other aspects of the graph, you will almost certainly be able to get what you want if you find the right options. A slightly more detailed syntax[6] for `graph twoway` is

`graph twoway` *plot*$_1$ $\big[$*plot*$_2$$\big]$...$\big[$*plot*$_N$$\big]$ $\big[$*if*$\big]$ $\big[$*in*$\big]$ $\big[$, *twoway_options*$\big]$

where *plot*$_i$ is defined to be

$[$(] *plottype varlist,* $\big[$<u>tit</u>le("*string*") <u>sub</u>title("*string*") <u>ytit</u>le("*string*")
   <u>xtit</u>le("*string*") <u>cap</u>tion("*string*") <u>xlab</u>el(*values*) <u>ylab</u>el(*values*)
   other_options$\big]$ $[$)]

This syntax highlights the fact that it is possible to put multiple plots on the same graph.[7] The plots can be of different plot types. For instance, suppose that we wanted the symbols in the plot corresponding to "No young children" to be connected. This plot type is called `connected`. For example,

---

6. The syntax presented here is incomplete. We wish only to explain the elements that we have found ourselves using in presenting analyses like those in this book. See the *Stata Graphics Reference Manual* for more information.

7. The parentheses are used to separate the different plots when there are multiple plots. When there is only one plot, the parentheses are not required.

```
. graph twoway (connected kid0p1 income)
>               (scatter kid1p1 kid2p1 income), ytitle(Probability)
```

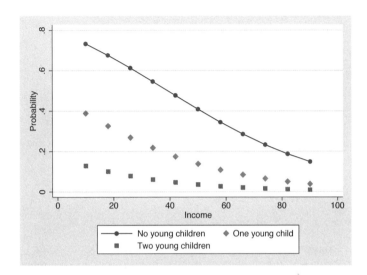

With the exception of the title on the *y*-axis, the default choices for the symbols, line styles, etc., all are all quite nice. Stata made these choices within the context of an overall look or scheme. For example, because our book is published in monochrome, we wanted our graphs to be drawn in monochrome. The *Graphics Reference Manual* describes how they could be changed. (Type `help schemes` in Stata for the latest information about the available schemes.) Users can choose the overall look of their graphs by setting the scheme. In writing this book, we simply included the line

```
. set scheme sj
```

at the top of our do-files.

Adding titles

Now we provide a quick introduction that shows how to set the five titles that we often wish to change: (1) overall title, (2) overall subtitle, (3) *y*-axis title, (4) *x*-axis title, and (5) graph caption. The options for setting each of these five titles are in the syntax diagram above. The command and graph below illustrate how we might use each of these titles.

```
. graph twoway (connected kid0p1 kid1p1 kid2p1 income),
>       ytitle("Probability")
>       title("Predicted Probability of Female LFP")
>       subtitle("(as predicted by logit model)")
>       xtitle("Family income, excluding wife's")
>       caption("Data from 1976 PSID-T Mroz")
```

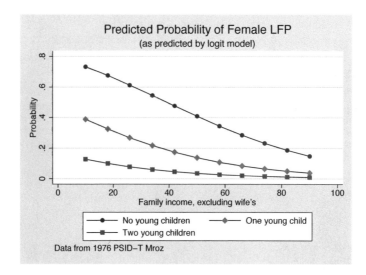

This graph is much more effective in illustrating that the probability of a woman being in the labor force declines as family income increases, and that the differences in predicted probabilities between women with no young children and those with one or two young children are greatest at the lowest levels of income.

Labeling the axes

Even though the defaults are nice, it is common to want to change the labeling of the ticks on the *x*-axis or *y*-axis. The `ylabel()` and `xlabel()` options allow users to specify either a rule or a set of values for the tick marks. A rule is simply a compact way to specify a list of values.

Let's consider specifying a list of values first. A common change is to alter the frequency or range of tick marks. This change can also be made with the `xlabel()` and `ylabel()` options. Suppose that we liked the frequency of the ticks on the *x*-axis but wanted to restrict the range to [10, 90]. We make this change in the command

```
. graph twoway (connected kid0p1 kid1p1 kid2p1 income),
>       ytitle("Probability")
>       title("Predicted Probability of Female LFP")
>       subtitle("(as predicted by logit model)")
>       xtitle("Family income, excluding wife's")
>       caption("Data from 1976 PSID-T Mroz")
>       xlabel(10 20 30 40 50 60 70 80 90)
```

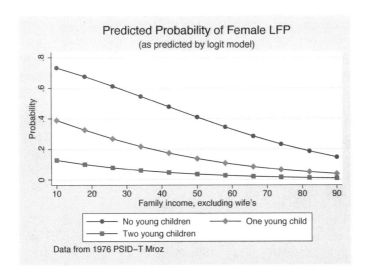

We could have obtained the same graph by specifying a rule for a new set of x-axis values. Although there are several ways to specify a rule,[8] we find the form $\#_1(\#_2)\#_3$ most useful. In this form, the user specifies three numbers: $\#_1$ specifies the beginning of the sequence of values, $\#_2$ specifies the increment between each value, and $\#_3$ specifies the maximum value. For instance, instead of specifying the option

```
xlabel(10 20 30 40 50 60 70 80 90)
```

in the previous graph, we could have specified

```
xlabel(10(10)90)
```

to obtain the same graph.

Naming graphs

When you create a graph, it is displayed in a Graph window and is also saved in memory. Accordingly, when you close the Graph window, you can redisplay the graph with the command **graph display**. By default, a graph is stored in memory with the name **Graph**, and this graph is overwritten whenever you generate a new graph. If you want to store more than one graph in memory (this is not the same as storing them to disk, which is discussed in the next section), you need to use the **name()** option. For example,

```
. scatter y x, name(example1)
```

stores the scatterplot for **y** against **x** in memory with the name **example1**. Then

```
. scatter z x, name(example2)
```

---

8. Type **help axis_label_options** for other ways to specify a rule.

will save the scatterplot for z against x with the name `example2`. Stata displays each named graph in its own window, and multiple Graph windows can be displayed simultaneously. Even if you have closed the Graph windows, you could redisplay the graphs with the commands

```
. graph display example1
. graph display example2
```

In do-files, you might want to use `name(example1, replace)` so that the program will overwrite graph `example1` if it exists.

Saving graphs

Graphs can be either saved in a file or stored in memory. When a graph is saved to a file, it remains there until the file is erased. When a graph is stored in memory, it remains there until you exit Stata or drop the graph from memory. Specifying `saving(`*filename*`, replace)` saves the graph to a file in Stata's proprietary format (indicated by the suffix `.gph`) in the working directory. Including `replace` tells Stata to overwrite a file with that name if it exists. Specifying `name(`*name*`, replace)` stores the graph in memory. The `replace` option tells Stata to replace any existing graphs stored under that name.

Graphs must be either saved to files or stored in memory before you can combine them. For example, if we were to later need the graph we just created, we could store it in memory under the name `graph1` with the command

```
. graph twoway (connected kid0p1 kid1p1 kid2p1 income),
>       ytitle("Probability")
>       title("Predicted Probability of Female LFP")
>       subtitle("(as predicted by logit model)")
>       xtitle("Family income, excluding wife's")
>       caption("Data from 1976 PSID-T Mroz")
>       xlabel(10(10)90) name(graph1, replace)
```

Tip: Exporting graphs to other programs If you are using Windows or Macintosh and want to export graphs to another program, such as a word processor, we find that it works best to save them as a Windows Enhanced Metafile (EMF) in Windows, or as a Macintosh PICT file in Mac OS X. In Windows with the graph currently displayed in the Graph window, you can export it in the EMF format with the command `graph export` *filename*`.emf`. In Mac OS X with the graph currently displayed in the Graph window, you can export it in the PICT format with the command `graph export` *filename*`.pct`. You cannot export a graph to the `.pct` format in Windows, nor can you export a graph to the `.emf` format in Mac OS X; the formats are exclusive to their respective operating systems. If the file is already saved in `.gph` format, you can export it to either `.emf` or `.pct` format in two steps. First, redisplay the graph with the command `graph use` *filename*. Then export the graph with the command `graph export` *filename*`.emf` or `graph export` *filename*`.pct`. The `replace` option can be used with `graph export` to automatically overwrite a graph of the same name, which is useful in do-files.

2.16.2   Displaying previously drawn graphs

There are several commands used for manipulating graphs that have been previously drawn and saved to memory or disk. `graph dir` lists graphs previously saved in memory or to a file in the current working directory. `graph use` copies a graph stored in a file into memory and displays it. `graph display` redisplays a graph stored in memory.

2.16.3   Printing graphs

It is easiest to print a graph once it is in the Graph window. When a graph is in the Graph window, you can print it by selecting File→Print→Graph(*graphname*) from the

menus or by clicking on . You can also print a graph in the Graph window with the command `graph print`. To print a graph saved to memory or disk, first use `graph use` or `graph display` to redisplay it, and then print it with the command `graph print`.

2.16.4   Combining graphs

Multiple graphs that have been saved can be combined. This is useful, for example, when you want to place two graphs side by side or stack them. In chapter 5, we will find it useful to combine two graphs. Here we use two of the graphs that we discuss in detail in section 5.8.6 to illustrate `graph combine`. When we originally drew the graphs, we saved them in memory under the names `graph1` and `graph2`. Now we use `graph combine` to put the two graphs side by side in one Graph window.

```
. graph combine graph1 graph2, imargin(small)
```

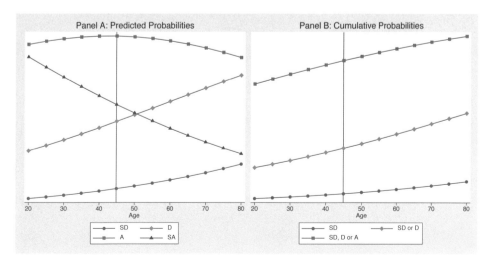

This combined graph is not as effective as one in which the graphs are stacked. The trick is to understand that when multiple graphs are combined, Stata divides the Graph window into an array. The `rows()` and `cols()` options can be used to set the number of rows and columns in the array. Of course, as with most aspects of a graph, the *Graphics Reference Manual* describes how almost any part of the combined graph can be changed.[9] By default, the individual graphs are allocated over the rows in the order in which the filenames are listed in the `graph combine` command.

To display the graphs stacked vertically, specify the `col()` option:

```
. graph combine graph1 graph2, iscale(*.9) imargin(small) col(1)
```

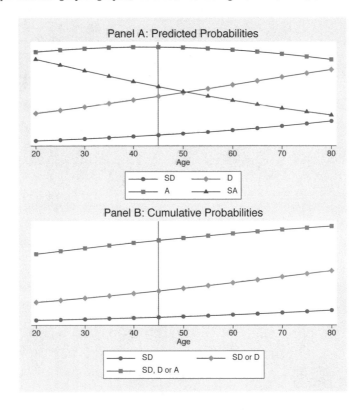

As we described earlier, `graph export` can be used to save the graph as a Windows Enhanced Metafile that can be imported to a word processor or other program. More details on combining graphs can be found in the *Stata Graphics Reference Manual*.

---

9. In particular, see *Advanced use* in [G] **graph combine** for a rather impressive example.

2.17 A brief tutorial

This tutorial uses the `science2.dta` dataset that is available from the book's web site. You can use your own dataset as you work through this tutorial, but you will need to change some of the commands to correspond to the variables in your data. In addition to our tutorial, the *User's Guide* provides a wealth of information for new users.

Opening a log

The first step is to open a log file for recording your results. Remember that all commands are case sensitive. The commands are listed with a period in front, but you do *not* type the period:

```
. capture log close
. log using tutorial, text

  log:  d:\spostdata\tutorial.log
  log type:  text
  opened on:  26 Sep 2005, 11:18:15
```

Loading the data

We assume that `science2.dta` is in your working directory. `clear` tells Stata to "clear out" any existing data from memory before loading the new dataset:

```
. use http://www.stata-press.com/data/lf2/science2, clear
(Note that some of the variables have been artificially constructed.)
```

The message after loading the data reflects that this dataset was created for teaching. Although most of the variables contain real information, some variables have been artificially constructed.

(Continued on next page)

Examining the dataset

`describe` gives information about the dataset.

```
. describe

Contains data from science2.dta
    obs:            308                    Note that some of the variables
                                           have been artificially
                                           constructed.
    vars:            35                    10 Mar 2001 05:51
    size:        17,556 (98.3% of memory free)  (_dta has notes)
```

variable name	storage type	display format	value label	variable label
id	float	%9.0g		ID Number.
cit1	int	%9.0g		Citations: PhD yr -1 to 1.
cit3	int	%9.0g		Citations: PhD yr 1 to 3.
cit6	int	%9.0g		Citations: PhD yr 4 to 6.
cit9	int	%9.0g		Citations: PhD yr 7 to 9.
enrol	byte	%9.0g		Years from BA to PhD.
fel	float	%9.0g		Fellow or PhD prestige.
felclass	byte	%9.0g	prstlb	* Fellow or PhD prestige class.
fellow	byte	%9.0g	fellbl	Postdoctoral fellow: 1=y,0=n.
female	byte	%9.0g	femlbl	Female: 1=female,0=male.
job	float	%9.0g		Prestige of 1st univ job.
jobclass	byte	%9.0g	prstlb	* Prestige class of 1st job.
mcit3	int	%9.0g		Mentor's 3 yr citation.
mcitt	int	%9.0g		Mentor's total citations.
mmale	byte	%9.0g	malelb	Mentor male: 1=male,0=female.
mnas	byte	%9.0g	naslb	Mentor NAS: 1=yes,0=no.
mpub3	byte	%9.0g		Mentor's 3 year publications.
nopub1	byte	%9.0g	nopublb	1=No pubs PhD yr -1 to 1.
nopub3	byte	%9.0g	nopublb	1=No pubs PhD yr 1 to 3.
nopub6	byte	%9.0g	nopublb	1=No pubs PhD yr 4 to 6.
nopub9	byte	%9.0g	nopublb	1=No pubs PhD yr 7 to 9.
phd	float	%9.0g		Prestige of Ph.D. department.
phdclass	byte	%9.0g	prstlb	* Prestige class of Ph.D. dept.
pub1	byte	%9.0g		Publications: PhD yr -1 to 1.
pub3	byte	%9.0g		Publications: PhD yr 1 to 3.
pub6	byte	%9.0g		Publications: PhD yr 4 to 6.
pub9	byte	%9.0g		Publications: PhD yr 7 to 9.
work	byte	%9.0g	worklbl	Type of first job.
workadmn	byte	%9.0g	wadmnlb	Admin: 1=yes; 0=no.
worktch	byte	%9.0g	wtchlb	* Teaching: 1=yes; 0=no.
workuniv	byte	%9.0g	wunivlb	* Univ Work: 1=yes; 0=no.
wt	byte	%9.0g		
faculty	byte	%9.0g	faclbl	1=Faculty in University
jobrank	byte	%9.0g	joblbl	Rankings of University Job.
totpub	byte	%9.0g		Total Pubs in 9 Yrs post-Ph.D.
				* indicated variables have notes

```
Sorted by:
```

Examining individual variables

A series of commands gives us information about individual variables. You can use whichever command you prefer, or all of them.

```
. summarize work
     Variable |       Obs        Mean    Std. Dev.       Min        Max
--------------+--------------------------------------------------------
         work |       302    2.062914     1.37829         1          5
. tabulate work, missing
   Type of |
first job. |      Freq.      Percent        Cum.
-----------+-----------------------------------
   FacUniv |        160        51.95       51.95
   ResUniv |         53        17.21       69.16
    ColTch |         26         8.44       77.60
    IndRes |         36        11.69       89.29
     Admin |         27         8.77       98.05
         . |          6         1.95      100.00
-----------+-----------------------------------
     Total |        308       100.00
. codebook work
```

```
--------------------------------------------------------------------------
work                                                        Type of first job.
--------------------------------------------------------------------------

                    type:   numeric (byte)
                   label:   worklbl

                   range:   [1,5]                        units:  1
           unique values:   5                        missing .:  6/308

               tabulation:   Freq.    Numeric  Label
                              160          1   FacUniv
                               53          2   ResUniv
                               26          3   ColTch
                               36          4   IndRes
                               27          5   Admin
                                6          .
```

Graphing variables

Graphs are also useful for examining data. The command

```
. dotplot work
```

creates the following graph:

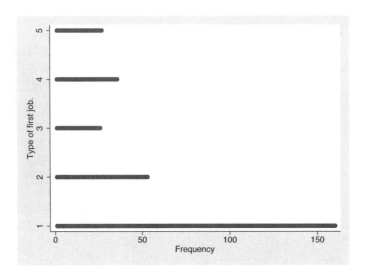

Saving graphs

To save the above graph as a Windows Enhanced Metafile, type

```
. graph export myname.emf, replace
(file d:\spostdata\myname.emf written in Windows Enhanced Metafile format)
```

Adding comments

To add comments to your output, which allows you to document your command files, type * at the beginning of each comment. The comments are listed in the log file:

```
. * saved graph as work.emf
```

Creating a dummy variable

Now let's make a dummy variable with faculty in universities coded 1 and all others coded 0. The command `gen isfac = (work==1) if work<.` generates `isfac` as a dummy variable where `isfac` equals 1 if `work` is 1, else 0. The statement `if work<.` makes sure that missing values are kept as missing in the new variable.

```
. generate isfac = (work==1) if work<.
(6 missing values generated)
```

Six missing values were generated because `work` contained six missing observations.

Checking transformations

One way to check transformations is with a table. In general, it is best to look at the missing values, which requires the `missing` option:

```
. tabulate isfac work, missing
```

			Type of first job.			
isfac	FacUniv	ResUniv	ColTch	IndRes	Admin	Total
0	0	53	26	36	27	142
1	160	0	0	0	0	160
.	0	0	0	0	0	6
Total	160	53	26	36	27	308

	Type of first job.	
isfac	.	Total
0	0	142
1	0	160
.	6	6
Total	6	308

Labeling variables and values

For many of the regression commands, value labels for the dependent variable are essential. We start by creating a variable label, then create `isfac` to store the value labels, and finally assign the value labels to the variable `isfac`:

```
. label variable isfac "1=Faculty in University"
. label define isfac 0 "NotFac" 1 "Faculty"
. label values isfac isfac
```

Then we can get labeled output:

```
. tabulate isfac
```

1=Faculty in University	Freq.	Percent	Cum.
NotFac	142	47.02	47.02
Faculty	160	52.98	100.00
Total	302	100.00	

Creating an ordinal variable

The prestige of graduate programs is often referred to using the categories of adequate, good, strong, and distinguished. Here we create such an ordinal variable from the continuous variable for the prestige of the first job. `missing` tells Stata to show cases with missing values.

```
. tab job, missing
Prestige of |
   1st univ |
       job. |     Freq.       Percent        Cum.
------------+-----------------------------------
       1.01 |         1          0.32        0.32
        1.2 |         1          0.32        0.65
       1.22 |         1          0.32        0.97
       1.32 |         1          0.32        1.30
       1.37 |         1          0.32        1.62
(output omitted )
       3.97 |         6          1.95       48.38
       4.18 |         2          0.65       49.03
       4.42 |         1          0.32       49.35
        4.5 |         6          1.95       51.30
       4.69 |         5          1.62       52.92
          . |       145         47.08      100.00
------------+-----------------------------------
      Total |       308        100.00
```

The `recode` command makes it easy to group the categories from `job`. Of course, we then label the variable:

```
. generate jobprst = job
(145 missing values generated)
. recode jobprst .=. 1/1.99=1 2/2.99=2 3/3.99=3 4/5=4
(jobprst: 162 changes made)
. label variable jobprst "Rankings of University Job"
. label define prstlbl 1 "Adeq" 2 "Good" 3 "Strong" 4 "Dist"
. label values jobprst prstlbl
```

Here is the new variable (we use the `missing` option so that missing values are included in the tabulation):

```
. tabulate jobprst, missing
 Rankings of |
  University |
        Job |     Freq.       Percent        Cum.
------------+-----------------------------------
       Adeq |        31         10.06       10.06
       Good |        47         15.26       25.32
     Strong |        71         23.05       48.38
       Dist |        14          4.55       52.92
          . |       145         47.08      100.00
------------+-----------------------------------
      Total |       308        100.00
```

Combining variables

Now we create a new variable by summing existing variables. If we add pub3, pub6, and pub9, we can obtain the scientist's total number of publications over the 9 years following receipt of the Ph.D.

```
. generate pubsum = pub3 + pub6 + pub9
. label variable pubsum "Total Pubs in 9 Yrs post-Ph.D."
. summarize pub3 pub6 pub9 pubsum
```

Variable	Obs	Mean	Std. Dev.	Min	Max
pub3	308	3.185065	3.908752	0	31
pub6	308	4.165584	4.780714	0	29
pub9	308	4.512987	5.315134	0	33
pubsum	308	11.86364	12.77623	0	84

A scatterplot matrix graph can be used to plot all pairs of variables simultaneously:

```
. graph matrix pub3 pub6 pub9 pubsum, half msymbol(smcircle_hollow)
```

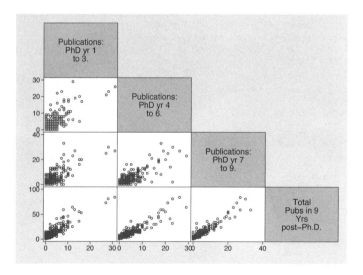

Saving the new data

After you make changes to your dataset, save the data with a new filename:

```
. save sciwork, replace
file sciwork.dta saved
```

Closing the log file

Last, we need to close the log file so that we can refer to it in the future.

```
. log close
    log:  d:\spostdata\tutorial.log
    log type:  text
    closed on:  26 Sep 2005, 11:18:27
```

A batch version

If you have read section 2.9, you know that a better idea is to create a batch (do-) file, perhaps called `tutorial.do`:[10]

```
// batch version of tutorial do-file
version 9
set scheme sj
set more off
capture log close
log using ch2tutorial, replace

// loading the data
use http://www.stata-press.com/data/lf2/science2, clear

// examining the dataset
describe

// examining individual variables
summarize work
tabulate work, missing
codebook work

// graphing variables
dotplot work

// saving graph
graph export 02dotplot2.emf, replace

// creating a dummy variable
gen isfac = (work==1) if work<.

// checking transformations
tabulate isfac work, missing

// labeling variables and values
label variable isfac "1=Faculty in University"
label define isfac 0 "NotFac" 1 "Faculty"
label values isfac isfac
tabulate isfac

// creating an ordinal variable
tabulate job, missing
generate jobprst=job
recode jobprst .=. 1/1.99=1 2/2.99=2 3/3.99=3 4/5=4
label variable jobprst "Rankings of University Job"
label define prstlbl 1 "Adeq" 2 "Good" 3 "Strong" 4 "Dist"
label values jobprst prstlbl
tabulate jobprst, missing
```

---

10. If you download this file from our web site, it is called `st9ch2tutorial.do`.

```
// combining variables
generate pubsum = pub3 + pub6 + pub9
label variable pubsum "Total Pubs in 9 Yrs post-Ph.D."
summarize pub3 pub6 pub9 pubsum

// graphing variables
graph matrix pub3 pub6 pub9 pubsum, half msymbol(smcircle_hollow)

// saving graph
graph export 02matrix.emf, replace

// saving the new data
note: temporary dataset for st9ch2tutorial.do
save sciwork, replace

// close the log
log close
```

Then type do `tutorial` in the Command window or select File→Do... from the menu.

3 Estimation, testing, fit, and interpretation

Our book deals with what we think are the most fundamental and useful cross-sectional regression models for categorical and count outcomes: binary logit and probit, ordinal logit and probit, multinomial and conditional logit, Poisson regression, negative binomial regression, and zero-inflated models for counts.[1] We also explore several less common models, such as the stereotype logistic regression model, rank-ordered logit, multinomial probit, and zero-truncated count models. Although these models differ in many respects, they share common features:

1. Each model is fitted by maximum likelihood.
2. The estimates can be tested with Wald and likelihood-ratio (LR) tests.
3. Measures of fit can be computed.
4. Models can be interpreted by examining predicted values of the outcome.

Because of these similarities, the same principles and commands can be applied to each model. Coefficients can be listed with `listcoef`. Wald and likelihood-ratio tests can be computed with `test` and `lrtest`. Measures of fit can be computed with `fitstat`. Stata's `estat` command can be used to display a variety of additional postestimation statistics. And, our SPost suite of postestimation commands can assist with the interpretation of predictions.

Building on the overview that this chapter provides, later chapters focus on the application of these principles and commands to exploit the unique features of each model. Also this chapter serves as a reference for the syntax and options for the SPost commands that we introduce here. Accordingly, we encourage you to read this chapter before proceeding to the chapter that covers the models of greatest interest to you.

(Continued on next page)

---

1. Although many of the principles and procedures discussed in this book apply to panel models, such as fitted by Stata's `xt` commands, or the multiple-equation systems, such as `biprobit` or `treatreg`, these models are not considered here.

3.1   Estimation

Each of the models that we consider is fitted by maximum likelihood (ML). [2] ML estimates are the values of the parameters that have the greatest likelihood (i.e., the *maximum likelihood*) of generating the observed sample of data if the assumptions of the model are true. To obtain the ML estimates, a *likelihood function* calculates how likely it is that we would observe the data we actually observed if a given set of parameter estimates were the true parameters. For example, in linear regression with one independent variable, we need to estimate both the slope, β, and the intercept, α (for simplicity, we are ignoring the parameter σ^2). For any combination of possible values for α and β, the likelihood function tells us how likely it is that we would have observed the data that we did observe if these values were the true population parameters. If we imagine a surface in which the range of possible values of α makes up one axis and the range of β makes up another axis, the resulting graph of the likelihood function would look like a hill, and the ML estimates would be the parameter values corresponding to the top of this hill. The variance of the estimates corresponds roughly to how quickly the slope is changing near the top of the hill.

For all but the simplest models, the only way to find the maximum of the likelihood function is by numerical methods. *Numerical methods* are the mathematical equivalent of how you would find the top of a hill if you were blindfolded and knew only the slope of the hill at the spot where you are standing and how the slope at that spot is changing (which you could figure out by poking your foot in each direction). The search begins with start values corresponding to your location as you start your climb. From the start position, the slope of the likelihood function and the rate of change in the slope determine the next guess for the parameters. The process continues to *iterate* until the maximum of the likelihood function is found, called *convergence*, and the resulting estimates are reported. Advances in numerical methods and computing hardware have made estimation by numerical methods routine.

3.1.1   Stata's output for ML estimation

The process of iteration is reflected in the initial lines of Stata's output. Consider the first lines of the output from the logit model of labor force participation that we use as an example in chapter 4:

```
. logit lfp k5 k618 age wc hc lwg inc

Iteration 0:    log likelihood =  -514.8732
Iteration 1:    log likelihood = -454.32339
Iteration 2:    log likelihood = -452.64187
Iteration 3:    log likelihood = -452.63296
Iteration 4:    log likelihood = -452.63296

Logistic regression                          Number of obs   =       753
     (output omitted )
```

---

2. Often there are convincing reasons for using Bayesian or exact methods for the estimation of these models. However, these methods are not generally available and hence are not considered here.

The output begins with the iteration log, where the first line reports the value of the *log* likelihood at the start values, reported as iteration 0. Whereas earlier we talked about maximizing the likelihood function, in practice, programs maximize the log of the likelihood, which simplifies the computations and yields the same result. For the probability models considered in this book, the log likelihood is always negative, because the likelihood itself is always between 0 and 1. Here the log likelihood at the start is -514.8732. The next four lines in this example show the progress in maximizing the log likelihood, converging to the value of -452.63296. The rest of the output is discussed later in this section.

3.1.2   ML and sample size

Under the usual assumptions (see Cramer 1986 or Eliason 1993 for specific details), the ML estimator is consistent, efficient, and asymptotically normal. These properties hold as the sample size approaches infinity. Although ML estimators are not necessarily bad estimators in small samples, the small-sample behavior of ML estimators for the models we consider is largely unknown. Except for the logit and Poisson regression, which can be fitted using exact permutation methods with LogXact (Cytel Corporation 2005), alternative estimators with known small-sample properties are generally not available. With this in mind, Long (1997, 54) proposed the following guidelines for the use of ML in small samples:

> It is risky to use ML with samples smaller than 100, while samples over 500 seem adequate. These values should be raised depending on characteristics of the model and the data. First, if there are many parameters, more observations are needed .... A rule of at least 10 observations per parameter seems reasonable .... This does not imply that a minimum of 100 is not needed if you have only two parameters. Second, if the data are ill-conditioned (e.g., independent variables are highly collinear) or if there is little variation in the dependent variable (e.g., nearly all the outcomes are 1), a larger sample is required. Third, some models seem to require more observations (such as the ordinal regression model or the zero-inflated count models).

3.1.3   Problems in obtaining ML estimates

Although the numerical methods used by Stata to compute ML estimates are highly refined and generally work extremely well, you can encounter problems. If your sample size is adequate, but you cannot get a solution or appear to get the wrong solution (i.e., the estimates do not make substantive sense), the most common cause is that the data have not been properly "cleaned". In addition to mistakes in constructing variables and selecting observations, the scaling of variables can cause problems. The larger the ratio between the largest and smallest standard deviations among variables in the model, the more problems you are likely to encounter with numerical methods due to rounding. For example, if income is measured in units of $1, income is likely to have a very large standard deviation relative to other variables. Recoding income to units of \$1,000 can solve the problem. For a more technical discussion of issues related to scaling, see Drukker and Wiggins (2004).

Overall, however, numerical methods for ML estimation work well when your model is appropriate for your data. Still, Cramer's (1986, 10) advice about the need for care in estimation should be taken seriously:

> Check the data, check their transfer into the computer, check the actual computations (preferably by repeating at least a sample by a rival program), and always remain suspicious of the results, regardless of the appeal.

3.1.4   Syntax of estimation commands

All single-equation estimation commands have the same syntax:[3]

command depvar [indepvars] [if] [in] [weight] [, options]

Elements in brackets [ ] are optional. Here are a few examples for a `logit` model with `lfp` as the dependent variable:

```
. logit lfp k5 k618 age wc lwg
  (output omitted )
. logit lfp k5 k618 age wc lwg if hc == 1
  (output omitted )
. logit lfp k5 k618 age wc lwg [pweight=wgtvar]
  (output omitted )
. logit lfp k5 k618 age wc lwg if hc == 1, level(90)
  (output omitted )
```

You can review the output from the last estimation by typing the command name again. For example, if the most recent model that you fitted was a logit model, you could have Stata replay the results by simply typing `logit`.

Variable lists

depvar is the dependent variable. *indepvars* is a list of the independent variables. If no independent variables are given, a model with only the intercept is fitted. Stata automatically corrects some mistakes in specifying independent variables. For example, if you include `wc` as an independent variable when the sample is restricted to a single value of `wc` (e.g., `logit lfp wc k5 k618 age hc if wc==1`), Stata drops `wc` from the list of variables. Or, suppose that you recode a k-category variable into a set of k dummy variables. Recall that one of the dummy variables must be excluded to avoid perfect

---

3. `mlogit` is a multiple-equation estimation command, but the syntax is the same as that for single-equation commands because the independent variables included in the model are the same for all equations. The zero-inflated count models `zip` and `zinb` are the only multiple-equation commands considered in our book where different sets of independent variables can be used in each equation. Details on the syntax for these models are given in chapter 8.

collinearity. If you included all k dummy variables in *indepvars*, Stata automatically excludes one of them.

Specifying the estimation sample

`if` and `in` restrictions can be used to define the estimation sample (i.e., the sample used to fit the model), where the syntax for `if` and `in` conditions follows the guidelines in chapter 2. For example, if you want to fit a logit model only for women who went to college, you could specify `logit lfp k5 k618 age hc lwg if wc==1`.

Missing data

Estimation commands use *listwise deletion* to exclude cases in which there are missing values for any of the variables *in the model*. Accordingly, if two models are fitted using the same dataset but have different sets of independent variables, it is possible to have different samples. The easiest way to understand this is with a simple example.[4] Suppose that among the 753 cases in the sample, 23 have missing data for at least one variable. If we fitted a model using all variables, we would get

```
. logit lfp k5 k618 age wc hc lwg inc, nolog
Logistic regression                          Number of obs   =        730
  (output omitted )
```

Suppose that seven of the missing cases were missing only for `k618` and that we fit a second model that excludes this variable:

```
. logit lfp k5 age wc hc lwg inc, nolog
Logistic regression                          Number of obs   =        737
  (output omitted )
```

The estimation sample for the second model has increased by seven cases. Thus you cannot compute a likelihood-ratio test comparing the two models (see section 3.3), and any changes in the estimates could be due either to changes in the model specification or to the use of different samples to fit the models. *When you compare coefficients across models, you want the samples to be exactly the same.* If they are not, you cannot compute likelihood-ratio tests, and any interpretations of why the coefficients have changed must take into account differences between the samples.

Although Stata uses listwise deletion when fitting models, *this does not mean that this is the only or the best way to handle missing data.* Even though the complex issues related to missing data are beyond the scope of our discussion (see Little and Rubin 1987; Schafer 1997; Allison 2001), we recommend that you make explicit decisions about which cases to include in your analyses, rather than let cases be dropped implicitly. We wish that Stata would issue an error rather than automatically dropping cases.

---

4. This example uses `binlfp2.dta`, which has no missing data. We have artificially created missing data. Remember that all our examples are available from
http://www.indiana.edu/~jslsoc/spost.htm or can be obtained by typing `net search spost`.

The `mark` and `markout` commands make it simple to explicitly exclude missing data. `mark` *markvar* generates a new variable *markvar* that equals 1 for all cases. `markout` *markvar varlist* changes the values of *markvar* from 1 to 0 for any cases in which values of any of the variables in *varlist* are missing. The following example illustrates how this works (missing data were artificially created):

```
. mark nomiss
. markout nomiss lfp k5 k618 age wc hc lwg inc
. tab nomiss
```

nomiss	Freq.	Percent	Cum.
0	23	3.05	3.05
1	730	96.95	100.00
Total	753	100.00	

```
. logit lfp k5 k618 age wc hc lwg inc if nomiss==1, nolog
```

Logistic regression	Number of obs	=	730

  (*output omitted*)

```
. logit lfp k5 age wc hc lwg inc if nomiss==1, nolog
```

Logistic regression	Number of obs	=	730

  (*output omitted*)

Because the `if` condition excludes the same cases from both equations, the sample size is the same for both models. After using `mark` and `markout`, we also could have used `drop if nomiss==0` to delete observations with missing values.

Missing data with misschk

Although `mark` and `markout` work well for determining which observations have missing values for a set of variables, these commands do not provide information on the patterns of missing data among these variables. For this, we have written the command `misschk` that quickly tells you which variables have missing data and how missing data are clustered among variables. For a specified set of variables, `misschk` provides the following information:

1. How many observations have no missing values? How many have missing data for one variable? How many have missing data for any given number of variables?

2. What percentage of cases are missing for each variable?

3. What patterns of missing data are there among the variables? For example, do missing values on one variable tend to occur when there are missing values for some other variable?

The easiest way to explain the command is with an example:

```
. use http://www.stata-press.com/data/lf2/gsskidvalue2, clear
(1993 and 1994 General Social Survey)
. misschk age anykids black degree female kidvalue othrrace year income91
> income, gen(m_) dummy help
```

The `misschk` command indicates that we want to look for missing data among the variables `age`, `anykids`, `black`, `degree`, `female`, `kidvalue`, `othrrace`, `year`, and `income91`. The `help` option requests detailed information explaining the output. The output begins like this:

```
Variables examined for missing values
    #  Variable      # Missing   % Missing
   ------------------------------------------
    1  age               0          0.0
    2  anykids          14          0.3
    3  black             0          0.0
    4  degree           14          0.3
    5  female            0          0.0
    6  kidvalue       1609         35.0
    7  othrrace          0          0.0
    8  year              0          0.0
    9  income91          0          0.0
   10  income          495         10.8
```

Each variable is assigned a number, with `age` assigned 1, `anykids` assigned 2, and so on. For each number, `misschk` reports the number and percentage of observations with missing values. For example, 495 cases representing 10.8% of the sample had missing data for `income`. The variable numbers are used to construct strings describing the combinations of missing values across all variables. Each column in a string represents the variable with the corresponding number. Column 1 corresponds to variable 1, column 2 to variable 2, and so on. An underscore (_) in a column indicates that the variable for that column has no missing data. A variable with missing data for at least one observation is indicated by the ones digit of the number assigned to that variable. For example, the string _2_4_ 6___0 indicates that data are missing for variables 2, 4, 6, and 10 (variable 10 is represented by 0), whereas there are no missing data for variables 1, 3, 5, 7, 8, and 9. For our example, we find

(Continued on next page)

```
The columns in the table below correspond to the # in the table above.
If a column is blank, there were no missing cases for that variable.
```

Missing for which variables?	Freq.	Percent	Cum.
_2_4_ 6___0	3	0.07	0.07
_2___ 6___0	6	0.13	0.20
_2___ 6____	3	0.07	0.26
_2___ ____0	1	0.02	0.28
_2___ _____	1	0.02	0.30
___4_ 6___0	1	0.02	0.33
___4_ 6____	1	0.02	0.35
___4_ ____0	5	0.11	0.46
___4_ _____	4	0.09	0.54
_____ 6___0	185	4.02	4.57
_____ 6____	1,410	30.67	35.23
_____ ____0	294	6.39	41.63
_____ _____	2,684	58.37	100.00
Total	4,598	100.00	

```
Table indicates the number of variables for which an observation
has missing data.
```

The first line in the table indicates that three observations have missing data for variables 2, 4, 6, and 10 (represented by string _2_4_ 6___0). Six cases have missing data for variables 2, 6, and 10 (represented by string _2___ 6___0). The last row in the table with the string _____ _____ indicates that 2,684 observations (58%) have no missing data. Next `misschk` tells you how many cases have missing data for how many variables:

Missing for how many variables?	Freq.	Percent	Cum.
0	2,684	58.37	58.37
1	1,709	37.17	95.54
2	195	4.24	99.78
3	7	0.15	99.93
4	3	0.07	100.00
Total	4,598	100.00	

This shows that 58% of the cases have no missing data, 37% have missing data on one variable, 4% have missing data for two variables, and so on.

`misschk` can also generate new variables that provide information about missing data for each observation. For example, the option `generate(m_)` creates two variables. `m_number` is the number of variables from the variable list for which a given observation has missing data. For example, if `m_number` is 3 for observation 1, it means that observation 1 has missing data for three of the variables from the variable list. The second variable that is created is `m_pattern`, which is a string variable that shows which variables have missing data for each observation. For example, if `m_pattern` contained

_2___ 6___0 for the first observation, it means that this observation has missing data for variables 2, 6, and 10.

The `dummy` option generates a dummy variable for each variable in the variable list. Here `misschk` would generate the variables `m_age`, `m_anykids`, ..., and `m_income`. `m_age` is 0 for all cases since `age` has no missing data. `m_anykids` equals 1 for those 14 cases where `anykids` is missing and equals 0 for other cases. We can use these dummy variables to informally examine whether missing data are associated with other variables. For example, the variable `m_income` is 1 when `income` is missing for a given observation, else 0. A binary logit can be estimated using basic demographics as predictors:

```
. logit m_income female black othrrace age, nolog
Logistic regression                             Number of obs   =       4598
                                                LR chi2(4)      =     100.88
                                                Prob > chi2     =     0.0000
Log likelihood = -1520.1691                     Pseudo R2       =     0.0321
```

m_income	Coef.	Std. Err.	z	P>|z|	[95% Conf. Interval]	
female	.4777436	.1023332	4.67	0.000	.2771742	.6783129
black	.5148555	.1308429	3.93	0.000	.2584081	.7713028
othrrace	.2518131	.2355403	1.07	0.285	-.2098374	.7134636
age	.0210619	.0026597	7.92	0.000	.0158489	.0262748
_cons	-3.521194	.1612841	-21.83	0.000	-3.837305	-3.205083

Although this is only an informal assessment of the missing data, it suggests that missing data on `income` are associated with being female, black, and older.

The full syntax for the command is

`misschk` *varlist* [*if*] [*in*] [, generate(*rootname*) replace dummy nosort help]

where *varlist* is the list of variables to be examined. By default, it is all the variables in the dataset. *if* and *in* restrict the cases to be examined.

The description of each option is as follows:

generate(*rootname*) creates two variables that describe the missing data for each observation. *rootname*number is the number of variables from the variable list for which a given observation has missing data. For example, if *rootname*number is 3 for observation 1, it means that observation 1 has missing data for three of the variables from the variable list. *rootname*pattern is a string variable that shows which variables have missing data for each observation. For example, _2___ 6___0 shows that there are missing data for variables 2, 6, and 10.

replace replaces existing variables that begin with *rootname* if they exist.

dummy generates a dummy variable corresponding to each variable in the variable list. Each dummy variable begins with the root specified with the generate() option and then adds the name of the variable (e.g., m_income). A value of 1 indicates missing data for that case; 0 indicates data are not missing. For example, with the options generate(m_) dummy, variables such as m_female and m_income would be generated.

nosort specifies that the list of patterns of missing data should not be sorted to list the most common patterns first.

help requests a description of each part of the output.

Postestimation commands and the estimation sample

Excepting predict, the postestimation commands for testing, assessing fit, and making predictions that are discussed below use only observations from the estimation sample, unless you specify otherwise. Accordingly, you do not need to worry about if and in conditions or cases deleted because of missing data when you use these commands. More details are given below.

Weights

Weights indicate that some observations should be given more weight than others when computing estimates. The syntax for specifying weights is [*type=varname*], where the brackets [ ] are part of the command, *type* is the abbreviation for the type of weight to be used, and *varname* is the weighting variable. Stata recognizes four types of weights:

1. fweights, or frequency weights, indicate that an observation represents multiple observations with *identical* values. For example, if an observation has an fweight of 5, this is equivalent to having 5 identical, duplicate observations. In large datasets, fweights can substantially reduce the size of the data file. If you do not include a weight option in your estimation command, this is equivalent to specifying fweight=1.

2. pweights, or sampling weights, denote the inverse of the probability that the observation is included because of the sampling design. For example, if a case has a pweight of 1200, that case represents 1200 observations in the population.

3. aweights, or analytic weights, are inversely proportional to the variance of an observation. The variance of the jth observation is assumed to be σ^2/w_j, where w_j is the analytic weight. Analytic weights are used most often when observations are averages and the weights are the number of elements that gave rise to the average. For example, if each observation is the cell mean from a larger dataset, the data are heteroskedastic, because we would expect the variance of these means to decrease as the number of observations used to calculate them increases. For some estimation problems, analytic weights can be used to transform the data to reinstate the homoskedasticity assumption.

4. `iweights`, or importance weights, have no formal statistical definition. They are used by programmers to facilitate certain types of computations under specific conditions.

The use of weights is a complex topic, and it is easy to apply weights incorrectly. If you need to use weights, we encourage you to read the detailed discussion in the *Stata User's Guide* ([U] **11.1.6 weight** and [U] **20.16 Weighted estimation**). Winship and Radbill (1994) also provide a useful introduction to weights in the linear regression model.

---

svy estimators For more complex sampling designs that include sampling weights, strata, and primary sampling unit (PSU) identifier variables, Stata allows you to fit many models using the prefix `svy`. For example, `svy: logit` fits a binary logit model with corrections for a complex sampling design. Type `help survey` for a current list of Stata's estimation commands that are supported by the `svy` prefix. Unfortunately, our SPost commands do not work when the `svy` prefix is specified. For our own work, when we want to use the postestimation techniques described in this book for models fitted using `svy`, we refit the model without the `svy` prefix and specify the appropriate `pweight` and `cluster()` variables.[5] The coefficients should be the same, but the standard errors will differ. You can then use `prvalue` and other SPost commands, but the confidence intervals for these predictions are not valid. For more details, type `help svy`. Other sources of information include Hosmer and Lemeshow (2000, chapter 6), Cameron and Trivedi (2005, chapter 24), Wooldridge (2002, chapter 17), the *Stata Survey Data Reference Manual*, and [U] **20.16.3 Sampling weights**.

Options

The following options apply to most regression models. Unique options for specific models are considered in later chapters.

`noconstant` constrains the intercept to equal 0. For example, in a linear regression the command `regress y x1 x2, noconstant` would fit the model $y = \beta_1 x_1 + \beta_2 x_2 + \varepsilon$.

`nolog` suppresses the iteration history. Although this option shortens the output, the iteration history might contain information that indicates problems with your model. If you use this option and you have problems obtaining estimates, it is a good idea to refit the model without this option and with the `trace` option.

`trace` lets you see the values of the parameters for each step of the iteration. This can be useful for determining which variables may be causing a problem if your model has difficulty converging.

---

5. `cluster()` does not affect the coefficient estimates, but it provides more conservative standard errors and confidence intervals.

`level(#)` specifies the level of the confidence interval. By default, Stata provides 95% confidence intervals for estimated coefficients, meaning that the interval around the estimated $\widehat{\beta}$ would capture the true value of β 95% of the time if repeated samples were drawn. `level` allows you to specify other intervals. For example, `level(90)` specifies a 90% interval. You can also change the default level with the command `set level 90` (for 90% confidence intervals).

`cluster(varname)` specifies that the observations are independent across the clusters that are defined by unique values of *varname* but are not necessarily independent within clusters. Specifying this option leads to robust standard errors, as discussed below, with one more correction for the effects of clustered data. See Hosmer and Lemeshow (2000, section 8.3) for a detailed discussion of logit models with clustered data.

Sometimes observations share similarities that violate the assumption of independent observations. For example, the same person might provide information at more than one point in time. Or, several members of the same family might be in the sample, again violating independence. In these examples, it is reasonable to assume that the observations within the groups, which are known as clusters, are not independent. With clustering, the usual standard errors will be incorrect.

`robust` replaces the traditional standard errors with robust standard errors, which are also known as Huber, White, or sandwich standard errors. These estimates are considered robust in the sense that they provide correct standard errors in the presence of violations of the assumptions of the model. For example, if the correct model is a binary logit model and a binary probit model is used, the model has been misspecified. The estimates obtained by fitting a logit model cannot be maximum likelihood estimates because an incorrect likelihood function is being used (i.e., a logistic probability density is used instead of the correct normal density). In this situation, the estimator is referred to by White (1982) as a *minimum ignorance estimator* because the estimators provide the best possible approximation to the true probability density function. When a model is misspecified in this way, the usual standard errors are incorrect. Arminger (1995) makes a compelling case for why robust standard errors should be used. He writes: "If one keeps in mind that most researchers misspecify the model ..., it is obvious that their estimated parameters can usually be interpreted only as minimum ignorance estimators and that the standard errors and test statistics may be far away from the correct asymptotic values, depending on the discrepancy between the assumed density and the actual density that generated the data". However, we have seen no information on the small-sample properties of robust standard errors for nonlinear models (i.e., how well these standard errors work in finite samples). Long and Ervin (2000) consider this problem in the context of the linear regression model, where they found that two small-sample versions of the robust standard error work quite well, whereas the asymptotic version often does *worse* than the usual standard errors.[6]

---

6. These versions can be computed by using the `hc2` or `hc3` options for `regress`. Long and Ervin (2000) recommend using `hc3`.

Robust estimators are automatically used with the `svy` commands and with the `cluster()` option. See [U] **20.14 Obtaining robust variance estimates** in the *Stata User's Guide*, and see Gould, Pitblado, and Sribney (2006, 1.2.5 Robust variance estimates) for a detailed discussion of how robust standard errors are computed in Stata; see Arminger (1995, 111–113) for a more mathematical treatment.

`vce(`*vcetype*`)` specifies how the variance–covariance matrix is to be estimated. One option, `vce(robust)`, is equivalent to specifying `robust`. The `robust` option cannot be used if a different *vcetype* is specified using `vce()`. Other options that you can specify for *vcetype* include `bootstrap`, which will estimate the variance–covariance matrix by bootstrap (which involves repeated re-estimation on samples drawn with replacement from the original estimation sample) or `jackknife` to estimate the variance–covariance matrix with the jackknife method (which involves refitting the model _N times, each time leaving out a single observation). Type `help vce_option` for further details.

3.1.5   Reading the output

We have already discussed the iteration log, so in the following example we suppress it with the `nolog` option. Here we consider other parts of the output from estimation commands. Although the sample output is from `logit`, our discussion generally applies to other regression models.

```
. use http://www.stata-press.com/data/lf2/binlfp2, clear
(Data from 1976 PSID-T Mroz)

. logit lfp k5 k618 age wc hc lwg inc, nolog
Logistic regression                             Number of obs   =        753
                                                LR chi2(7)      =     124.48
                                                Prob > chi2     =     0.0000
Log likelihood = -452.63296                     Pseudo R2       =     0.1209
```

lfp	Coef.	Std. Err.	z	P>|z|	[95% Conf. Interval]	
k5	-1.462913	.1970006	-7.43	0.000	-1.849027	-1.076799
k618	-.0645707	.0680008	-0.95	0.342	-.1978499	.0687085
age	-.0628706	.0127831	-4.92	0.000	-.0879249	-.0378162
wc	.8072738	.2299799	3.51	0.000	.3565215	1.258026
hc	.1117336	.2060397	0.54	0.588	-.2920969	.515564
lwg	.6046931	.1508176	4.01	0.000	.3090961	.9002901
inc	-.0344464	.0082084	-4.20	0.000	-.0505346	-.0183583
_cons	3.18214	.6443751	4.94	0.000	1.919188	4.445092

Header

1. `Log likelihood = -452.63296` corresponds to the value of the log likelihood at convergence.

2. `Number of obs` is the number of observations, excluding those with missing values and after any `if` or `in` conditions have been applied.

3. `LR chi2(7)` is the value of a likelihood-ratio chi-squared for the test of the null hypothesis that all the coefficients associated with independent variables are simultaneously equal to zero. The p-value is indicated by `Prob > chi2`. The number in parentheses is the number of coefficients being tested.

4. `Pseudo R2` is the measure of fit also known as McFadden's R^2. Details on how this measure is computed are given below, along with a discussion of alternative measures of fit.

Estimates and standard errors

1. The left column lists the variables in the model, with the dependent variable listed at the top. The independent variables are always listed in the same order as they were entered on the command line. The constant, labeled `_cons`, is last.

2. Column `Coef.` contains the ML-point estimates.

3. Column `Std. Err.` contains the standard errors of the estimates.

4. The resulting z-test, equal to the estimate divided by its standard error, is labeled z with the two-tailed significance level listed as `P > | z |`. A significance level listed as 0.000 means that $P < .0005$ (for example, .0006 is rounded to .001, whereas .00049 is rounded to .000).

5. The end points of the confidence interval for each estimate are listed under `[95% Conf. Interval]`.

Confidence intervals

Instead of testing a specific hypothesis (e.g., $H_0: \pi = 0$) about a point estimate (e.g., $\hat{\pi}$), we can estimate a confidence interval that contains the true parameter with some chosen probability, known as the confidence level. For a given confidence level, the estimated upper and lower bounds define the confidence interval. Similar to point estimators, the upper and lower bounds of a confidence interval are random variables whose estimates vary from sample to sample.

Consider the problem of estimating the probability of labor force participation for an average woman who went to college. Suppose that the estimated lower and upper bounds of the 95% confidence interval are .633 and .786, respectively. To understand what this means, imagine that we take repeated samples from our population and that, for each sample, we estimate the upper and lower bounds of the confidence interval for that probability. About 95% of these confidence intervals would contain the true probability of labor force participation for an average woman who went to college.

Confidence intervals are reported in many ways, almost always accompanying a point estimate. For example, the predicted probability of labor force participation is .71 and a 95% confidence interval is (.63, .79). Sometimes an author might simply report the point estimate followed by the confidence interval at the 95% level. For instance, the estimated probability of labor force participation is .71 (.63, .79). Or, we might say:

the results suggest that the predicted probability of labor force participation is .71 and the probability could be as small as .63 or as large as .79 with 95% confidence.

3.1.6 Storing estimation results

Stata's `estimates` command provides the facility to store estimation results in memory under a given name. Storing the estimation results allows postestimation commands to use them as input. For instance, to perform a likelihood-ratio test, the Stata command `lrtest` needs information from both the constrained and the unconstrained estimation results. Here we discuss how to use the `estimates` command to store results from a given model and sample. In the next section, we will discuss how to perform postestimation analysis on stored estimation results.

After running any estimation command in Stata, the syntax of the `estimates` command for storing results is

estimates store *name*

Suppose that we know that we will want to perform postestimation analysis on the results from the above regression. To store the estimation results in memory under the name `reg1`, we could type

```
. regress job female phd mcit3 fellow pub1 cit1
  (output omitted )
. estimates store reg1
```

We could then perform postestimation analysis on the estimation results stored as `reg1` as described in the next section.

3.1.7 Reformatting output with estimates table

`estimates table` can be used to reformat from an estimation command to look more like the tables that are seen in articles that use regression models. `estimates table` also makes it easier to move estimation results into a word processor or spreadsheet to make a presentation-quality table there. We strongly recommend using this command or some other automated procedure (such as `estout`, which we discuss in the next section) rather than retyping the results. Not only is this much less tedious, but it also diminishes the possibility of errors. Also if you revise your model and have used `estimates table` or `estout` in your do-file, you can easily generate new tables. The syntax for `estimates table` is

estimates table [*modelname* [*modelname* ...]] [, *options*]

where *modelname* is the name of a model whose results have been saved using `estimates store` or is the previously fitted model if none was specified. As we will show later, the command can be used to place the results from multiple models side by side.

After fitting the logit model we presented above, we could run `estimates table` as follows:

```
. logit lfp k5 k618 age wc hc lwg, nolog
  (output omitted)
. estimates table, b(%9.3f) t label varwidth(30)
```

Variable	active
# kids < 6	-1.439
	-7.44
# kids 6-18	-0.087
	-1.31
Wife's age in years	-0.069
	-5.49
Wife College: 1=yes 0=no	0.693
	3.10
Husband College: 1=yes 0=no	-0.142
	-0.73
Log of wife's estimated wages	0.561
	3.77
Constant	2.939
	4.67

legend: b/t

`estimates table` provides much flexibility for what you include in your table. Although you should check the *Stata Base Reference Manual* or type `help estimates` for complete information, here are some of the most helpful options:

se, t, p, and star specify whether and how standard errors are to be included in the table. se tells Stata to print standard errors along with the coefficients, t specifies t or z statistics, and p tells Stata to print the p-values. By contrast, star tells Stata to print one star by the coefficient if the p-value is $< .05$, two if $< .01$, and three if $< .001$. At least as of this writing, the star option cannot be used in conjunction with the se (or t or p) option.

b(*format*) specifies the format in which the coefficients are printed, e.g., the number of decimal places shown. Also formats can be specified in parentheses after se, t, or p. We use the format %9.3f for many of our tables; the 3 in this format means three decimal places. For more information on formats, see `help format` or the *Stata User's Guide*.

keep(*varlist*) or drop(*varlist*) can be used to specify which of the independent variables from the regression you wish to include in the table, if you do not wish to include them all.

label indicates that variable labels should be used instead of variable names in the rows of the table.

varwidth(*#*) specifies the width of the column that includes variable names and labels, which is useful when these are long.

stats(*list*) indicates that the scalar statistics included in the *list* should also be included
in the model. N is one such statistic that can be specified here. Others, including
several goodness-of-fit statistics, are also available.

3.1.8   Reformatting output with estout

The command estout is a powerful, user-written alternative to estimates table (Jann
2005).[7] To install estout, type findit estout and follow the links. estout allows
great flexibility in the way that estimation results are presented and allows you to
export tables in several formats that can be imported into other programs (e.g., a plain-
text file, a tab-delimited text file, an HTML table, or a LATEX file). The cost of estout's
flexibility is that the command has many options, although once you master a few basic
options, you can create tables easily. For full details on estout, enter help estout or
see Jann (2005). For an introduction to estout, enter help estout_intro. Here we
show you only a few options that we use to make a table for the logit model that has
been our running example. In later chapters, we use estout to construct more complex
tables.

First, we fit the model:

```
. logit lfp k5 k618 age wc hc lwg inc
  (output omitted)
```

(*Continued on next page*)

---

7. We thank Ben Jann for his help with this section.

Using `estout`, we create the following table:

```
Model of women's labor force participation

                                        Coef
------------------------------------------------
# kids < 6                              -1.463
                                       (0.20)
# kids 6-18                             -0.065
                                       (0.07)
Wife's age in years                    -0.063
                                       (0.01)
Wife College: 1=yes 0=no                0.807
                                       (0.23)
Husband College: 1=yes 0=no             0.112
                                       (0.21)
Log of wife's estimated wages           0.605
                                       (0.15)
Family income excluding wife's         -0.034
                                       (0.01)
Constant                                3.182
                                       (0.64)
------------------------------------------------
N                                          753
ll                                    -452.633
Note: Standard errors in parentheses
```

This is the `estout` command we used,

```
estout using table1.txt, replace style(fixed)
    prehead("Model of women's labor force participation")   ///
    posthead("---------------------------------------------")  ///
    collabels("Coef")                                       ///
    cells(b(fmt(%9.3f)) se(par fmt(%9.2f)))                 ///
    label varwidth(30) varlabels(_cons "Constant")         ///
    prefoot("---------------------------------------------")  ///
    stats(N ll, fmt(%9.0g))                                 ///
    postfoot("Note: Standard errors in parentheses")
```

where `///` is used to continue the command across multiple lines. Here is what the options are doing.

`using table1.txt, replace` indicates that the results should be written to the output file `table1.txt` in the working directory, overwriting the file if it exists.

`style(fixed)` specifies that the file is a simple text file, where spaces are used to line up the columns (as opposed to, for example, tab characters).

`prehead("Model of women's labor force participation")` adds the label at the top of the table.

`posthead("---------------------------------------------")` adds a dashed line under the labels at the top of the table.

`collabels("Coef")` adds the label "Coef" at the top of the column.

cells(b(fmt(%9.3f)) se(par fmt(%9.2f))) specifies what the cells of the table should
contain and in what format. Specifying b and se causes the $\hat{\beta}$s and the standard
errors to be shown. fmt(%9.3f) and fmt(%9.2f) specify the numeric format for the
$\hat{\beta}$s (nine columns with three decimal digits) and standard errors (nine columns with
two decimal digits); see help format for a full discussion of formatting options. The
par option puts the standard errors in parentheses.

label, varwidth(30), and varlabels(_cons "Constant") affect the labels on the left
of the table. label indicates that variable labels as opposed to variable names should
be used. varwidth() indicates the number of columns to be used for the variable
labels. varlabels() allows us to customize the label for the intercept term.

prefoot("--") adds a dashed line
at the bottom of the table.

stats(N ll, fmt(%9.0g)) indicates that we want some model statistics listed at the
bottom—here the sample size and the value of the log likelihood.

postfoot("Note: Standard errors in parentheses") adds a label at the bottom
of the table.

One of the advantages of estout is that you can quickly change the format of your
table. Suppose that the journal publishing your paper requires tables to be formatted
with betas along with stars indicating significance in one column and standard errors
in a second column. For example,

```
Model of women's labor force participation

                                        Coef        Std Err
-------------------------------------------------------------
# kids < 6                            -1.463***      0.197
# kids 6-18                           -0.065         0.068
Wife's age in years                   -0.063***      0.013
Wife College: 1=yes 0=no              0.807***       0.230
Husband College: 1=yes 0=no           0.112          0.206
Log of wife's estimated wages         0.605***       0.151
Family income excluding wife's        -0.034***      0.008
Constant                              3.182***       0.644
-------------------------------------------------------------
N                                       753
ll                                    -452.633
```

We can change the estout command to

```
estout using table2.txt, replace style(fixed)                             ///
    prehead("Model of women's labor force participation")                 ///
    posthead("------------------------------------------------------") /// 
    cells("b(fmt(%9.3f) star label(Coef)) se(fmt(%9.3f) label(Std Err))") ///
    label varwidth(30) varlabels(_cons "Constant")                        ///
    prefoot("-------------------------------------------------------") /// 
    stats(N ll, fmt(%9.0g))
```

where the new option is

cells("b(fmt(%9.3f) star label(Coef)) se(fmt(%9.3f) label(Std Err))"), in
    which we request two cells, one for the betas and the second for the standard
    errors with labels specified by the label option. The star option requests that
    stars be used to indicate significance.

We have only touched on the power of estout. We encourage readers who generate
tables to invest some time in studying this command.

Tip: Copy Table and Copy Table as HTML Here is a quick way to move tables
    into a word processor, spreadsheet, or email. Highlight the results and right-click
    the mouse, and select Copy Table; you will copy the results using tabs instead of
    spaces to separate columns and will remove the lines used to format the table on
    screen. If you then paste the results into a spreadsheet, the columns will be pre-
    served, and most word processors make it very easy to convert tab-delimited text
    into a table. If you select Copy Table as HTML, the results can be directly pasted
    as a table into most word processors or spreadsheets. If your mailer supports
    HTML or rich-text formatting, you can paste a formatted table into your message.

3.1.9 Alternative output with listcoef

The interpretation of regression models often involves transformations of the usually
estimated parameters. For some official Stata estimation commands, there are options
to list transformations of the parameters, such as the or option to list odds ratios in
logit or the beta option to list standardized coefficients for regress. Although Stata is
commendably clear in explaining the meaning of the estimated parameters, in practice
it is easy to be confused about proper interpretations. For example, the zip model
(discussed in chapter 8) simultaneously fits a binary and count model, and it is easy to
be confused regarding the direction of the effects.

For the estimation commands considered in this book (plus some not considered
here), our command listcoef lists estimated coefficients in ways that facilitate interpretation. You can list coefficients by name or significance level, list transformations of
the coefficients, and request help on proper interpretation. In fact, often you will not
even need the normal output from the estimation. You could suppress this output with
the prefix quietly (e.g., quietly logit lfp k5 wc hc) and then use the listcoef
command. An abbreviated syntax is

listcoef [*varlist*] [, [factor|percent|std] lt gt adjacent pvalue(#)

    nolabel constant help]

where *varlist* indicates that coefficients for only these variables are to be listed. If no
varlist is given, coefficients for all variables are listed.

Options for types of coefficients

Depending on the model fitted and the specified options, listcoef computes standard-ized coefficients, factor changes in the odds or expected counts, or percent changes in the odds or expected counts. More information on these different types of coefficients is provided below, as well as in the chapters that deal with specific types of outcomes. The table below lists which options (details on these options are given below) are avail-able for each estimation command. If an option is the default, it does not need to be specified.

	Option		
	std	factor	percent
Type 1: cloglog, cnreg, intreg, mprobit, oprobit, probit, regress, tobit	Default	No	No
Type 2: logistic, logit, ologit	Yes	Default	Yes
Type 3: clogit, mlogit, nbreg, poisson, rologit, slogit, zinb, zip, ztnb, ztp	No	Default	Yes

factor requests factor change coefficients.

percent requests percent change coefficients instead of factor change coefficients.

std indicates that coefficients are to be standardized to a unit variance for the inde-pendent and dependent variables. For models with a latent-dependent variable, the variance of the latent outcome is estimated.

Options for mlogit, mprobit, and slogit

For mlogit, mprobit, and slogit, listcoef can show the coefficients for each pair of outcome categories. When these models are used with ordinal outcomes, it is helpful to look at a subset of these coefficients. The following options are for this purpose:

lt specifies that only the coefficients from comparisons in which the first category has a smaller value than the second will be printed (e.g., comparing outcome 1 versus 2, but not 2 versus 1).

gt specifies that only the coefficients from comparisons in which the first category has a larger value than the second will be printed (e.g., comparing outcome 2 versus 1, but not 1 versus 2).

adjacent specifies that only the coefficients from comparisons in which the two category values are adjacent will be printed (e.g., comparing outcome 1 versus 2 and 2 versus 1, but not 1 versus 3). This option can be combined with gt or lt.

Other options

pvalue(#) specifies that only coefficients significant at the # significance level or smaller will be printed. For example, pvalue(.05) specifies that only coefficients significant at the .05 level of less should be listed. If pvalue is not given, all coefficients are listed.

nolabel requests that category numbers rather than value labels be used in the output.

constant includes the constants in the output. By default, they are not listed.

help gives details for interpreting each coefficient.

Standardized coefficients

std requests coefficients after some or all the variables have been standardized to a unit variance. Standardized coefficients are computed as follows:

x-standardized coefficients

The linear regression model can be expressed as

$$y = \beta_0 + \beta_1 x_1 + \beta_2 x_2 + \varepsilon \tag{3.1}$$

The independent variables can be standardized with simple algebra. Let σ_k be the standard deviation of x_k. Then dividing each x_k by σ_k and multiplying the corresponding β_k by σ_k

$$y = \beta_0 + (\sigma_1 \beta_1)\frac{x_1}{\sigma_1} + (\sigma_2 \beta_2)\frac{x_2}{\sigma_2} + \varepsilon$$

$\beta_k^{S_x} = \sigma_k \beta_k$ is an x-standardized coefficient. For a continuous variable, $\beta_k^{S_x}$ can be interpreted as follows:

For a standard deviation increase in x_k, y is expected to change by $\beta_k^{S_x}$ units, holding all other variables constant.

The same method of standardization can be used in all the other models we consider in this book.

y- and y^*-standardized coefficients

To standardize for the dependent variable, let σ_y be the standard deviation of y. We can standardize y by dividing (3.1) by σ_y:

$$\frac{y}{\sigma_y} = \frac{\beta_0}{\sigma_y} + \frac{\beta_1}{\sigma_y}x_1 + \frac{\beta_2}{\sigma_y}x_2 + \frac{\varepsilon}{\sigma_y}$$

Then $\beta_k^{S_y} = \beta_k/\sigma_y$ is a *y-standardized coefficient* that can be interpreted as follows:

> For a unit increase in x_k, y is expected to change by $\beta_k^{S_y}$ standard deviations, holding all other variables constant.

For a dummy variable,

> Having characteristic x_k (as opposed to not having the characteristic) results in an expected change in y of $\beta_k^{S_y}$ standard deviations, holding all other variables constant.

In models with a latent-dependent variable, the equation $y^* = \beta_0 + \beta_1 x_1 + \beta_2 x_2 + \varepsilon$ can be divided by $\hat{\sigma}_{y^*}$. To estimate the variance of the latent variable, the quadratic form is used:

$$\widehat{\mathrm{Var}}(y^*) = \widehat{\boldsymbol{\beta}}' \, \widehat{\mathrm{Var}}(\mathbf{x}) \, \widehat{\boldsymbol{\beta}} + \mathrm{Var}(\varepsilon)$$

where $\widehat{\boldsymbol{\beta}}$ is a vector of estimated coefficients and $\widehat{\mathrm{Var}}(\mathbf{x})$ is the covariance matrix for the x's computed from the observed data. By assumption, $\mathrm{Var}(\varepsilon) = 1$ in probit models, and $\mathrm{Var}(\varepsilon) = \pi^2/3$ in logit models.

Fully standardized coefficients

In the linear regression model, it is possible to standardize both y and the x's:

$$\frac{y}{\sigma_y} = \frac{\beta_0}{\sigma_y} + \left(\frac{\sigma_1 \beta_1}{\sigma_y}\right)\frac{x_1}{\sigma_1} + \left(\frac{\sigma_2 \beta_2}{\sigma_y}\right)\frac{x_2}{\sigma_2} + \frac{\varepsilon}{\sigma_y}$$

Then $\beta_k^S = (\sigma_k \beta_k)/\sigma_y$ is a *fully standardized coefficient* that can be interpreted as follows:

> For a standard deviation increase in x_k, y is expected to change by β_k^S standard deviations, holding all other variables constant.

The same approach can be used in models with a latent-dependent variable y^*.

Example of listcoef for standardized coefficients

Here we illustrate the computation of standardized coefficients for the regression model. Examples for other models are given in later chapters. The standard output from `regress` is

```
. use http://www.stata-press.com/data/lf2/science2, clear
(Note that some of the variables have been artificially constructed.)
. regress job female phd mcit3 fellow pub1 cit1
```

Source	SS	df	MS
Model	28.8930452	6	4.81550754
Residual	95.7559074	154	.621791607
Total	124.648953	160	.779055954

```
Number of obs =     161
F(  6,   154) =    7.74
Prob > F      =  0.0000
R-squared     =  0.2318
Adj R-squared =  0.2019
Root MSE      =  .78854
```

job	Coef.	Std. Err.	t	P>|t|	[95% Conf. Interval]	
female	-.1243218	.1573559	-0.79	0.431	-.4351765	.1865329
phd	.2898888	.0732633	3.96	0.000	.145158	.4346196
mcit3	.0021852	.0023485	0.93	0.354	-.0024542	.0068247
fellow	.1839757	.133502	1.38	0.170	-.0797559	.4477073
pub1	-.0068635	.0255761	-0.27	0.789	-.0573889	.0436618
cit1	.0080916	.0041173	1.97	0.051	-.0000421	.0162253
_cons	1.763224	.2361352	7.47	0.000	1.296741	2.229706

Now we use listcoef:

```
. listcoef female cit1, help
regress (N=161): Unstandardized and Standardized Estimates
  Observed SD: .88264146
  SD of Error: .78853764
```

job	b	t	P>|t|	bStdX	bStdY	bStdXY	SDofX
female	-0.12432	-0.790	0.431	-0.0534	-0.1409	-0.0605	0.4298
cit1	0.00809	1.965	0.051	0.1719	0.0092	0.1947	21.2422

```
      b = raw coefficient
      t = t-score for test of b=0
  P>|t| = p-value for t-test
  bStdX = x-standardized coefficient
  bStdY = y-standardized coefficient
 bStdXY = fully standardized coefficient
  SDofX = standard deviation of X
```

By default for regress, listcoef lists the standardized coefficients. We requested information only on two of the variables.

Factor and percent change

In logit-based models and models for counts, coefficients can be expressed as (1) a factor or multiplicative change in the odds or the expected count (requested in listcoef by the factor option), or (2) the percent change in the odds or expected count (requested with the percent option). Although these can be computed with options to some estimation commands, listcoef provides a single method to compute these. Details on these coefficients are given in later chapters for each specific model.

3.2 Postestimation analysis

We consider three types of postestimation analysis in the rest of this chapter. The first is statistical testing that goes beyond routine tests of a single coefficient. This is done with Stata's powerful `test` and `lrtest` commands. In later chapters, we present other tests of interest for a given model (e.g., tests of the parallel regression assumption for ordered regression models). The second task is assessing the fit of a model using scalar measures computed by our `fitstat` command and Stata's `estat` command. The third task, and the focus of much of this book, is interpreting the predictions from nonlinear models. We begin by discussing general issues that apply to all nonlinear models. We then discuss our SPost commands and various Stata commands that implement these methods. In chapter 4, we discuss the important step of examining outliers and residuals.

3.3 Testing

Coefficients estimated by ML can be tested with Wald tests using `test` and likelihood-ratio (LR) tests using `lrtest`. For both types of tests, there is a null hypothesis, H_0, that implies constraints on the model's parameters. For example, H_0: $\beta_{wc} = \beta_{hc} = 0$ hypothesizes that two of the parameters are zero in the population.

The Wald test assesses H_0 by considering two pieces of information. First, all else being equal, the greater the distance between the estimated coefficients and the hypothesized values, the less support we have for H_0. Second, the greater the curvature of the log-likelihood function, the more certainty we have about our estimates. This means that smaller differences between the estimates and hypothesized values are required to reject H_0. The LR test assesses a hypothesis by comparing the log likelihood from the full model (i.e., the model that does not include the constraints implied by H_0) and a restricted model that imposes the constraints. If the constraints significantly reduce the log likelihood, then H_0 is rejected. Thus the LR test requires fitting two models. Even though the LR and Wald tests are asymptotically equivalent, in finite samples they give different answers, particularly for small samples. In general, it is unclear which test is to be preferred. Rothenberg (1984) and Greene (2003) suggest that neither test is uniformly superior, although many statisticians prefer the LR.

3.3.1 Wald tests

`test` computes Wald tests for linear hypotheses about parameters from the last model fitted. Here we consider the most useful features of this command for regression models. Information on features for multiple-equation models, such as `mlogit`, `zip`, and `zinb`, are discussed in chapters 6 and 8. Use `help test` for more features and `help testnl` for testing nonlinear hypotheses.

The first syntax for test allows you to specify that one or more coefficients from the last estimation are simultaneously equal to 0:

<u>test</u> *varlist* [, <u>a</u>ccumulate]

where *varlist* contains names of one or more independent variables from the last estimation. The accumulate option will be discussed shortly.

Some examples should make this first syntax clear. With one variable listed, here k5, we are testing H_0: $\beta_{k5} = 0$.

```
. use http://www.stata-press.com/data/lf2/binlfp2, clear
(Data from 1976 PSID-T Mroz)
. logit lfp k5 k618 age wc hc lwg inc, nolog
  (output omitted)
. test k5
 ( 1)  k5 = 0

          chi2( 1) =    55.14
        Prob > chi2 =    0.0000
```

The resulting chi-squared test with 1 degree of freedom equals the square of the z-test in the output from the estimation command, and we can reject the null hypothesis.

With two variables listed, we are testing H_0: $\beta_{k5} = \beta_{k618} = 0$:

```
. test k5 k618
 ( 1)  k5 = 0
 ( 2)  k618 = 0

          chi2( 2) =    55.16
        Prob > chi2 =    0.0000
```

We can reject the hypothesis that the effects of young and older children are simultaneously zero.

In our last example, we include all the independent variables.

```
. test k5 k618 age wc hc lwg inc
 ( 1)  k5 = 0
 ( 2)  k618 = 0
 ( 3)  age = 0
 ( 4)  wc = 0
 ( 5)  hc = 0
 ( 6)  lwg = 0
 ( 7)  inc = 0

          chi2( 7) =    94.98
        Prob > chi2 =    0.0000
```

This is a test of the hypothesis that all the coefficients except the intercept are simultaneously equal to zero. As noted above, a likelihood-ratio test of this same hypothesis is part of the standard output of estimation commands (e.g., LR chi2(7)=124.48 from the earlier logit output).

The second syntax for `test` allows you to test hypotheses about linear combinations of coefficients:

<u>test</u> [*exp=exp*] [, <u>accumulate</u>]

For example, to test that two coefficients are equal, for example H_0: $\beta_{k5} = \beta_{k618}$:

```
. test k5=k618
( 1)   k5 - k618 = 0
          chi2(  1) =    49.48
        Prob > chi2 =    0.0000
```

Because the test statistic is significant, we can reject the null hypothesis that the effect of having young children on labor force participation is equal to the effect of having older children.

The accumulate option

The `accumulate` option allows you to build more complex hypotheses based on the prior use of the `test` command. For example, you might begin with a test of H_0: $\beta_{k5} = \beta_{k618}$:

```
. test k5=k618
( 1)   k5 - k618 = 0
          chi2(  1) =    49.48
        Prob > chi2 =    0.0000
```

Then add the constraint that $\beta_{wc} = \beta_{hc}$:

```
. test wc=hc, accumulate
( 1)   k5 - k618 = 0
( 2)   wc - hc = 0
          chi2(  2) =    52.16
        Prob > chi2 =    0.0000
```

This results in a test of H_0: $\beta_{k5} = \beta_{k618}$, $\beta_{wc} = \beta_{hc}$.

3.3.2   LR tests

`lrtest` compares nested models using an LR test. The syntax is

`lrtest` *model₁* [*model₂*]

where *model₁*, and *model₂* if specified, are the names of estimation results saved using `estimates store`. When *model₂* is not specified, the most recent estimation results are used in its place.

We prefer to save the results of both models before running lrtest. Typically, we begin by fitting the full or unconstrained model and then save the results using estimates store. For example,

```
. logit lfp k5 k618 age wc hc lwg inc, nolog
(output omitted)
. estimates store fmodel
```

where fmodel is the name of the estimation results from the full model. Although any name up to 27 characters can be used, we recommend keeping the names short but informative. After you save the results, you fit a model that is *nested* in the full model. A nested model is one that can be created by imposing constraints on the coefficients in the prior model. Most commonly, one excludes some of the variables that were included in the first model, which in effect constrains the coefficients on these variables to be zero. For example, if we drop k5 and k618 from the last model, this gives

```
. logit lfp age wc hc lwg inc, nolog
(output omitted)
. estimates store nmodel
. lrtest fmodel nmodel
Likelihood-ratio test                       LR chi2(2)   =     66.49
(Assumption: nmodel nested in fmodel)       Prob > chi2 =    0.0000
```

We stored the results for the nested models as nmodel. The output indicates that the test assumes that nmodel is nested in fmodel. It is up to the user to ensure that the models are nested. Here the models are nested. The result is an LR test of the hypothesis $H_0: \beta_{k5} = \beta_{k618} = 0$. The significant chi-squared statistic means that we reject the null hypothesis that these two coefficients are simultaneously equal to zero.

The syntax of lrtest allows us to fit the constrained model first, as in the following example.

```
. logit lfp age wc hc lwg inc, nolog
(output omitted)
. estimates store nmodel
. logit lfp k5 k618 age wc hc lwg inc, nolog
(output omitted)
. estimates store fmodel
. lrtest nmodel fmodel
Likelihood-ratio test                       LR chi2(2)   =     66.49
(Assumption: nmodel nested in fmodel)       Prob > chi2 =    0.0000
```

The output from lrtest states that, in this case, the nmodel estimation results are assumed to be based on a model that nests in fmodel.

Avoiding invalid LR tests

lrtest does *not* always prevent you from computing an invalid test. There are two things that you must check. First, the two models must be nested. Second, the two models must be fitted on the same sample. Although `lrtest` will exit with an error message if the number of observations differs over the models, this check does not catch those cases in which the number of observations is the same but the samples are different. If either of these conditions is violated, the results of `lrtest` are meaningless. For details on ensuring the same sample size, see our discussion of `mark` and `markout` in section 3.1.4.

3.4   estat command

`estat`, added in Stata 9, is a postestimation command that lists a variety of statistics and can be useful for interpreting a model. Here we provide a general overview, whereas specific uses of the command are illustrated throughout the book. `estat` has a series of subcommands that we describe briefly.

estat summarize

`estat summarize` provides descriptive statistics for the estimation sample from the last model. For example,

```
. estat summarize
    Estimation sample logit                 Number of obs =     753
```

Variable	Mean	Std. Dev.	Min	Max
lfp	.5683931	.4956295	0	1
k5	.2377158	.523959	0	3
k618	1.353254	1.319874	0	8
age	42.53785	8.072574	30	60
wc	.2815405	.4500494	0	1
hc	.3917663	.4884694	0	1
lwg	1.097115	.5875564	-2.05412	3.21888
inc	20.12897	11.6348	-.029	96

The output is equivalent to that from `summarize` *modelvars* `if e(sample)`, where *modelvars* is the list of all variables included in your model. Several options to `estat summarize` are useful

`labels` displays variable labels rather than the names of the variables.

`noheader` suppresses the header.

`equation` specifies that the table of statistics be divided with the dependent variable first, then the independent variables for the first equation in the model, then those for the second equation. This could be useful, for example, with the ZIP model that is discussed in chapter 8.

`noweights` ignores the weights if they have been used in estimation.

estat ic

`estat ic` lists the AIC and BIC statistics for the last model fitted (see page 112 for details).

estat vce

`estat vce` lists the variance–covariance matrix for the coefficient estimates. For further details, see `help estat vce`.

Other variations of `estat` depend on the model being fitted and will be discussed along with those models. If you are writing programs, `estat` is handy since it leaves each of the statistics in `r()`. To see what is saved, type `return list` after the `estat` command.

3.5   Measures of fit

Assessing fit involves both the analysis of the fit of individual observations and the evaluation of scalar measures of fit for the model as a whole. Regarding the former, Pregibon (1981) extended methods of residual and outlier analysis from the linear regression model to the case of binary logit and probit (see also Cook and Weisberg 1999, part IV). These measures are considered in chapter 4. Measures for many count models are also available (Cameron and Trivedi 1998). Unfortunately, similar methods for ordinal and nominal outcomes are not available. Many scalar measures have been developed to summarize the overall goodness of fit for regression models of continuous, count, or categorical dependent variables. A scalar measure can be useful in comparing competing models and ultimately in selecting a final model. Within a substantive area, measures of fit can provide a *rough* index of whether a model is adequate. However, *there is no convincing evidence that selecting a model that maximizes the value of a given measure results in a model that is optimal in any sense other than the model's having a larger (or, in some instances, smaller) value of that measure.* Although measures of fit provide some information, it is only partial information that must be assessed within the context of the theory motivating the analysis, past research, and the estimated parameters of the model being considered.

Syntax of fitstat

Our command `fitstat` calculates many fit statistics for the estimation commands we consider in this book. With its `saving()` and `using()` options, the command also allows the comparison of fit measures across two models. Although `fitstat` duplicates some measures computed by other commands (e.g., the pseudo-R^2 in standard Stata output or `lfit`), `fitstat` adds many more measures and makes it convenient to compare measures across models. The syntax is fitstat $\big[$, <u>s</u>aving(*name*) <u>u</u>sing(*name*) <u>b</u>ic force save <u>d</u>iff $\big]$

Although many of the calculated measures are based on values returned by the estimation command, for some measures it is necessary to compute more statistics from the estimation sample. This is done automatically using the estimation sample from the last estimation command. fitstat can also be used when models are fitted with weighted data, with two limitations. First, some measures cannot be computed with some types of weights. Second, with pweights, we use the "pseudolikelihoods" rather than the likelihood to compute our measures of fit. Given the heuristic nature of the various measures of fit, we see no reason why the resulting measures would be inappropriate.

fitstat terminates with an error if the last estimation command does not return a value for the log-likelihood function with only an intercept (i.e., if e(ll_0)=.). This occurs, for example, if the noconstant option is used to fit a model.

Options

saving(*name*) saves the computed measures in a matrix for later comparisons. *name* must be four characters or shorter.

using(*name*) compares the fit measures for the current model with those of the model saved as *name*. *name* cannot be longer than four characters.

bic presents only the Bayesian information criterion (BIC) and other information measures. When comparing two models, fitstat reports Raftery's (1996) guidelines for assessing the strength of one model over another, which are detailed at the end of this section.

force is required to compare two models when the number of observations or the estimation method varies between the two models.

save and diff are equivalent to saving(0) and using(0), respectively.

Models and measures

Details on the measures of fit are given below. Here we only summarize which measures are computed for which models. ■ indicates that a measure is computed, and □ indicates that the measure is not computed.

(Continued on next page)

	regress	logit / probit	cloglog	ologit / oprobit	clogit / mlogit	cnreg / intreg / tobit	nbreg / poisson / zinb / zip	slogit	mprobit	ztnb / ztp	rologit
Log likelihood	■	■	■[1]	■	■	■	■[2]	■	■	■	■
Deviance and LR χ^2	□	■	■	■	■	□	■	□	■	■	□
Deviance and Wald χ^2	□	□	□	□	□	□	□	■	□	□	■
AIC, AIC*n, BIC, BIC′, BIC^S	■	■	■	■	■	■	■	■	■	■	■
R^2 and adjusted R^2	■	□	□	□	□	□	□	□	□	□	□
Efron's R^2	□	■	■	□	□	□	■	□	□	■	□
McFadden's, ML, Cragg and Uhler's R^2	□	■	■	■	■	■	■	□	■	■	■
Count and adjusted count R^2	□	■	■	■	■[3]	□	□	□	■	□	□
Var(e), Var(y^*), and McKelvey and Zavoina's R^2	□	■	■	■	□	■	□	■	■	□	□

1: For cloglog, the log likelihood for the intercept-only model does not correspond to the first step in the iterations.

2: For zinb and zip, the log likelihood for the intercepts-only model is calculated by estimating zinb | zip *depvar* , inf(_cons).

3: The adjusted count R^2 is not defined for clogit.

Example of fitstat

To compute fit statistics for one model, we first fit the model and then run `fitstat`:

```
. logit lfp k5 k618 age wc hc lwg inc, nolog
  (output omitted)
. fitstat
Measures of Fit for logit of lfp
Log-Lik Intercept Only:      -514.873   Log-Lik Full Model:          -452.633
D(745):                       905.266   LR(7):                        124.480
                                        Prob > LR:                      0.000
McFadden's R2:                  0.121   McFadden's Adj R2:              0.105
ML (Cox-Snell) R2:              0.152   Cragg-Uhler(Nagelkerke) R2:     0.204
McKelvey & Zavoina's R2:        0.217   Efron's R2:                     0.155
Variance of y*:                 4.203   Variance of error:              3.290
Count R2:                       0.693   Adj Count R2:                   0.289
AIC:                            1.223   AIC*n:                        921.266
BIC:                        -4029.663   BIC':                         -78.112
BIC used by Stata:            958.258   AIC used by Stata:            921.266
```

`fitstat` is particularly useful for comparing two models. To do this, you fit a model and then save the results from `fitstat`. Here we use `quietly` to suppress the output from `fitstat` because we list those results in the next step:

```
. logit lfp k5 k618 age wc hc lwg inc, nolog
  (output omitted)
. quietly fitstat, saving(mod1)
```

Next we generate `agesq`, which is the square of `age`. The new model adds `agesq` and drops `k618`, `hc`, and `lwg`. To compare the saved model with the current model, type

```
. generate agesq = age*age
. logit lfp k5 age agesq wc inc, nolog
  (output omitted)
```

(Continued on next page)

```
. fitstat, using(mod1)

Measures of Fit for logit of lfp
                                    Current        Saved       Difference
Model:                              logit          logit
N:                                  753            753              0
Log-Lik Intercept Only              -514.873       -514.873         0.000
Log-Lik Full Model                  -461.653       -452.633        -9.020
D                                   923.306(747)   905.266(745)    18.040(2)
LR                                  106.441(5)     124.480(7)      18.040(2)
Prob > LR                           0.000          0.000            0.000
McFadden's R2                       0.103          0.121           -0.018
McFadden's Adj R2                   0.092          0.105           -0.014
ML (Cox-Snell) R2                   0.132          0.152           -0.021
Cragg-Uhler(Nagelkerke) R2          0.177          0.204           -0.028
McKelvey & Zavoina's R2             0.182          0.217           -0.035
Efron's R2                          0.135          0.155           -0.020
Variance of y*                      4.023          4.203           -0.180
Variance of error                   3.290          3.290            0.000
Count R2                            0.677          0.693           -0.016
Adj Count R2                        0.252          0.289           -0.037
AIC                                 1.242          1.223            0.019
AIC*n                               935.306        921.266         14.040
BIC                                 -4024.871      -4029.663        4.791
BIC'                                -73.321        -78.112          4.791
BIC used by Stata                   963.050        958.258          4.791
AIC used by Stata                   935.306        921.266         14.040

Difference of    4.791 in BIC' provides positive support for saved model.

Note: p-value for difference in LR is only valid if models are nested.
```

Methods and formulas for fitstat

This section provides brief descriptions of each measure computed by fitstat. Full details along with citations to original sources are found in Long (1997). The measures are listed in the same order as the output above.

Log-likelihood based measures

Stata begins maximum likelihood iterations by computing the log likelihood of the model with all parameters but the intercept constrained to zero, referred to as $L(M_{\text{Intercept}})$. The log likelihood upon convergence, referred to as M_{Full}, is also listed. This information is usually presented as the first step of the iteration log and in the header for the estimation results.[8]

---

8. In cloglog, the value at iteration 0 is not the log likelihood with only the intercept. For zip and zinb, the "intercept-only" model can be defined in different ways. These commands return as e(ll_0), the value of the log likelihood with the binary portion of the model unrestricted, whereas only the intercept is free for the Poisson or negative binomial portion of the model. fitstat returns the value of the log likelihood from the model with only an intercept in both the binary and the count portion of the model.

Chi-squared test of all coefficients An LR test of the hypothesis that all coefficients except the intercepts are zero can be computed by comparing the log likelihoods: $LR = 2 \ln L(M_{\text{Full}}) - 2 \ln L(M_{\text{Intercept}})$. This statistic is sometimes designated as G^2. LR is reported by Stata as `LR chi2(7) = 124.48`, where the degrees of freedom, (7), are the number of constrained parameters. `fitstat` reports this statistic as `LR(7): 124.48`. For `zip` and `zinb`, LR tests that the coefficients in the count portion (not the binary portion) of the model are zero.

Deviance The *deviance* compares a given model with a model that has one parameter for each observation so that the model reproduces the observed data perfectly. The deviance is defined as $D = -2 \ln L(M_{\text{Full}})$, where the degrees of freedom equal N minus the number of parameters. D does not have a chi-squared distribution.

R^2 in the LRM

For `regress`, `fitstat` reports the standard coefficient of determination, which can be defined variously as

$$R^2 = 1 - \frac{\sum_{i=1}^{N}(y_i - \widehat{y}_i)^2}{\sum_{i=1}^{N}(y_i - \overline{y})^2} = \frac{\widehat{\text{Var}}(\widehat{y})}{\widehat{\text{Var}}(\widehat{y}) + \widehat{\text{Var}}(\widehat{\varepsilon})} = 1 - \left\{ \frac{L(M_{\text{Intercept}})}{L(M_{\text{Full}})} \right\}^{2/N} \tag{3.2}$$

The adjusted R^2 is defined as

$$\overline{R}^2 = \left(R^2 - \frac{K}{N-1} \right) \left(\frac{N-1}{N-K-1} \right)$$

where K is the number of independent variables.

Pseudo-R^2s

Although each of the definitions of R^2 in (3.2) give the same numeric value in the LRM, they give different answers and thus provide different measures of fit when applied to the other models evaluated by `fitstat`.

McFadden's R^2 McFadden's R^2, also known as the "likelihood-ratio index", compares a model with just the intercept to a model with all parameters. It is defined as

$$R^2_{\text{McF}} = 1 - \frac{\ln \widehat{L}(M_{\text{Full}})}{\ln \widehat{L}(M_{\text{Intercept}})}$$

If model $M_{\text{Intercept}} = M_{\text{Full}}$, R^2_{McF} equals 0, but R^2_{McF} can never exactly equal 1. This measure, which is computed by Stata as `Pseudo R2 = 0.1209`, is listed in `fitstat` as `McFadden's R2: 0.121`. Because R^2_{McF} always increases as new variables are added, an adjusted version is also available:

$$\overline{R}^2_{\text{McF}} = 1 - \frac{\ln \widehat{L}(M_{\text{Full}}) - K^*}{\ln \widehat{L}(M_{\text{Intercept}})}$$

where K^* is the number of parameters (not independent variables).

Maximum likelihood R^2 Another analogy to R^2 in the LRM was suggested by Maddala:

$$R^2_{\text{ML}} = 1 - \left\{ \frac{L(M_{\text{Intercept}})}{L(M_{\text{Full}})} \right\}^{2/N} = 1 - \exp(-G^2/N)$$

This R^2 is also known as the Cox–Snell R^2.

Cragg and Uhler's R^2 Since R^2_{ML} reaches a maximum of only $1 - L(M_{\text{Intercept}})^{2/N}$, Cragg and Uhler suggested a normed measure:

$$R^2_{\text{C\&U}} = \frac{R^2_{\text{ML}}}{\max R^2_{\text{ML}}} = \frac{1 - \left\{ L(M_{\text{Intercept}}) / L(M_{\text{Full}}) \right\}^{2/N}}{1 - L(M_{\text{Intercept}})^{2/N}}$$

This R^2 is also known as the Nagelkerke R^2.

Efron's R^2 For binary outcomes, Efron's pseudo-R^2 defines $\widehat{y} = \widehat{\pi} = \widehat{\Pr}(y = 1 \mid \mathbf{x})$ and equals

$$R^2_{\text{Efron}} = 1 - \frac{\sum_{i=1}^{N} (y_i - \widehat{\pi}_i)^2}{\sum_{i=1}^{N} (y_i - \overline{y})^2}$$

$V(y^*)$, $V(\varepsilon)$, and McKelvey and Zavoina's R^2 Some models can be defined in terms of a latent variable y^*. This includes the models for binary or ordinal outcomes: `logit`, `probit`, `ologit`, and `oprobit`, as well as some models with censoring: `tobit`, `cnreg`, and `intreg`. Each model is defined in terms of a regression on a latent variable y^*:

$$y^* = \mathbf{x}\boldsymbol{\beta} + \varepsilon$$

Using $\widehat{\text{Var}}(\widehat{y}^*) = \widehat{\boldsymbol{\beta}}' \widehat{\text{Var}}(\mathbf{x}) \widehat{\boldsymbol{\beta}}$, McKelvey and Zavoina proposed

$$R^2_{M\&Z} = \frac{\widehat{\text{Var}}(\widehat{y}^*)}{\widehat{\text{Var}}(y^*)} = \frac{\widehat{\text{Var}}(\widehat{y}^*)}{\widehat{\text{Var}}(\widehat{y}^*) + \text{Var}(\varepsilon)}$$

In models for categorical outcomes, $\text{Var}(\varepsilon)$ is assumed to identify the model.

Count and adjusted count R^2 Observed and predicted values can be used in models with categorical outcomes to compute what is known as the count R^2. Consider the binary case where the observed y is 0 or 1 and $\pi_i = \widehat{\Pr}(y = 1 \mid \mathbf{x}_i)$. Define the expected outcome as

$$\widehat{y}_i = \begin{cases} 0 & \text{if } \widehat{\pi}_i \leq 0.5 \\ 1 & \text{if } \widehat{\pi}_i > 0.5 \end{cases}$$

This allows us to construct a table of observed and predicted values, such as that produced for the logit model by the command `estat classification`, which we abbreviate as

```
. estat class

Logistic model for lfp
                    ———— True ————
Classified |        D           ~D    |     Total

     +              342          145         487
     -               86          180         266

  Total  |          428          325   |      753
Classified + if predicted Pr(D) >= .5
True D defined as lfp != 0

Sensitivity                     Pr( +| D)    79.91%
Specificity                     Pr( -|~D)    55.38%
Positive predictive value       Pr( D| +)    70.23%
Negative predictive value       Pr(~D| -)    67.67%

False + rate for true ~D        Pr( +|~D)    44.62%
False - rate for true D         Pr( -| D)    20.09%
False + rate for classified +   Pr(~D| +)    29.77%
False - rate for classified -   Pr( D| -)    32.33%

Correctly classified                         69.32%
```

From this output, we can see that positive responses were predicted for 487 observations, of which 342 were correctly classified because the observed response was positive $(y = 1)$, whereas the other 145 were incorrectly classified because the observed response was negative $(y = 0)$. Likewise, of the 266 observations for which a negative response was predicted, 180 were correctly classified and 86 were incorrectly classified.

A seemingly appealing measure is the proportion of correct predictions, referred to as the *count* R^2,

$$R^2_{\text{Count}} = \frac{1}{N} \sum_j n_{jj}$$

where the n_{jj}'s are the number of correct predictions for outcome j. The count R^2 can give the faulty impression that the model is predicting very well. In a binary model *without* knowledge about the independent variables, it is possible to correctly predict at least 50% of the cases by choosing the outcome category with the largest percentage of observed cases. To adjust for the largest row marginal,

$$R^2_{\text{AdjCount}} = \frac{\sum_j n_{jj} - \max_r (n_{r+})}{N - \max_r (n_{r+})}$$

where n_{r+} is the marginal for row r. The *adjusted count* R^2 is the proportion of correct guesses beyond the number that would be correctly guessed by choosing the largest marginal.

Information measures

Information measures can be used to compare both nested and nonnested models.

AIC Akaike's (1973) information criterion is defined as

$$\text{AIC} = \frac{\{-2\ln \widehat{L}(M_k) + 2P_k\}}{N}$$

where $\widehat{L}(M_k)$ is the likelihood of the model and P_k is the number of parameters in the model (e.g., $K + 1$ in the binary regression model, where K is the number of regressors). All else being equal, the model with the smaller AIC is considered the better-fitting model. Another definition of AIC is equal to N times the value we report (e.g., Tobias and Campbell [1998]). This is the quantity that Stata reports with the command `estat ic` (where ic stands for information criteria) and in `estimates table` with the `stats(aic)` option is used. Because this question comes up often, `fitstat` now reports this quantity labeled both as `AIC*n` and `AIC used by Stata`.

BIC and BIC′ The Bayesian information criterion (BIC) has been proposed by Raftery (1996 and literature cited therein) and others as a means to compare nested and nonnested models. There are at least three ways in which the BIC statistic is defined. Although this can be confusing, the differences are not important, as we will show after presenting the various definitions.

Consider the model M_k with deviance $D(M_k)$. BIC is defined as

$$\text{BIC}_k = D(M_k) - df_k \ln N$$

where df_k is the degrees of freedom associated with the deviance. The more negative the BIC_k, the better the fit. A second version of BIC is based on the LR chi-squared with df_k' equal to the number of regressors (not parameters) in the model. Then

$$\text{BIC}_k' = -G^2(M_k) + df_k' \ln N$$

Again the more negative the BIC_k', the better the fit. A third definition, the one that is included with `estimates table` with the `stats(bic)` option and in `estat ic`, is

$$\text{BIC}_k^S = -2\ln N\widehat{L}(M_k) + df_k^S \ln N$$

where df_k^2 is the number of parameters in the model (including auxiliary parameters such as α in the negative binomial regression model).

The difference in the BICs from two models indicates which model is preferred, at least according to the BIC. Since $BIC_1 - BIC_2 = BIC_1' - BIC_2' = BIC_1^S - BIC_2^S$, the choice of which form of BIC to use is a matter of convenience. `fitstat` shows you all three:

```
Measures of Fit for logit of lfp
   (output omitted)
BIC                        -4024.871         -4029.663           4.791
BIC'                          -73.321           -78.112           4.791
BIC used by Stata             963.050           958.258           4.791
Difference of    4.791 in BIC' provides positive support for saved model.
```

When $BIC_1 - BIC_2 < 0$, the first model is preferred. When $BIC_1 - BIC_2 > 0$, the second model is preferred. Raftery (1996) suggested guidelines for the strength of evidence favoring M_2 against M_1 based on a difference in BIC, BIC', or BIC^S:

Absolute difference	Evidence
0–2	Weak
2–6	Positive
6–10	Strong
> 10	Very strong

Thus for our example, there is positive support for the saved model.

3.6   Interpretation

Models for categorical outcomes are nonlinear. Understanding the implications of nonlinearity is fundamental to the proper interpretation of these models. Here we begin with a heuristic discussion of the idea of nonlinearity and the implications of nonlinearity for the proper interpretation of these models. We then introduce a set of commands that facilitate proper interpretation. Later chapters contain the details for specific models.

(Continued on next page)

Linear models

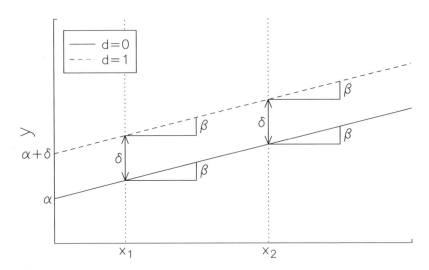

Figure 3.1: A simple linear model.

Figure 3.1 shows a simple, linear regression model, where y is the dependent variable, x is a continuous independent variable, and d is a binary independent variable. The model being fitted is

$$y = \alpha + \beta x + \delta d$$

where for simplicity we assume that there is no error term. The solid line plots y as x changes holding $d = 0$; that is, $y = \alpha + \beta x$. The dashed line plots y as x changes when $d = 1$, which has the effect of changing the intercept: $y = \alpha + \beta x + \delta 1 = (\alpha + \delta) + \beta x$.

The effect of x on y can be computed as the partial derivative or slope of the line relating x to y, often called the *marginal effect* or *marginal change*:

$$\frac{\partial y}{\partial x} = \frac{\partial (\alpha + \beta x + \delta d)}{\partial x} = \beta$$

This equation is the ratio of the change in y to the change in x, when the change in x is infinitely small, holding d constant. In a linear model, the marginal is the same at *all* values of x and d. Consequently, when x increases by one unit, y increases by β units regardless of the current values for x and d. This is shown by the four small triangles with bases of length 1 and heights of β.

The effect of d cannot be computed with a partial derivative because d is discrete. Instead, we measure the *discrete change* in y as d changes from 0 to 1, holding x constant:

$$\frac{\Delta y}{\Delta d} = (\alpha + \beta x + \delta \, 1) - (\alpha + \beta x + \delta \, 0) = \delta$$

When d changes from 0 to 1, y changes by δ units regardless of the level of x. This is shown by the two arrows marking the distance between the solid and dashed lines. Because of the linearity of the model, the discrete change equals the partial change in linear models.

The distinguishing feature of interpretation in the LRM is that the effect of a given change in an independent variable is the same regardless of the value of that variable at the start of its change and regardless of the level of the other variables in the model. That is, interpretation needs only to specify which variable is changing and by how much, and specify that all other variables are being held constant.

Given the simple structure of linear models, such as `regress`, most interpretations require only reporting the estimates. Sometimes it is useful to standardize the coefficients, which can be obtained with `listcoef` as discussed earlier.

Nonlinear models

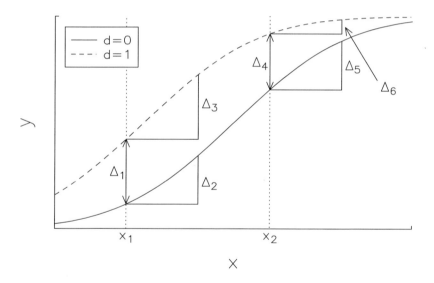

Figure 3.2: A simple nonlinear model.

Figure 3.2 plots a logit model where $y = 1$ if the outcome event occurs, say, if a person is in the labor force, else $y = 0$. The curves are from the logit equation[9]:

$$\Pr(y = 1) = \frac{\exp(\alpha + \beta x + \delta d)}{1 + \exp(\alpha + \beta x + \delta d)} \tag{3.3}$$

Once again, x is continuous and d is binary.

---

9. The α, β, and δ parameters in this equation are unrelated to those in figure 3.1.

The nonlinearity of the model makes it more difficult to interpret the effects of x and d on the probability of an event occurring. For example, neither the marginal nor the discrete change with respect to x is constant:

$$\frac{\partial \Pr(y=1)}{\partial x} \neq \beta$$

$$\frac{\Delta \Pr(y=1)}{\Delta d} \neq \delta$$

This is illustrated by the triangles. Because the solid curve for $d=0$ and the dashed curve for $d=1$ are not parallel, $\Delta_1 \neq \Delta_4$. And, the effect of a unit change in x differs according to the level of both d and x: $\Delta_2 \neq \Delta_3 \neq \Delta_5 \neq \Delta_6$. *In nonlinear models, the effect of a change in a variable depends on the values of all variables in the model and is no longer simply equal to one of the parameters of the model.*

3.6.1 Approaches to interpretation

There are several general approaches for interpreting nonlinear models:

1. Predictions can be computed for each observation in the sample using `predict`.
2. The marginal or discrete change in the outcome can be computed at a representative value of the independent variables using `prchange`.
3. Predicted values for substantively meaningful "profiles" of the independent variables can be compared using `prvalue`, `prtab`, or `prgen`.
4. The nonlinear model can be transformed to a model that is linear in some other outcome. As we discuss in chapter 4, the logit model in (3.3) can be written as

$$\ln\left\{\frac{\Pr(y=1)}{1-\Pr(y=1)}\right\} = \alpha + \beta x + \delta d$$

   which can then be interpreted with methods for linear models, or the exponential of the coefficients can be interpreted in terms of factor changes in the odds.

We now consider the first three of these methods. Details on using these approaches for specific models are given in later chapters.

3.6.2 Predictions using predict

`predict` can be used to compute predicted values for each observation in the current dataset. Although `predict` is a powerful command with many options, we consider only the simplest form of the command that provides all the details that we need. For more options, you can enter `help predict`. The simplest syntax for `predict` is

`predict` *newvarname*

where *newvarname* is the name or names of the new variables that are being generated. The quantity computed for *newvarname* depends on the model that was fitted, and the number of new variables created depends on the model. The defaults are listed in the following table.

Estimation quantity	Quantity computed
regress	Predicted value $\widehat{y} = \mathbf{x}\widehat{\boldsymbol{\beta}}$
cloglog, logistic, logit, probit	Predicted probability $\widehat{\Pr}(y = 1)$
asmprobit, clogit, mlogit, mprobit, ologit, oprobit, rologit, slogit	Predicted probabilities $\widehat{\Pr}(y = k)$
nbreg, poisson, zinb, zip, ztnb, ztp	Predicted count or rate

Here we generate predicted probabilities for a logit model of women's labor force participation. The values of `pr1` generated by `predict` are the probabilities of a woman being in the labor force for each of the observations in the dataset:

```
. logit lfp k5 k618 age wc hc lwg inc
  (output omitted)
. predict pr1
(option p assumed; Pr(lfp))
. summarize pr1
    Variable |       Obs        Mean    Std. Dev.       Min        Max
-------------+--------------------------------------------------------
         pr1 |       753    .5683931    .1944213    .0139875    .9621198
```

The summary statistics show that the predicted probabilities in the sample range from .014 to .962, with an average probability of .568. More discussion of predicted probabilities for the logit model is provided in chapter 4.

For models for ordinal or nominal categories, `predict` computes the predicted probability of an observation falling into each of the outcome categories. So, instead of specifying one variable for predictions, you specify as many names as there are categories. For example, after fitting a model for a nominal dependent variable with four categories, you can type `predict pr1 pr2 pr3 pr4`. These new variables contain the predicted probabilities of being in the first, second, third, and fourth categories, as ordered from the lowest value of the dependent variable to the highest.

For count models, `predict` computes predicted counts for each observation. Our command `prcounts` computes the predicted probabilities of observing specific counts, e.g., $\Pr(y = 3)$ and the cumulative probabilities, e.g., $\Pr(y \leq 3)$. More details are given in chapter 8.

3.6.3   Overview of prvalue, prchange, prtab, and prgen

We have written the postestimation commands `prvalue`, `prchange`, `prtab`, and `prgen` to make it simple to compute specific predictions for interpreting models for categorical and count outcomes. Details on installing these programs as part of SPost are given in chapter 1. These commands are used throughout the book. Here we provide an overview of what the commands do and their basic options.

`prvalue` computes predicted values of the outcomes for specified values of the independent variables and can compute differences in predictions for two sets of values. `prvalue` is the most basic command. Indeed, it can be used to compute all the quantities except marginal change from the next three commands. `asprvalue` works like `prvalue` but handles models with alternative-specific data; see chapter 7.

`prchange` computes discrete and marginal changes in the predicted outcomes.

`prtab` creates a table of predicted outcomes for a cross-classification of up to four categorical independent variables, while other independent variables are held at specified values.

`prgen` computes predicted outcomes as one independent variable changes over a specified range, holding other variables constant. New variables containing these values are generated and can then be plotted. `prgen` is limited in that it cannot handle complex model specifications in which a change in the value of the key independent variable implies a change in another independent variable, such as in models that include terms for both age and age-squared. For these models, we have created the more general, but more complex, command `praccum`, which we describe in chapter 9.

The most effective interpretation of nonlinear regression models involves using all these commands to discover the most convincing way to convey the predictions from the model.

We now discuss the syntax and options for these commands. Many of the specific details might be clear only after you read the more detailed discussions in later chapters.

Specifying the levels of variables

Each command computes predicted values for the last regression model that was fitted. To compute predicted values, you must specify values for all the independent variables in the regression. By default, all variables are set to their means in the estimation sample.[10] Using the `x()` and `rest()` options, variables can be assigned to specific values or to a sample statistic computed from the data in memory.

---

10. The estimation sample includes only those cases that were used in fitting a model. Cases that were dropped due to missing values or `if` and `in` conditions are not part of the estimation sample.

x(*variable1*=*value1* [...]) assigns *variable1* to *value1*, *variable2* to *value2*, and so on. Although equal signs are optional, they make the commands easier to read. You can assign values to as many or as few variables as you want. The assigned value is either a specific *number* (e.g., female=1) or a *mnemonic* specifying the descriptive statistic (e.g., phd=mean to set variable phd to the mean; pub3=max to assign pub3 to the maximum value). Details on the mnemonics that can be used are given below.

rest(*stat*) sets the values of all variables not specified in x() to the sample statistic indicated by *stat*. For example, rest(mean) sets all variables to their mean. If x() is not specified, all variables are set to *stat*. The value of *stat* can be calculated for the whole sample or can be conditional based on the values specified by x(). For example, if x(female=1) is specified, rest(grmean) specifies that all other variables should equal their mean in the sample defined by female=1. This is referred to as a group statistic (i.e., statistics that begin with *gr*). If you specify a group statistic for rest(), only numeric values can be used for x(). For example, x(female=mean) rest(grmean) is not allowed. If rest() is not specified, it is assumed to be rest(mean).

The statistics that can be used with x() and rest() are

mean, median, min, and max specify the unconditional mean, median, minimum, and maximum. By default, the estimation sample is used to compute these statistics. If the option all is specified, all cases in memory are used for computing descriptive statistics, regardless of whether they were used in the estimation. if or in conditions can also be used. For example, adding if female==1 to any of these commands restricts the computations of descriptive statistics to women, even if the estimation sample included men and women.

previous sets values to what they were the last time the command was called; this option can be used only if the set of independent variables is the same in both cases. This can be useful if you want to change the value of only one variable from the last time the command was used.

upper and lower set values to those that yield the maximum or minimum predicted values, respectively. These options can be used only for binary models.

grmean, grmedian, grmin, and grmax compute statistics that are conditional on the group specified in x(). For example, x(female=0) rest(grmean) sets female to 0 and all other variables to the means *of the subsample* in which female is 0 (i.e., the means of these other variables for male respondents).

Options controlling output

nobase suppresses printing of the base values of the independent variables.

brief provides only minimal output.

all specifies that any calculations of means, medians, etc., should use the entire sample instead of the sample used to fit the model.

3.6.4 Syntax for prvalue

prvalue computes the predicted outcome for one set of values of the independent variables. For some commands, prvalue provides confidence intervals for the predictions; see section A.18. The syntax is

prvalue $\big[$ *if* $\big]$ $\big[$ *in* $\big]$ $\big[$, x(*variable1=value1* $[...]$) rest(*stat*) maxcnt(*#*) save
     diff ystar nobase nolabel brief all level(*#*) delta ept bootstrap
     reps(*#*) dots match size(*#*) saving(*filename*, *save_options*)
     $\big[$ biascorrected | percentile | normal $\big]$ $\big]$

Options

Options x(), rest(), nobase, brief, and all are documented earlier in the chapter.

maxcnt(*#*) is the maximum count value for which a predicted probability is computed for count models. The default is 9.

save preserves predicted values computed by prvalue for later comparison.

diff compares predicted values computed by prvalue with those previously preserved with the save option.

ystar prints the predicted value of y^* for binary and ordinal models.

nolabel uses values rather than value labels in output.

Options for confidence intervals

prvalue computes confidence intervals for many predictions. When used with the save and diff options, it computes confidence intervals for discrete changes in predictions. Here we list the options related to confidence intervals. Details on each method for computing intervals are given in section 3.7.

level(*#*) specifies the confidence level, as a percentage, for confidence intervals. For example, level(95) requests a 95% confidence interval.

delta calculates confidence intervals by the delta method using analytic derivatives.

ept computes confidence intervals for predicted probabilities for cloglog, logit, and probit by endpoint transformation. This method cannot be used for changes in predictions.

bootstrap computes confidence intervals using the bootstrap method. This method
    takes roughly 1,000 times longer to compute than other methods, but it provides
    closer to nominal coverage in small samples.

Options used for bootstrapped confidence intervals

reps(#) specifies the number of bootstrap replications to be performed. The default
    is 1,000. The accuracy of a bootstrap estimate depends critically on the number of
    replications. Although sources differ on the recommended number of replications,
    Efron and Tibshirani (1993, 188) suggest 1,000 replications for confidence intervals.
    You can use the user-written command bssize (Poi 2004) to calculate the number
    of bootstrap replications to be used; to install bssize, type findit bssize. In our
    experience, this method often suggests more than 1,000 replications.

saving(*filename, save_options*) creates a data file with the estimates from each of the
    bootstrapped samples (i.e., one case for each replication). This option is useful when
    you need to examine the distribution of bootstrapped estimates. For example, this
    option is required if you plan to use bssize to calculate the number of replications
    to be used (Poi 2004).

dots is used with bootstrap to write a dot (.) at the beginning of each replication
    and periodically prints the percentage of total replications that has been completed.
    If computations appear to be stuck (i.e., new dots do not appear), the estimation is
    probably not converging for the current bootstrap sample. We have found this to be
    most common with zip and zinb (see chapter 8 for a detailed discussion of what to do
    when nonconvergence is a problem). When this happens, you can click on the break
    symbol to stop computations for the current sample or wait until the maximum
    number of iterations have been computed (by default, the maximum number of
    iterations is 16,000). When a model does not converge for a given bootstrap sample,
    that sample is dropped.

match specifies that the bootstrap will resample within each category of the dependent
    variable in proportion to the distribution of the outcome categories in the estimation
    sample. If match is not specified, the proportions in each category of the bootstrap
    sample are determined entirely by the random draw, and it is possible to have
    samples with no cases in some categories. This option does not apply to cnreg,
    intreg, nbreg, poisson, regress, tobit, zinb, and zip.

size(#) specifies the number of cases to be sampled when bootstrapping. The default
    is the size of the estimation sample. If size() is specified, # must be less than or
    equal to the size of the estimation sample. In general, it is best to not specify size()
    (see *http://www.stata.com/support/faqs/stat/reps.html* for more information).

biascorrected computes the bootstrapped confidence interval using the bias-corrected
    method.

`percentile` computes the bootstrapped confidence interval using the percentile method. This is the default method.

`normal` computes the bootstrapped confidence interval using the normal approximation method.

3.6.5   Syntax for prchange

`prchange` computes marginal and discrete change coefficients. An abbreviated syntax is

`prchange` [*varlist*] [*if*] [*in*] [, x(*variable1*=*value1*[...]) rest(*stat*) outcome(#) fromto brief nobase nolabel help all uncentered delta(#)]

varlist specifies that changes are to be listed only for these variables. By default, changes are listed for all variables.

Options

`outcome(#)` specifies that changes will be printed only for the outcome indicated. For example, if `ologit` was run with outcome categories 1, 2, and 3, `outcome(1)` requests that only changes in the probability of outcome 1 be listed. For `ologit`, `oprobit`, and `mlogit`, the default is to provide results for all outcomes. For the count models, the default is to present results with respect to the predicted rate; specifying an outcome number will provide changes in the probability of that outcome.

`fromto` specifies that the starting and ending probabilities from which the discrete change is calculated for `prchange` should also be displayed.

`nolabel` uses values rather than value labels in the output.

`help` provides information explaining the output.

`uncentered` specifies that the uncentered discrete change, rather than the centered discrete change, is to be computed. By default, the change in an independent variable is centered on its value.

`delta(#)` specifies the amount of the discrete change in the independent variable. The default is a 1-unit change (i.e., `delta(1)`).

3.6.6   Syntax for prtab

`prtab` constructs a table of predicted values for all combinations of up to three variables. An abbreviated syntax is

prtab *rowvar* [*colvar* [*supercolvar*]] [*if*] [*in*] [, by(*superrowvar*)
    x(*variable1=value1* [...]) rest(*stat*) outcome(*string*) nobase nolabel
    novarlbl brief all]

Options

rowvar, *colvar*, *supercolvar*, and *superrowvar* are independent variables from the previously fitted model. These define the table that is constructed.

by(*superrowvar*) specifies the categorical independent variable that is to be used to form the superrows of the table.

outcome(*string*) specifies that changes be printed only for the outcome indicated. The default for ologit, oprobit, and mlogit is to provide results for all outcomes. For the count models, the default is to present results with respect to the predicted rate; specifying an outcome number provides changes in the probability of that outcome.

nolabel uses a variable's numerical values rather than value labels in the output. Sometimes this is more readable.

novarlbl uses a variable's name rather than the variable label in the output. Sometimes this is more readable.

3.6.7 Syntax for prgen

prgen computes a variable containing predicted values as one variable changes over a range of values, which is useful for constructing plots. You can add confidence intervals to these plots and can plot the marginal changes in the outcome. An abbreviated syntax is

prgen *varname* [*if*] [*in*], generate(*prefix*) [from(#) to(#) ncases(#)
    gap(#) x(*variable1=value1* [...]) rest(*stat*) maxcnt(#) brief all noisily
    marginal ci *prvalueci_options*]

varname is the name of the variable that changes while all other variables are held at specified values.

Options

x(), rest(), maxcnt(), brief, and all work the same way as for prvalue; see section 3.6.4.

generate(*prefix*) sets the prefix for the new variables created by prgen. Choosing a prefix that is different from the beginning letters of any of the variables in your dataset makes it easier to examine the results. For example, if you choose the prefix abcd, you can use the command summarize abcd* to examine all newly created variables.

from(#) and to(#) are the start and end values for *varname*. The default is for
varname to range from the observed minimum to the observed maximum of *varname*.

ncases(#) specifies the number of predicted values prgen computes as *varname* varies
from the start value to the end value. The default is ncases(11).

gap(#) is an alternative to ncases(). You specify the gap or size between tic marks and
prgen determines if the specified value divides evenly into the range specified with
from() and to(). If it does, prgen determines the appropriate value for ncases().

maxcnt(#) is the maximum count value for which a predicted probability is computed
for count models. The default is maxcnt(9).

Options for confidence intervals and marginals

marginal requests that a variable be created containing the marginal change in the
outcome relative to *varname*, holding all other variables constant.

ci generates confidence intervals for the predictions. All the other options for constructing confidence intervals listed for prvalue can be used with prgen. Indeed,
prgen simply passes the options along to prvalue—the command that is doing all
the work.

prvalueci_options are any options available with prvalue for computing confidence intervals; see section 3.6.4.

Variables generated

prgen constructs variables that can be graphed. The observations contain predicted
values and probabilities for a range of values for the variable *varname*, holding the
other variables at the specified values. *n* observations are created, where *n* is 11 by
default or specified by ncases(). The new variables all start with the *prefix* specified
by gen(). The variables created are

For which model	Name	Content
All models	*prefix*x	Values of *varname* from `from(#)` to `to(#)`
logit, probit	*prefix*p0 *prefix*p1	Predicted probability $\Pr(y=0)$ Predicted probability $\Pr(y=1)$

mlogit, mprobit,
  ologit, oprobit,
  slogit
	*prefix*pk
	
	*prefix*sk
		Predicted probability $\Pr(y=k)$,
		  for all outcomes
		Cumulative probability $\Pr(y \le k)$,
		  for all outcomes

nbreg, poisson, zinb,
  zip, ztnb, ztp
	*prefix*mu
	*prefix*pk
	
	*prefix*sk
		Predicted rate μ
		Predicted probability $\Pr(y=k)$,
		  for $0 \le k \le$ `maxcnt()`
		Cumulative probability $\Pr(y \le k)$,
		  for $0 \le k \le$ `maxcnt()`

zinb, zip	*prefix*inf	Predicted probability $\Pr(\text{Always } 0 = 1) =$ $\Pr(\text{inflate})$
cnreg, intreg, regress, tobit	*prefix*xb	Predicted value of y

If `ci` is specified as an option for `prgen`, variables are created containing the upper and lower bounds of the confidence interval for the outcome. These variables have the same names as those in the table above, except for adding `ub` at the end for the variable with the upper bound and `lb` for the lower bound. If `marginal` is specified, variables are created that contain the marginal change in the outcome with respect to *varname*, holding all other variables constant. The variables containing marginal changes have the same names as those in the table above, except for adding a `D` prior to the outcome abbreviation and D*varname* after. For example, the marginal for *prefix*p0 is named *prefix*Dp0D*varname*. These are computed only for those models for which `prchange` computes the marginal change.

3.6.8 Computing marginal effects using mfx

`mfx` is the Stata command for calculating marginal effects. Recall that the marginal effect is the partial derivative of y with respect to x_k. For nonlinear models, the value of the marginal effect depends on the specific values of all the independent variables. After fitting a model, `mfx` will compute the marginal effects for all the independent variables, evaluated at values that are specified using the `at()` option. `at()` is similar to the `x()` and `rest()` syntax used in our commands. To compute the marginal effects while holding `age` at 40 and `female` at 0, the command is `mfx, at(age=40 female=0)`. As with our commands for working with predicted values, unspecified independent variables are held at their mean by default.

`mfx` has several features that make it worth exploring. For one, it works after many different estimation commands. For dummy independent variables, `mfx` computes the discrete change rather than the marginal effect. Of particular interest for economists, the command optionally computes elasticities instead of marginal effects. And, `mfx` computes standard errors for the effects. The derivatives are calculated numerically, which means that the command can take a long time to execute when there are many independent variables and observations, especially when used with `mlogit`. We encourage readers who are interested in learning more about this command to examine its entry in the *Base Reference Manual.*

3.7   Confidence intervals for predictions

Depending on the model and options chosen, `prvalue` computes predictions for different outcomes. For models with categorical outcomes, the probability of each outcome is computed. When these models can be derived from a latent-variable model, the predicted value for the latent variable can also be computed. For count models, both the predicted rate and the probability of each count are computed. The values computed are summarized here:[11]

Model	$\Pr(y)$	y or y^*	Rate μ	$\Pr(\text{count})$
`cloglog, logistic, logit, probit`	Default	Yes[1]	No	No
`ologit, oprobit`	Default	Yes[1]	No	No
`mlogit`	Default	No	No	No
`nbreg, poisson, ztnb, ztp`	No	No	Default	Default
`cnreg, intreg, regress, tobit`	No	Default	No	No

1. When `ystar` is specified as an option.

Although technical details on each method for computing confidence intervals are provided in Xu and Long (2005), here we provide general information about these methods and why you might use each. `prvalue` uses four methods: maximum likelihood, endpoint transformation, the bootstrap method, and the delta method.

1. **Maximum likelihood** For models such as the linear regression model, the standard method of computing confidence intervals using maximum likelihood theory is used.

2. **Endpoint transformation** With the `ept` option, confidence intervals are computed using endpoint transformations. This method is appropriate for bounded monotonic functions only, such as predicted probabilities in binary regression models. One advantage of this method is that the bounds cannot be smaller than 0 or greater than 1. This method cannot be used for computing confidence intervals for changes in predictions.

---

11. We have not added confidence intervals for all estimation commands discussed in this edition.

3. **Delta method** With the `delta` option, confidence intervals are computed using the delta method. This method takes a function that is too complex for analytically computing the variance (for example, the change in the predicted probabilities in a multinomial logit model) and creates a linear approximation of that function. The variance of the simpler approximation is used for constructing the confidence interval. Because `prvalue` uses analytic formulas for the derivatives, rather than numerical estimation, the computation of confidence intervals is extremely fast. Unlike the method of endpoint transformation, the bounds computed by the delta method can include values that exceed the range of the statistic being estimated (e.g., a bound for a predicted probability could be negative or greater than one).

4. **Bootstrap method** The `bootstrap` option computes bootstrapped confidence intervals (see Guan [2003] for an introduction to the bootstrap using Stata). The idea of the bootstrap is that by repeatedly taking samples from the sample used to fit your model, you can estimate the sampling variability that would occur if you took repeated samples from the population. This is done by taking a random sample from the estimation sample, computing the statistics of interest, and repeating this process for some number of replications. The variation in the estimates across the replications is used to estimate the standard deviation of the sampling distribution. Although the bootstrap method frequently provides better estimates of the confidence interval bounds, it is computationally intensive. For example, computations of the confidence intervals for a multinomial logit with five outcomes, three variables, and 337 cases by the delta method took .15 seconds, whereas computing the confidence intervals by bootstrap took 141 seconds for 1,000 replications. Computations of confidence intervals for a multinomial logit with six outcomes, four variables and 7,357 cases using the delta method took .61 seconds, whereas computing the confidence intervals by bootstrap took 13 minutes 2 seconds for 1,000 replications.[12] Roughly speaking, each replication in a bootstrap takes as long as the entire computation for the delta method. The `zip` and `zinb` models are too complex for computing the derivatives necessary for the delta method, so only the bootstrap method can be used.

At present, `prvalue` and `prchange` only provide confidence intervals for some of the commands for which they can be used to generate predictions. The models with which these command currently work are listed below, along with the methods used to compute the intervals.

---

12. Computations were made using Stata/SE 8.2 (born 10 Jan 2005) under Windows XP on a Dell XPS with an Intel® Pentium® 4 running at 3.4 GHz. Estimates of times for the delta method were based on the average of 1,000 computations. We expect that the timings would be similar with Stata 9.

Model	Maximum likelihood	Endpoint transfor- mations	Delta method	Bootstrap method	Default method
cloglog, logistic, logit, probit	Yes[1]	Yes[2]	Yes	Yes	Delta
ologit, oprobit	Yes[1]	No	Yes	Yes	Delta
mlogit, nbreg, poisson	No	No	Yes	Yes	Delta
zinb, zip	No	No	No	Yes	No ci
cnreg, intreg, regress tobit	ystar	No	No	No	ML

1. When ystar is specified.
2. This does not work for changes in predictions.

3.8   Next steps

This concludes our discussion of the basic commands and options that are used for fitting, testing, assessing fit, and interpreting regression models. In the next four chapters, we show how each of the commands can be applied for models relevant to one particular type of outcome. Although chapter 4 has somewhat more detail than later chapters, readers should be able to proceed from here to any of the chapters that follow.

Part II

Models for Specific Kinds of Outcomes

In part II, we provide information on the models appropriate for different kinds of dependent outcomes.

- **Chapter 4** considers *binary outcomes*. Models for binary outcomes are the most basic type that we consider and, to a large extent, they provide a foundation for the models in later chapters. For this reason, chapter 4 has more detailed explanations, and we recommend that all readers review this chapter, even if they are interested mainly in other types of outcomes. We show how to fit the binary regression model, how to test hypotheses, how to compute residuals and influence statistics, and how to calculate scalar measures of model fit. Then we describe how these models can be interpreted using predicted probabilities, discrete and marginal change in these probabilities, and odds ratios.

Chapters 5, 6, and 7 can be read in any combination or order, depending on your interests. Each chapter provides information on fitting the relevant models, testing hypotheses about the coefficients, and interpretation in terms of predicted probabilities. Also

- **Chapter 5** on *ordinal outcomes* describes the parallel regression assumption that is made by the ordered logit and probit models and shows how this assumption can be tested. We also discuss interpretation in terms of the underlying latent variable and odds ratios.

- **Chapter 6** on *nominal outcomes with case-specific data* introduces the multinomial logit model. We discuss the the assumption of the independence of irrelevant alternatives (IIA) and present two graphical methods of interpretation. We then consider the multinomial probit model without correlated errors and the stereotype logistic regression model.

- **Chapter 7** on *nominal outcomes with alternative-specific data* examines models in which at least some variables vary over the alternatives for each individual. After showing you how to set up date for such models, we begin with the conditional logit model. We then consider the alternative-specific multinomial probit model, which relaxes the assumption of the independence of irrelevant alternatives. Finally, we present the rank-ordered logistic regression model, in which the outcome is the ranking of a set of alternatives.

- **Chapter 8** on *count outcomes* presents the Poisson and negative binomial regression models. We show how to test the Poisson model's assumption of equidispersion and how to incorporate differences in exposure time into the models. The next two models, the zero-truncated Poisson and negative binomial models, deal with the common problem of having no zeros in your data. We combine these models with the logit model to construct the hurdle model for counts. We conclude by considering two zero-inflated models that are designed for data with zero counts.

- **Chapter 9** covers more topics that extend material presented earlier. We discuss the use and interpretation of categorical independent variables, interactions, and nonlinear terms. We also provide tips on how to use Stata more efficiently and effectively.

4 Models for binary outcomes

Regression models for binary outcomes are the foundation from which more complex models for ordinal, nominal, and count models can be derived. Ordinal and nominal regression models are equivalent to the simultaneous estimation of a series of binary outcomes. Although the link is less direct in count models, the Poisson distribution can be derived as the outcome of many binary trials. More importantly for our purposes, the zero-inflated count models that we discuss in chapter 8 merge a binary logit or probit with a standard Poisson or negative binomial model. Consequently, the principles of fitting, testing, and interpreting binary models provide tools that can be readily adapted to models in later chapters. Thus although each chapter is largely self-contained, this chapter provides somewhat more detailed explanations than later chapters. As a result, even if your interests are in models for ordinal, nominal, or count outcomes, you should benefit from reading this chapter.

Binary dependent variables have two values, typically coded as 0 for a negative outcome (i.e., the event did not occur) and 1 as a positive outcome (i.e., the event did occur). Binary outcomes are ubiquitous, and examples come easily to mind. Did a person vote? Is a manufacturing firm unionized? Is someone a feminist or nonfeminist? Did a startup company go bankrupt? Five years after a person was diagnosed with cancer, is he or she still alive? Was a purchased item returned to the store or kept?

Regression models for binary outcomes allow a researcher to explore how each explanatory variable affects the probability of the event occurring. We focus on the two most often used models, the binary logit and binary probit models, referred to jointly as the *binary regression model* (BRM). Because the model is nonlinear, the magnitude of the change in the outcome probability that is associated with a given change in one of the independent variables depends on the levels of all the independent variables. The challenge of interpretation is to find a summary of the way in which changes in the independent variables are associated with changes in the outcome that best reflect the key substantive processes without overwhelming yourself or your readers with distracting detail.

The chapter begins by reviewing the mathematical structure of binary models. We then examine statistical testing and fit, and finally, methods of interpretation. These discussions are intended as a review for those who are familiar with the models. For a complete discussion, see Long (1997). You can obtain sample do-files and data files that reproduce the examples in this chapter by downloading the spost9_do and spost9_ado packages (see chapter 1 for details).

4.1   The statistical model

There are three ways to derive the BRM, with each method leading to the same mathematical model. First, an unobserved or latent variable can be hypothesized along with a measurement model relating the latent variable to the observed, binary outcome. Second, the model can be constructed as a probability model. Third, the model can be generated as a random utility or discrete-choice model. This last approach is not considered in our review; see Long (1997, 155–156) for an introduction or Pudney (1989) for a detailed discussion.

4.1.1   A latent-variable model

Assume a *latent* or unobserved variable y^* ranging from $-\infty$ to ∞ that is related to the observed independent variables by the structural equation

$$y_i^* = \mathbf{x}_i \boldsymbol{\beta} + \varepsilon_i$$

where i indicates the observation and ε is a random error. For one independent variable, we can simplify the notation to

$$y_i^* = \alpha + \beta x_i + \varepsilon_i$$

These equations are identical to those for the linear regression model except that the dependent variable is unobserved.

The link between the observed binary y and the latent y^* is made with a simple measurement equation:

$$y_i = \begin{cases} 1 & \text{if } y_i^* > 0 \\ 0 & \text{if } y_i^* \leq 0 \end{cases}$$

Cases with positive values of y^* are observed as $y = 1$, whereas cases with negative or zero values of y^* are observed as $y = 0$.

Imagine a survey item that asks respondents if they agree or disagree with the proposition that "a working mother can establish just as warm and secure a relationship with her children as a mother who does not work". Obviously, respondents vary greatly in their opinions on this issue. Some people adamantly agree with the proposition, some adamantly disagree, and still others have only weak opinions one way or the other. We can imagine an underlying continuum of possible responses to this item, with every respondent having some value on this continuum (i.e., some value of y^*). Those respondents whose value of y^* is positive answer "agree" to the survey question ($y = 1$), and those whose value of y^* is 0 or negative answer "disagree" ($y = 0$). A shift in a respondent's opinion might move them from agreeing strongly with the position to agreeing weakly with the position, which would not change the response we observe. Or, the respondent might move from weakly agreeing to weakly disagreeing, in which case we would observe a change from $y = 1$ to $y = 0$.

Consider a second example, which we use throughout this chapter. Let $y = 1$ if a woman is in the paid labor force and $y = 0$ if she is not. The independent variables include variables such as number of children, education, and expected wages. Not all women in the labor force ($y = 1$) are there with the same certainty. One woman might be close to leaving the labor force, whereas another woman could be firm in her decision to work. In both cases, we observe $y = 1$. The idea of a latent y^* is that an underlying *propensity to work* generates the observed state. Again although we cannot directly observe the propensity, at some point a change in y^* results in a change in what we observe, namely, whether the woman is in the labor force.

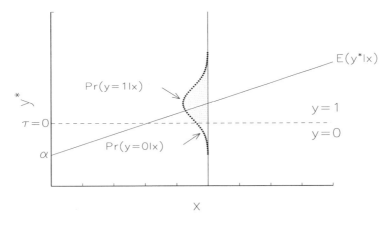

Figure 4.1: Relationship between latent variable y^* and $\Pr(y = 1)$ for the BRM.

The latent-variable model for binary outcomes is shown in figure 4.1 for one independent variable. For a given value of x, we see that

$$\Pr(y = 1 \mid x) = \Pr(y^* > 0 \mid x)$$

Substituting the structural model and rearranging terms,

$$\Pr(y = 1 \mid x) = \Pr(\varepsilon > - [\alpha + \beta x] \mid x) \tag{4.1}$$

This equation shows that the probability depends on the distribution of the error, ε.

Two distributions of ε are commonly assumed, both with an assumed mean of 0. First, ε is assumed to be distributed normally with $\mathrm{Var}(\varepsilon) = 1$. This leads to the binary probit model, in which (4.1) becomes

$$\Pr(y = 1 \mid x) = \int_{-\infty}^{\alpha+\beta x} \frac{1}{\sqrt{2\pi}} \exp\left(-\frac{t^2}{2}\right) dt$$

Alternatively, ε is assumed to be distributed logistically with $\mathrm{Var}(\varepsilon) = \pi^2/3$, leading to the binary logit model with the simpler equation

$$\Pr(y = 1 \mid x) = \frac{\exp(\alpha + \beta x)}{1 + \exp(\alpha + \beta x)} \tag{4.2}$$

The peculiar value assumed for $\text{Var}(\varepsilon)$ in the logit model illustrates a basic point about the identification of models with latent outcomes. In the LRM, $\text{Var}(\varepsilon)$ can be estimated because y is observed. For the BRM, the value of $\text{Var}(\varepsilon)$ must be assumed because the dependent variable is unobserved. The model is unidentified unless an assumption is made about the variance of the errors. For probit, we assume $\text{Var}(\varepsilon) = 1$ because this leads to a simple form of the model. If a different value were assumed, this would simply change the values of the structural coefficients uniformly. In the logit model, the variance is set to $\pi^2/3$ because this leads to the simple form in (4.2). Although the value assumed for $\text{Var}(\varepsilon)$ is arbitrary, the value chosen does *not* affect the computed value of the probability (see Long 1997, 49–50 for a simple proof). In effect, changing the assumed variance affects the spread of the distribution but not the proportion of the distribution above or below the threshold.

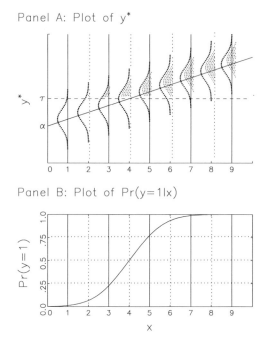

Figure 4.2: Relationship between the linear model $y^* = \alpha + \beta x + \varepsilon$ and the nonlinear probability model $\Pr(y = 1 \mid x) = F(\alpha + \beta x)$.

For both models, the probability of the event occurring is the cumulative density function (cdf) of ε evaluated at given values of the independent variables:

$$\Pr(y = 1 \mid \mathbf{x}) = F(\mathbf{x}\boldsymbol{\beta})$$

where F is the normal cdf Φ for the probit model and the logistic cdf Λ for the logit model. The relationship between the linear latent-variable model and the resulting nonlinear probability model is shown in figure 4.2 for a model with one independent variable. Panel A shows the error distribution for nine values of x, which we have labeled 1, 2, ..., 9. The area where $y^* > 0$ corresponds to $\Pr(y = 1 \mid x)$ and has been shaded. Panel B plots $\Pr(y = 1 \mid x)$ corresponding to the shaded regions in panel A. As we move from 1 to 2, only a portion of the thin tail crosses the threshold in panel A, resulting in a small change in $\Pr(y = 1 \mid x)$ in panel B. As we move from 2 to 3 to 4, thicker regions of the error distribution slide over the threshold, and the increase in $\Pr(y = 1 \mid x)$ becomes larger. The resulting curve is the well-known S-curve associated with the BRM.

4.1.2 A nonlinear probability model

Can all binary dependent variables be conceptualized as observed manifestations of some underlying latent propensity? Although philosophically interesting, perhaps, the question is of little practical importance, as the BRM can also be derived without appealing to a latent variable. This is done by specifying a nonlinear model relating the x's to the probability of an event. Following Theil (1970), the logit model can be derived by constructing a model in which the predicted $\Pr(y = 1 \mid \mathbf{x})$ is forced to be within the range 0 to 1. For example, in the linear probability model,

$$\Pr(y = 1 \mid \mathbf{x}) = \mathbf{x}\boldsymbol{\beta} + \varepsilon$$

the predicted probabilities can be greater than 1 and less than 0. To constrain the predictions to the range 0 to 1, we first transform the probability into the *odds*,

$$\Omega(\mathbf{x}) = \frac{\Pr(y = 1 \mid \mathbf{x})}{\Pr(y = 0 \mid \mathbf{x})} = \frac{\Pr(y = 1 \mid \mathbf{x})}{1 - \Pr(y = 1 \mid \mathbf{x})}$$

which indicate how often something happens ($y = 1$) relative to how often it does not happen ($y = 0$), and range from 0 when $\Pr(y = 1 \mid \mathbf{x}) = 0$ to ∞ when $\Pr(y = 1 \mid \mathbf{x}) = 1$. The log of the odds, or *logit*, ranges from $-\infty$ to ∞. This range suggests a model that is *linear in the logit*:

$$\ln \Omega(\mathbf{x}) = \mathbf{x}\boldsymbol{\beta}$$

This equation can be shown to be equivalent to the logit model from (4.2). Interpretation of this form of the logit model often focuses on factor changes in the odds, which are discussed below.

Other binary regression models are created by choosing functions of $\mathbf{x}\boldsymbol{\beta}$ that range from 0 to 1. Cumulative distribution functions have this property and readily provide several examples. For example, the cdf for the standard normal distribution results in the probit model.

4.2   Estimation using logit and probit

Logit and probit can be fitted with the following commands and their basic options:

logit *depvar* [*indepvars*] [*if*] [*in*] [*weight*] [, <u>noco</u>nstant <u>l</u>evel(*#*) or
    <u>r</u>obust <u>cl</u>uster(*varname*) <u>nolog</u>]

<u>p</u>robit *depvar* [*indepvars*] [*if*] [*in*] [*weight*] [, <u>noco</u>nstant <u>l</u>evel(*#*)
    <u>r</u>obust <u>cl</u>uster(*varname*) <u>nolog</u>]

We have never had a problem with either of these models converging, even with small samples and data with wide variation in scaling.

Variable lists

depvar is the dependent variable. *indepvars* is a list of independent variables. If *indepvars* is not included, Stata fits a model with only an intercept.

Warning For binary models, Stata defines observations in which *depvar* = 0 as negative outcomes and observations in which *depvar* equals *any* other nonmissing value (including negative values) as positive outcomes. To avoid possible confusion, we urge you to explicitly create a 0/1 variable for use as *depvar*.

Specifying the estimation sample

if and in qualifiers can be used to restrict the estimation sample. For example, if you wanted to fit a logit model for only women who went to college (as indicated by the variable wc), you could specify logit lfp k5 k618 age hc lwg if wc==1.

Listwise deletion Stata excludes cases in which there are missing values for any of the variables in the model. Accordingly, if two models are fitted using the same dataset but have different sets of independent variables, it is possible to have different samples. We recommend that you use mark and markout (discussed in chapter 3) to explicitly remove cases with missing data.

Weights

Both logit and probit can be used with fweights, pweights, and iweights. In chapter 3, we provide a brief discussion of the different types of weights and how weighting variables are specified.

Options

noconstant specifies that the model should not have a constant term. This would rarely be used for these models.

level(*#*) specifies the level of the confidence interval. By default, Stata provides 95% confidence intervals for estimated coefficients. You can also change the default level, say to a 90% interval, with the command set level 90.

or (logit only) reports the "odds ratios" defined as $\exp(\widehat{\beta})$. Standard errors and confidence intervals are similarly transformed. Alternatively, our listcoef command can be used.

robust indicates that robust variance estimates are to be used. When cluster() is specified, robust standard errors are used automatically. We provide a brief general discussion of these options in chapter 3.

cluster(*varname*) specifies that the observations are independent across the groups specified by unique values of *varname* but not necessarily within the groups.

nolog suppresses the iteration history.

Example

Our example is from Mroz's (1987) study of the labor force participation of women, using data from the 1976 Panel Study of Income Dynamics.[1] The sample consists of 753 white, married women between the ages of 30 and 60 years. The dependent variable lfp equals 1 if a woman is employed and otherwise equals 0. Because we have assigned variable labels, a complete description of the data can be obtained using describe and summarize:

```
. use http://www.stata-press.com/data/lf2/binlfp2, clear
(Data from 1976 PSID-T Mroz)

. describe lfp k5 k618 age wc hc lwg inc

              storage  display   value
variable name  type    format    label     variable label

lfp            byte    %9.0g     lfplbl    Paid Labor Force: 1=yes 0=no
k5             byte    %9.0g               # kids < 6
k618           byte    %9.0g               # kids 6-18
age            byte    %9.0g               Wife's age in years
wc             byte    %9.0g     collbl    Wife College: 1=yes 0=no
hc             byte    %9.0g     collbl    Husband College: 1=yes 0=no
lwg            float   %9.0g               Log of wife's estimated wages
inc            float   %9.0g               Family income excluding wife's
```

---

1. These data were generously made available by Thomas Mroz.

```
. summarize lfp k5 k618 age wc hc lwg inc
```

Variable	Obs	Mean	Std. Dev.	Min	Max
lfp	753	.5683931	.4956295	0	1
k5	753	.2377158	.523959	0	3
k618	753	1.353254	1.319874	0	8
age	753	42.53785	8.072574	30	60
wc	753	.2815405	.4500494	0	1
hc	753	.3917663	.4884694	0	1
lwg	753	1.097115	.5875564	-2.054124	3.218876
inc	753	20.12897	11.6348	-.0290001	96

Using these data, we fitted the model

$$\Pr\left(\texttt{lfp}=1\right)=F(\beta_0+\beta_{\texttt{k5}}\texttt{k5}+\beta_{\texttt{k618}}\texttt{k618}+\beta_{\texttt{age}}\texttt{age}$$
$$+\beta_{\texttt{wc}}\texttt{wc}+\beta_{\texttt{hc}}\texttt{hc}+\beta_{\texttt{lwg}}\texttt{lwg}+\beta_{\texttt{inc}}\texttt{inc})$$

with both the `logit` and `probit` commands, and then we created a table of results with `estimates table`:

```
. logit lfp k5 k618 age wc hc lwg inc, nolog
```

Logistic regression				Number of obs	=	753
				LR chi2(7)	=	124.48
				Prob > chi2	=	0.0000
Log likelihood = -452.63296				Pseudo R2	=	0.1209

lfp	Coef.	Std. Err.	z	P>|z|	[95% Conf.	Interval]
k5	-1.462913	.1970006	-7.43	0.000	-1.849027	-1.076799
k618	-.0645707	.0680008	-0.95	0.342	-.1978499	.0687085
age	-.0628706	.0127831	-4.92	0.000	-.0879249	-.0378162
wc	.8072738	.2299799	3.51	0.000	.3565215	1.258026
hc	.1117336	.2060397	0.54	0.588	-.2920969	.515564
lwg	.6046931	.1508176	4.01	0.000	.3090961	.9002901
inc	-.0344464	.0082084	-4.20	0.000	-.0505346	-.0183583
_cons	3.18214	.6443751	4.94	0.000	1.919188	4.445092

```
. estimates store logit
```

```
. probit lfp k5 k618 age wc hc lwg inc, nolog
Probit regression                              Number of obs   =      753
                                               LR chi2(7)      =   124.36
                                               Prob > chi2     =   0.0000
Log likelihood = -452.69496                    Pseudo R2       =   0.1208
```

lfp	Coef.	Std. Err.	z	P>|z|	[95% Conf. Interval]	
k5	-.8747112	.1135583	-7.70	0.000	-1.097281	-.6521411
k618	-.0385945	.0404893	-0.95	0.340	-.117952	.0407631
age	-.0378235	.0076093	-4.97	0.000	-.0527375	-.0229095
wc	.4883144	.1354873	3.60	0.000	.2227642	.7538645
hc	.0571704	.1240052	0.46	0.645	-.1858754	.3002161
lwg	.3656287	.0877792	4.17	0.000	.1935847	.5376727
inc	-.020525	.0047769	-4.30	0.000	-.0298875	-.0111626
_cons	1.918422	.3806536	5.04	0.000	1.172355	2.66449

```
. estimates store probit
```

Although the iteration log was suppressed by the `nolog` option, the value of the log likelihood at convergence is listed as `Log likelihood`. The information in the header and table of coefficients is in the same form as discussed in chapter 3.

We can use `estimates table` to create a table that combines the results:

```
. estimates table logit probit, b(%9.3f) t label varwidth(30)
```

Variable	logit	probit
# kids < 6	-1.463	-0.875
	-7.43	-7.70
# kids 6-18	-0.065	-0.039
	-0.95	-0.95
Wife's age in years	-0.063	-0.038
	-4.92	-4.97
Wife College: 1=yes 0=no	0.807	0.488
	3.51	3.60
Husband College: 1=yes 0=no	0.112	0.057
	0.54	0.46
Log of wife's estimated wages	0.605	0.366
	4.01	4.17
Family income excluding wife's	-0.034	-0.021
	-4.20	-4.30
Constant	3.182	1.918
	4.94	5.04

```
                            legend: b/t
```

The estimated coefficients differ from logit to probit by a factor of about 1.7. For example, the ratio of the logit to probit coefficient for k5 is 1.67 and for inc is 1.68. This illustrates how the magnitudes of the coefficients are affected by the assumed $Var(\varepsilon)$. The exception to the ratio of 1.7 is the coefficient for hc. This estimate has a great deal of sampling variability (i.e., a large standard error), and in such cases, the 1.7 rule often does not hold. Values of the z-tests are quite similar because they are not affected by the assumed $\text{Var}(\varepsilon)$. The z-test statistics are not exactly the same because the two models assume different distributions of the errors.

4.2.1 Observations predicted perfectly

ML estimation is not possible when the dependent variable does not vary within one of the categories of an independent variable. Say that you are fitting a logit model predicting whether a person voted in the last election, vote, and that one of the independent variables is whether the person is enrolled in college, college. If you had a small number of college students in your sample, it is possible that none of them voted in the last election. That is, vote==0 every time college==1. The model cannot be fitted because the coefficient for college is effectively negative infinity. Stata's solution is to drop the variable college along with all observations where college==1. For example,

```
. logit vote college phd, nolog
Note: college!=0 predicts failure perfectly
      college dropped and 4 obs not used

Logistic regression                        Number of obs   =         299
   (output omitted)
```

4.3 Hypothesis testing with test and lrtest

Hypothesis tests of regression coefficients can be conducted with the z-statistics in the estimation output, with test for Wald tests of simple and complex hypotheses, and with lrtest for the corresponding likelihood-ratio tests. We consider the use of each of these to test hypotheses involving only one coefficient, and then we show you how both test and lrtest can be used to test hypotheses involving multiple coefficients.

4.3.1 Testing individual coefficients

If the assumptions of the model hold, the ML estimators (e.g., the estimates produced by logit or probit) are distributed asymptotically normally:

$$\widehat{\beta}_k \overset{a}{\sim} \mathcal{N}\left(\beta_k, \sigma^2_{\widehat{\beta}_k}\right)$$

The hypothesis $H_0: \beta_k = \beta^*$ can be tested with the z-statistic:

$$z = \frac{\widehat{\beta}_k - \beta^*}{\widehat{\sigma}_{\widehat{\beta}_k}}$$

z is included in the output from logit and probit. Under the assumptions justifying ML, if H_0 is true, then z is distributed approximately normally with a mean of zero and a variance of one for large samples. This is shown in the following figure, where the shading shows the rejection region for a two-tailed test at the .05 level:

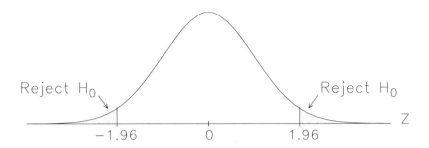

For example, consider the results for variable k5 from the logit output generated in
section 4.2:

lfp	Coef.	Std. Err.	z	P>|z|	[95% Conf. Interval]
k5	-1.462913	.1970006	-7.43	0.000	-1.849027 -1.076799

(*output omitted*)

We conclude that having young children has a significant effect on the probability of
working ($z = -7.43$, $p < 0.01$ for a two-tailed test).

One- and two-tailed tests

The probability levels in the output for estimation commands are for two-tailed tests.
That is, the result corresponds to the area of the curve that is either greater than $|z|$
or less than $-|z|$. When past research or theory suggests the sign of the coefficient, a
one-tailed test can be used, and H_0 is rejected only when z is in the *expected* tail. For
example, assume that my theory proposes that having children can have only a negative
effect on labor force participation. For k618, $z = -0.95$ and P > | z | is .342. This
is the proportion of the sampling distribution for z that is less than -0.95 or greater
than 0.95. Because we want a one-tailed test, and the coefficient is in the expected
direction, we want only the proportion of the distribution that is less than -0.95, which
is $.342/2 = .171$. We conclude that having older children does not significantly affect a
woman's probability of working ($z = -0.95$, $p = .17$ for a one-tailed test).

You should divide P > | z | by 2 only when the estimated coefficient is in the
expected direction. Suppose I am testing a theory that having a husband who went to
college has a negative effect on labor force participation, but the estimated coefficient is
positive with $z = 0.542$ and P > | z | is .588. The one-tailed significance level would be
the percentage of the distribution less than .542 (not the percentage less than $-.542$),
which is equal to $1 - (.588/2) = .706$, not $.588/2 = .294$. We conclude that having
a husband who attends college does not significantly affect a woman's probability of
working ($z = 0.542$, $p = .71$ for a one-tailed test).

Testing single coefficients using test

The z-test included in the output of estimation commands is a Wald test, which can also be computed using test. For example, to test $H_0: \beta_{k5} = 0$,

```
. test k5
 ( 1)  k5 = 0

           chi2( 1) =    55.14
         Prob > chi2 =    0.0000
```

We can conclude that the effect of having young children on the probability of entering the labor force is significant at the .01 level ($X^2 = 55.14$, $df = 1$, $p < .01$).

The value of a chi-squared test with 1 degree of freedom is identical to the square of the corresponding z-test. For example, using Stata's display as a calculator

```
. display sqrt(55.14)
7.4256313
```

This corresponds to -7.43 from the logit output. Some packages, such as SAS, present chi-squared tests rather than the corresponding z-test.

Testing single coefficients using lrtest

An LR test is computed by comparing the log likelihood from a full model with that of a restricted model. To test a single coefficient, we begin by fitting the full model and storing the results:

```
. logit lfp k5 k618 age wc hc lwg inc, nolog
Logistic regression                        Number of obs   =        753
                                           LR chi2(7)      =     124.48
                                           Prob > chi2     =     0.0000
Log likelihood = -452.63296                Pseudo R2       =     0.1209
   (output omitted )
. estimates store fmodel
```

Then we fit the model without k5 and run lrtest:

```
. logit lfp k618 age wc hc lwg inc, nolog
Logistic regression                        Number of obs   =        753
                                           LR chi2(6)      =      58.00
                                           Prob > chi2     =     0.0000
Log likelihood = -485.87503                Pseudo R2       =     0.0563
   (output omitted )
. estimates store nmodel
. lrtest fmodel nmodel
Likelihood-ratio test                      LR chi2(1)   =      66.48
(Assumption: nmodel nested in fmodel)      Prob > chi2 =     0.0000
```

The resulting LR test can be interpreted as indicating that the effect of having young children is significant at the .01 level ($LRX^2 = 66.48$, $df = 1$, $p < .01$).

4.3.2 Testing multiple coefficients

Often you may wish to test complex hypotheses that involve more than one coefficient. For example, we have two variables that reflect education in the family, hc and wc. The conclusion that education has (or does not have) a significant effect on labor force participation cannot be based on a pair of tests of single coefficients. But a joint hypothesis can be tested using either test or lrtest.

Testing multiple coefficients using test

To test that the effect of the wife attending college and of the husband attending college on labor force participation are both equal to 0, H_0: $\beta_{wc} = \beta_{hc} = 0$, we fit the full model and then

```
. test hc wc
 ( 1)   hc = 0
 ( 2)   wc = 0
           chi2(  2) =    17.66
         Prob > chi2 =    0.0001
```

We conclude that the hypothesis that the effects of the husband's and the wife's education are simultaneously equal to zero can be rejected at the .01 level ($X^2 = 17.66$, $df = 2$, $p < .01$).

This form of the test command can be readily extended to hypotheses regarding more than two independent variables by listing more variables; for example, test wc hc k5.

test can also be used to test the equality of coefficients. For example, to test that the effect of the wife attending college on labor force participation is equal to the effect of the husband attending college, H_0: $\beta_{wc} = \beta_{hc}$:

```
. test hc=wc
 ( 1)  - wc + hc = 0
           chi2(  1) =     3.54
         Prob > chi2 =    0.0600
```

Here test has translated $\beta_{wc} = \beta_{hc}$ into the equivalent expression $-\beta_{wc} + \beta_{hc} = 0$. We conclude that the null hypothesis that the effects of husband's and wife's education are equal is marginally significant at the .05 level ($X^2 = 3.54$, $df = 1$, $p = .06$). This result suggests that we have weak evidence that the effects are not equal.

Testing multiple coefficients using lrtest

To compute an LR test of multiple coefficients, we first fit the full model and then save the results using the command `estimates store`. Then to test the hypothesis that the effect of the wife attending college and of the husband attending college on labor force participation are both equal to zero, H_0: $\beta_{\mathtt{wc}} = \beta_{\mathtt{hc}} = 0$, we fit the model that excludes these two variables and then run `lrtest`:

```
. logit lfp k5 k618 age wc hc lwg inc, nolog
  (output omitted )
. estimates store fmodel
. logit lfp k5 k618 age lwg inc, nolog
  (output omitted )
. estimates store nmodel
. lrtest fmodel nmodel
Likelihood-ratio test                                LR chi2(2)  =      18.50
(Assumption: nmodel nested in fmodel)                Prob > chi2 =     0.0001
```

We conclude that the hypothesis that the effects of the husband's and the wife's education are simultaneously equal to zero can be rejected at the .01 level ($LRX^2 = 18.50$, $df = 2$, $p < .01$).

This logic can be extended to exclude other variables. Say that we wish to test the null hypothesis that all the effects of the independent variables are simultaneously equal to zero. We do not need to fit the full model again because the results are still saved from our use of `estimates store fmodel` above. We fit the model with no independent variables and run `lrtest`:

```
. logit lfp, nolog
  (output omitted )
. estimates store intercept_only
. lrtest fmodel intercept_only
Likelihood-ratio test                                LR chi2(7)  =     124.48
(Assumption: intercept_only  nested in fmodel)       Prob > chi2 =     0.0000
```

We can reject the hypothesis that all coefficients except the intercept are zero at the .01 level ($LRX^2 = 124.48$, $df = 7$, $p < .01$). This test is identical to the test in the header of the `logit` output:
`LR chi2(7) = 124.48`.

4.3.3   Comparing LR and Wald tests

Although the LR and Wald tests are *asymptotically* equivalent, their values differ in finite samples. For example,

Hypothesis	df	LR test G^2	p	Wald test W	p
$\beta_{k5} = 0$	1	66.48	<.01	55.14	<.01
$\beta_{wc} = \beta_{hc} = 0$	2	18.50	<.01	17.66	<.01
All slopes $= 0$	7	124.48	<.01	95.0	< .01

Statistical theory is unclear on whether the LR or Wald test is to be preferred in models for categorical outcomes, although many statisticians, ourselves included, prefer the LR test. The choice of which test to use is often determined by convenience, personal preference, and convention within an area of research.

4.4   Residuals and influence using predict

Examining residuals and outliers is an important way to assess the fit of a regression model. *Residuals* are the difference between a model's predicted and observed outcome for each observation in the sample. Cases that fit poorly (i.e., have large residuals) are known as *outliers*. When an observation has a large effect on the estimated parameters, it is said to be *influential*.

Not all outliers are influential, as figure 4.3 shows. In the top panel, we show a scatterplot of some simulated data, and we have drawn the line that results from the linear regression of y on x. The residual of any observation is its vertical distance from the regression line. The observation highlighted by the box has a very large residual and so is an outlier. Even so, it is not very influential on the slope of the regression line. In the bottom panel, the only observation whose value has changed is the highlighted one. Now the magnitude of the residual for this observation is much smaller, but it is very influential; its presence is entirely responsible for the slope of the new regression line being positive instead of negative.

(Continued on next page)

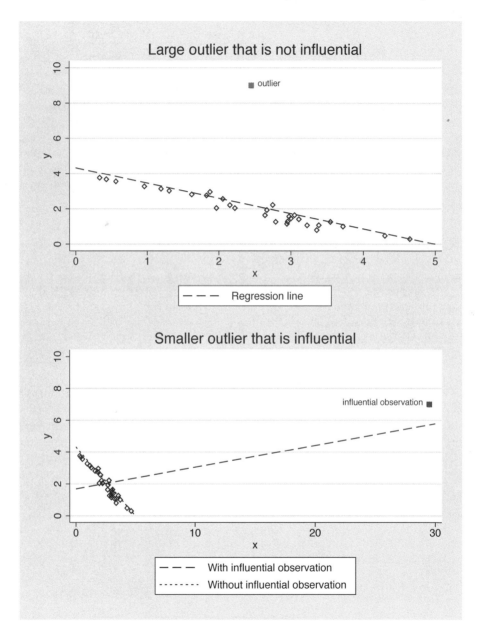

Figure 4.3: The distinction between an outlier and an influential observation.

Building on the analysis of residuals and influence in the linear regression model (see Fox 1991 and Weisberg 1980, chapter 5 for details), Pregibon (1981) extended these ideas to the BRM.

4.4.1   Residuals

If we define the predicted probability for a given set of independent variables as

$$\pi_i = \Pr\left(y_i = 1 \mid \mathbf{x}_i\right)$$

then the deviations $y_i - \pi_i$ are heteroskedastic, with

$$\mathrm{Var}\left(y_i - \pi_i \mid \mathbf{x}_i\right) = \pi_i\left(1 - \pi_i\right)$$

This implies that the variance in a binary outcome is greatest when $\pi_i = .5$ and least as π_i approaches 0 or 1. For example, $.5\left(1 - .5\right) = .25$ and $.01\left(1 - .01\right) = .0099$. In other words, there is heteroskedasticity that depends on the probability of a positive outcome. This suggests the *Pearson residual*, which divides the residual $y - \widehat{\pi}$ by its standard deviation:

$$r_i = \frac{y_i - \widehat{\pi}_i}{\sqrt{\widehat{\pi}_i\left(1 - \widehat{\pi}_i\right)}}$$

Large values of r suggest a failure of the model to fit a given observation. Pregibon (1981) showed that the variance of r is not 1, as $\mathrm{Var}(y_i - \widehat{\pi}_i) \neq \widehat{\pi}_i\left(1 - \widehat{\pi}_i\right)$, and proposed the *standardized Pearson residual*

$$r_i^{\mathrm{Std}} = \frac{r_i}{\sqrt{1 - h_{ii}}}$$

where

$$h_{ii} = \widehat{\pi}_i\left(1 - \widehat{\pi}_i\right)\mathbf{x}_i\, \widehat{\mathrm{Var}}\left(\widehat{\boldsymbol{\beta}}\right)\mathbf{x}_i' \tag{4.3}$$

Although r^{Std} is preferred over r because of its constant variance, we find that the two residuals are often similar in practice. But, because r^{Std} is simple to compute in Stata, we recommend that you use this measure.

Example

An *index plot* is a useful way to examine residuals by simply plotting them against the observation number. The standardized residuals can be computed by specifying the `rs` option with `predict`. For example,

```
. logit lfp k5 k618 age wc hc lwg inc, nolog
  (output omitted)
. predict rstd, rs
. label var rstd "Standardized Residual"
. sort inc, stable
. generate index = _n
. label var index "Observation Number"
```

Here we first fit the logit model. Second, we use the `rs` option for `predict` to specify that we want standardized residuals, which are placed in a new variable that we have named `rstd`. Third, we sort the cases by `income`, so that observations are ordered from lowest to highest incomes. This results in a plot of residuals in which cases are ordered from low income to high income. The next line creates a new variable `index`, whose value for each observation is that observation's number (i.e., row) in the dataset. Note that `_n` on the right side of `generate` inserts the observation number. All that remains is to plot the residuals against the index using the commands[2]

```
. graph twoway scatter rstd index, xlabel(0(200)800) ylabel(-4(2)4) ///
> xtitle("Observation Number") yline(0) msymbol(Oh)
```

which produces the following index plot of standardized Pearson residuals:

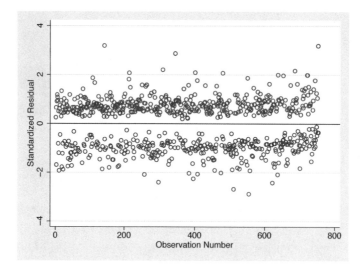

There is no hard-and-fast rule for what counts as a "large" residual. Indeed, in their detailed discussion of residuals and outliers in the binary regression model, Hosmer and Lemeshow (2000, 176) sagely caution that it is impossible to provide any absolute standard: "In practice, an assessment of 'large' is, of necessity, a judgment call based on experience and the particular set of data being analyzed".

One way to search for problematic residuals is to sort the residuals by the value of a variable that you think may be a problem for the model. Here we sorted the data by `income` before plotting. If this variable had been primarily responsible for the lack of fit of some observations, the plot would show a disproportionate number of cases with large residuals among either the low-income or the high-income observations in our model. However, this does not appear to be the case for these data.

Still, in our plot, several residuals stand out as being large relative to the others. In such cases, it is important to identify the specific observations with large residuals for further inspection. We can do this by instructing `graph` to use the observation number to label each point in our plot. Recall that we just created a new variable called `index`

---

2. The `///` is just a way of executing long lines in do-files. You should not type these characters if you are working from the Command window.

whose value is equal to the observation number for each observation. We want the values of this index variable to be the marker symbols. We do this by labeling the marker with the index value and then placing the label over an invisible marker. In the command below, `msymbol(none)` makes the marker symbol invisible, `mlabel(index)` specifies that the variable `index` contains the labels, and `mlabposition(0)` causes the label to be positioned where the marker would have appeared. For example,

```
. graph twoway scatter rstd index, xlabel(0(200)800) ylabel(-4(2)4) ///
> xtitle("Observation Number") yline(0)                            ///
> msymbol(none) mlabel(index) mlabposition(0)
```

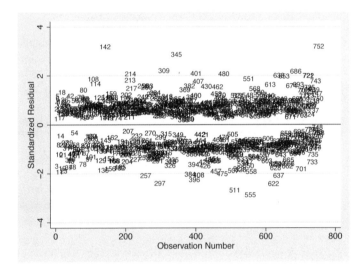

Although labeling points with observations leads to chaos where there are many points, it effectively highlights and identifies the isolated cases. You can then easily list these cases. For example, observation 142 stands out and should be examined:

```
. list in 142, noobs
```

lfp	k5	k618	age	wc	hc	lwg	inc	rstd	index
inLF	1	2	36	NoCol	NoCol	-2.054124	11.2	3.191524	142

(Continued on next page)

We can also use `list` to list all observations with large residuals:

```
. list rstd index if rstd>2.5 | rstd<-2.5
```

	rstd	index
142.	3.191524	142
345.	2.873378	345
511.	-2.677243	511
555.	-2.871972	555
752.	3.192648	752

We can then check the listed cases to see if there are problems.

An initial review of the cases with large positive or negative residuals suggests that these could be related to the number of children a scientist has. To explore this further, we modify the last graph to plot only large residuals and to label these residuals with the number of young children a scientist has. To do this,

```
. graph twoway scatter rstd index if (rstd>1.7) | (rstd<-1.7), ///
> msymbol(none) mlabel(k5) mlabposition(0)                     ///
> caption("Values indicate # of young children")               ///
> xlabel(0(200)800) xtitle("Observation Number")               ///
> ylabel(-4(2)4) yline(0)
```

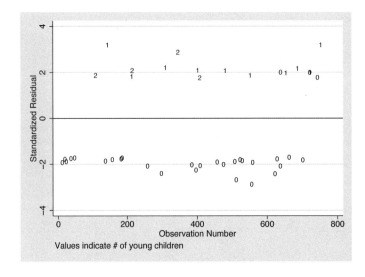

Here all the cases with negative residuals correspond to individuals without young children, whereas most cases with positive residuals have children. This suggests that we should further consider the way in which the number of children has been modeled.

Regardless of which method is used, further analyses of the highlighted cases might reveal either incorrectly coded data or some inadequacy in the specification of the model.

Cases with large positive or negative residuals should *not* simply be discarded from the analysis but rather should be examined to determine why they fit so poorly.

4.4.2 Influential cases

As shown in figure 4.3, large residuals do not necessarily have a strong influence on the estimated parameters, and observations with relatively small residuals can have a large influence. *Influential* points are also sometimes called *high-leverage* points. These can be determined by examining the change in the estimated $\widehat{\boldsymbol{\beta}}$ that occurs when the ith observation is deleted. Although estimating a new logit for each case is usually impractical (although as the speed of computers increases, this may soon no longer be so), Pregibon (1981) derived an approximation that requires fitting the model only once. This measure summarizes the effect of removing the ith observation on the entire vector $\widehat{\boldsymbol{\beta}}$, which is the counterpart to Cook's distance for the linear regression model. The measure is defined as

$$P_i = \frac{r_i^2 h_{ii}}{(1 - h_{ii})^2}$$

where h_{ii} was defined in (4.3). In Stata, which refers to Cook's distance as `dbeta`, we can compute and plot Cook's distance as follows:

```
. predict cook, dbeta
. label var cook "Cook's Statistic"
. graph twoway scatter cook index, xlabel(0(200)800) ylabel(0(.1).3) ///
> xtitle("Observation Number") yline(.1 .2)                        ///
> msymbol(none) mlabel(index) mlabposition(0)
```

These commands produce the following plot, which shows that cases 142, 309, and 752 merit further examination:

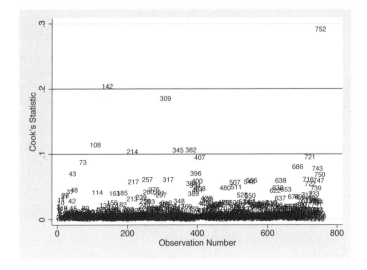

Methods for plotting residuals and outliers can be extended in many ways, including plots of different diagnostics against one another. Details of these plots are found in Cook and Weisberg (1999), Hosmer and Lemeshow (2000), and Landwehr, Pregibon, and Shoemaker (1984).

4.4.3   Least likely observations

A common motivation for examining residuals in the linear regression model is to uncover the largest residuals and to check if there is some reason why the model fits these observations so poorly. Observations with large residuals are those for which the observed values of the dependent variable are most "surprising" given the regression coefficients and the values of the independent variables. Maximum likelihood estimates maximize the probability of observing the outcomes that were actually observed. In this context, we can think of the most surprising outcomes as those that have the smallest predicted probabilities of observing that outcome. These cases may warrant closer inspection precisely because observations with large residuals do in the more familiar linear regression model. Our command `leastlikely` (Freese 2002) will list the least likely observations. For example, for a binary model, `leastlikely` will list both the observations with the smallest $\widehat{\Pr}(y = 0)$ among cases in which $y = 0$ and the smallest $\widehat{\Pr}(y = 1)$ for cases in which $y = 1$. In addition to `logit` and `probit`, `leastlikely` can be used after most binary models in which the option `p` for `predict` generates the predicted probabilities of a positive outcome (e.g., `cloglog`, `scobit`, `hetprob`) and after many models for ordinal or nominal outcomes in which the option `outcome(#)` for `predict` generates the predicted probability of outcome # (e.g., `ologit`, `oprobit`, `mlogit`, `mprobit`, `slogit`). `leastlikely` is not appropriate for models in which the probabilities produced by `predict` are probabilities within groups or panels, such as `clogit`, `nlogit`, or `asmprobit`.

Syntax

The syntax for `leastlikely` is as follows:

`leastlikely` [*varlist*] [*if*] [*in*] [, n(#) generate(*varname*) [no]display nolabel noobs]

where *varlist* contains any variables whose values are to be listed in addition to the observation numbers and probabilities.

Options

n(#) specifies the number of observations to be listed for each level of the outcome variable. The default is n(5). For multiple observations with identical predicted probabilities, all observations will be listed.

generate(*varname*) specifies that the probabilities of observing the outcome that was observed should be stored in *varname*.

Options controlling the list of values

leastlikely can also include any of the options available after list. These include the following:

[no]display forces the format into display or tabular (nodisplay) format. If you do not specify one of these two options, Stata chooses the one it decides will be most readable.

nolabel causes numeric values rather than labels to be displayed.

noobs suppresses printing of the observation numbers.

For example, we can use leastlikely to identify the least likely observations for our model of labor force participation and to list the values of the variables k5, k618, and wc for these observations.

```
. use http://www.stata-press.com/data/lf2/binlfp2
(Data from 1976 PSID-T Mroz)
. logit lfp k5 k618 age wc hc lwg inc
  (output omitted)
. leastlikely k5 k618 wc
Outcome: 0 (NotInLF)
```

	Prob	k5	k618	wc
60.	.1231792	0	1	College
172.	.1490344	0	2	College
221.	.1470691	0	2	College
235.	.1666356	0	4	College
252.	.1088271	0	0	College

```
Outcome: 1 (inLF)
```

	Prob	k5	k618	wc
338.	.1760865	1	2	College
534.	.0910262	1	2	NoCol
568.	.178205	1	5	NoCol
635.	.0916614	1	3	College
662.	.1092709	2	0	NoCol

Among women not in the labor force (lfp is 0), we find that the lowest predicted probability of not being in the labor force occurs for those who attended college and have young children. For women in the labor force (lfp is 1) with the lowest probabilities of being in the labor force, all individuals have young children and most have more than one older child. This suggests further consideration of how labor force participation is affected by having children in the family.

4.5   Measuring fit

As discussed in chapter 3, a scalar measure of fit can be useful in comparing competing models. Within a substantive area, measures of fit provide a *rough* index of whether a model is adequate. For example, if prior models of labor force participation routinely have values of .4 for a given measure of fit, you would expect that new analyses with a different sample or with revised measures of the variables would result in a similar value for that measure of fit. Remember: there is *no convincing evidence that selecting a model that maximizes the value of a given measure of fit results in a model that is optimal in any sense other than the model's having a larger value of that measure.* Details on these measures are presented in chapter 3.

4.5.1   Scalar measures of fit using fitstat

To illustrate the use of scalar measures of fit, consider two models. M_1 contains our original specification of independent variables k5, k618, age, wc, hc, lwg, and inc. M_2 drops the variables k618, hc, and lwg and adds agesq, which is the square of age. These models are fitted, and measures of fit are computed:

```
. quietly logit lfp k5 k618 age wc hc lwg inc, nolog
. estimates store model1
. quietly fitstat, save
. gen agesq = age*age
. quietly logit lfp k5 age agesq wc inc, nolog
. estimates store model2
```

We used `quietly` to suppress the output from `logit` and now use `estimates table` to combine the results from the two logits:

```
. estimates table model1 model2, b(%9.3f) t
```

Variable	model1	model2
k5	-1.463	-1.380
	-7.43	-7.06
k618	-0.065	
	-0.95	
age	-0.063	0.057
	-4.92	0.50
wc	0.807	1.094
	3.51	5.50
hc	0.112	
	0.54	
lwg	0.605	
	4.01	
inc	-0.034	-0.032
	-4.20	-4.18
agesq		-0.001
		-1.00
_cons	3.182	0.979
	4.94	0.40

legend: b/t

The output from `fitstat` for M_1 was suppressed, but the results were saved to be listed by a second call to `fitstat` using the `diff` option:

```
. fitstat, diff
Measures of Fit for logit of lfp
                                Current            Saved         Difference
Model:                          logit              logit
N:                                 753                753                   0
Log-Lik Intercept Only        -514.873           -514.873               0.000
Log-Lik Full Model            -461.653           -452.633              -9.020
D                          923.306(747)       905.266(745)          18.040(2)
LR                          106.441(5)         124.480(7)            18.040(2)
Prob > LR                        0.000              0.000               0.000
McFadden's R2                    0.103              0.121              -0.018
McFadden's Adj R2                0.092              0.105              -0.014
ML (Cox-Snell) R2                0.132              0.152              -0.021
Cragg-Uhler(Nagelkerke) R2       0.177              0.204              -0.028
McKelvey & Zavoina's R2          0.182              0.217              -0.035
Efron's R2                       0.135              0.155              -0.020
Variance of y*                   4.023              4.203              -0.180
Variance of error                3.290              3.290               0.000
Count R2                         0.677              0.693              -0.016
Adj Count R2                     0.252              0.289              -0.037
AIC                              1.242              1.223               0.019
AIC*n                          935.306            921.266              14.040
BIC                          -4024.871          -4029.663               4.791
BIC'                           -73.321            -78.112               4.791
BIC used by Stata              963.050            958.258               4.791
AIC used by Stata             935.306            921.266              14.040
Difference of    4.791 in BIC' provides positive support for saved model.
Note: p-value for difference in LR is only valid if models are nested.
```

These results illustrate the limitations inherent in scalar measures of fit. M_2 deleted two variables that were not significant and one that was from M_1. It added a new variable that was not significant in the new model. Because the models are not nested, they cannot be compared using a difference of chi-squares test.[3] What do the fit statistics show? First, the values of the pseudo-R^2s are slightly larger for M_1. If you take the pseudo-R^2s as evidence for the best model, which we do not, there is some evidence preferring M_1. Second, the BIC statistic is smaller for M_1, which provides support for that model. Following Raftery's (1996) guidelines, one would say that there is positive (neither weak nor strong) support for M_1.

4.5.2   Hosmer–Lemeshow statistic

Earlier we showed how to use `predict` to compute the predicted probabilities for each observation in the sample. The idea of the Hosmer–Lemeshow (HL) test statistic is to compare these predicted probabilities with the observed data (Lemeshow and Hosmer 1982; Hosmer and Lemeshow 1980). To explain what this statistic is doing, we go through the steps that are used to compute HL.

---

3. `fitstat, diff` computes the difference between all measures, even if the models are not nested. As with the Stata command `lrtest`, it is up to the user to determine if it makes sense to interpret the computed difference.

1. Fit the model.

2. Compute the predicted probabilities $\hat{\pi}_i$.

3. Sort the data from the smallest value of $\hat{\pi}_i$ to the largest.

4. Divide the observations into G groups, where 10 groups are often used. Each group will have $n_g \approx \frac{N}{G}$ cases (if G does not divide equally into N, the group sizes will differ slightly). The first group will have the n_1 smallest values of $\hat{\pi}_i$, and so on.

5. Within each group, compute the mean prediction $\overline{\pi}_g = \sum\limits_{\text{Group g}} \hat{\pi}_i / n_g$ and the mean number of observed ones, $\overline{y}_g = \sum\limits_{\text{Group g}} y_i / n_g$.

6. HL is a Pearson χ^2 statistic with $G - 2$ degrees of freedom:

$$\text{HL} = \sum_{g=1}^{G} \frac{\left(n_g \overline{y}_g - n_g \overline{\pi}_g\right)^2}{n_g \overline{\pi}_g \left(1 - \overline{\pi}_g\right)}$$

Hosmer and Lemeshow (2000) ran extensive simulations that showed that HL is approximately distributed as χ^2 if the model is correct. But, since the value of HL depends on the number of groups chosen, it is better to think of this statistic as a guide to assessing the fit of a model rather than a formal test. When using this statistic, keep in mind what Hosmer and Lemeshow (2000) wrote: "The advantage of a summary goodness-of-fit statistics like [HL] is that it provides a single, easily interpretable value that can be used to assess fit. The great disadvantage is that in the process of grouping we may miss an important deviation from fit due to the small number of individual data points. Hence, we advocate that, before finally accepting that a model fits, an analysis of the individual residuals and relevant diagnostic statistics be performed."

For our example of labor force participation, we computed the HL statistic using the command

```
. estat gof, group(10)
Logistic model for lfp, goodness-of-fit test

  (Table collapsed on quantiles of estimated probabilities)
        number of observations =        753
            number of groups =         10
    Hosmer-Lemeshow chi2(8) =      23.79
              Prob > chi2 =       0.0025
```

The HL statistic suggests that the model does not fit well. However, if you experiment with different values for group(), you will see that the p-value is sensitive to the number of groups used. Still, the results suggests that we should explore the fit of the model. One way to do this is to make a lowess graph comparing predicted probabilities to a moving average of the proportion of cases that are one. This can be done with the following commands:

```
. predict p1
(option p assumed; Pr(lfp))

. lowess lfp p1, ylabel(0(.2)1, grid) xlabel(0(.2)1, grid) ///
> addplot(function y = x, legend(off))
```

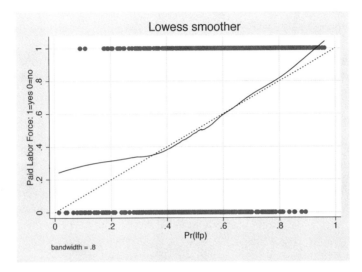

Roughly speaking, the solid line shows the fraction of observed cases that equal 1 at each level of the model's predicted probability of observing a 1. The closer the solid line to the diagonal, dashed line, the better the fit of the model. The graph suggests that the model fails in predicting the lower probabilities of being in the labor force, where the fractions of observed cases exceeds the predicted probabilities.

4.6   Interpretation using predicted values

Because the BRM is nonlinear, no approach to interpretation can fully describe the relationship between a variable and the outcome. We suggest that you try a variety of methods, with the goal of finding an elegant way to present the results that does justice to the complexities of the nonlinear model.

In general, the estimated parameters from the BRM do not provide directly useful information for understanding the relationship between the independent variables and the outcome. With the exception of the rarely used method of interpreting the latent variable (which we discuss in our treatment of ordinal models in chapter 5), substantively meaningful interpretations are based on predicted probabilities and functions of those probabilities (e.g., ratios, differences). As shown in figure 4.1, for a given set of values of the independent variables, the predicted probability in BRMs is defined as

$$\text{Logit: } \widehat{\Pr}(y=1 \mid \mathbf{x}) = \Lambda\left(\mathbf{x}\widehat{\boldsymbol{\beta}}\right) \qquad \text{Probit: } \widehat{\Pr}(y=1 \mid \mathbf{x}) = \Phi\left(\mathbf{x}\widehat{\boldsymbol{\beta}}\right)$$

where Λ is the cdf for the logistic distribution with variance $\pi^2/3$, and Φ is the cdf for the normal distribution with variance 1. For any set of values of the independent variables, the predicted probability can be computed. Several commands in Stata and our pr* commands make it simple to work with these predicted probabilities.

4.6.1 Predicted probabilities with predict

After running logit or probit,

predict *newvarname* $\left[\,if\,\right]$ $\left[\,in\,\right]$

can be used to compute the predicted probability of a positive outcome for each observation, given the values on the independent variables for that observation. The predicted probabilities are stored in the new variable *newvarname*. The predictions are computed for all cases in memory that do not have missing values for the variables in the model, regardless of whether if and in had been used to restrict the estimation sample. For example, if you estimate logit lfp k5 age if wc==1, only 212 cases are used. But predict *newvarname* computes predictions for the entire dataset, 753 cases. If you want predictions only for the estimation sample, you can use the command predict *newvarname* if e(sample)==1.[4]

predict can be used to examine the range of predicted probabilities from your model. For example,

```
. predict prlogit
(option p assumed; Pr(lfp))

. summarize prlogit
```

Variable	Obs	Mean	Std. Dev.	Min	Max
prlogit	753	.5683931	.1944213	.0139875	.9621198

The message (option p assumed; Pr(lfp)) reflects that predict can compute many different quantities. Because we did not specify an option indicating which quantity to predict, option p for predicted probabilities was assumed, and the new variable prlogit was given the variable label Pr(lfp). summarize computes summary statistics for the new variable and shows that the predicted probabilities in the sample range from .014 to .962, with a mean predicted probability of being in the labor force of .568.

---

4. Stata estimation commands create the variable e(sample), indicating whether a case was used when fitting a model. Accordingly, the condition if e(sample)==1 selects only cases used in the last estimation.

We can use `dotplot` to plot the predicted probabilities for our sample:

```
. label var prlogit "Logit: Pr(lfp)"
. dotplot prlogit, ylabel(0(.2)1)
```

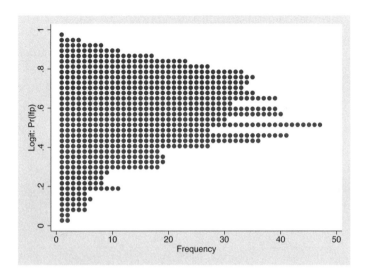

The plot clearly shows that the predicted probabilities for individual observations span almost the entire range from 0 to 1 but that roughly two-thirds of the observations have predicted probabilities between .40 and .80.

`predict` can also be used to demonstrate that the predictions from logit and probit models are essentially identical. Even though the two models make different assumptions about Var(ε), these differences are absorbed in the relative magnitudes of the estimated coefficients. To see this, we first fit the two models and compute their predicted probabilities:

```
. use http://www.stata-press.com/data/lf2/binlfp2, clear
(Data from 1976 PSID-T Mroz)
. logit lfp k5 k618 age wc hc lwg inc, nolog
  (output omitted)
. predict prlogit
(option p assumed; Pr(lfp))
. label var prlogit "Logit: Pr(lfp)"
. probit lfp k5 k618 age wc hc lwg inc, nolog
  (output omitted)
. predict prprobit
(option p assumed; Pr(lfp))
. label var prprobit "Probit: Pr(lfp)"
```

Next we check the correlation between the two sets of predicted values:

```
. pwcorr prlogit prprobit
             |  prlogit prprobit
   ----------+------------------
     prlogit |  1.0000
    prprobit |  0.9998   1.0000
```

The extremely high correlation is confirmed by plotting them against one another:

```
. graph twoway scatter prlogit prprobit,   ///
> xlabel(0(.25)1) ylabel(0(.25)1)          ///
> xline(.25(.25)1) yline(.25(.25)1)        ///
> plotregion(margin(zero)) msymbol(Oh)
```

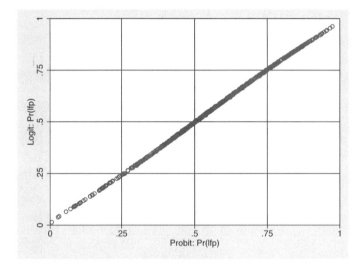

For predictions, there is little reason to prefer either logit or probit. If your substantive findings turn on whether you used logit or probit, *we would not place much confidence in either result.* In our own research, we tend to use logit, primarily because of the availability of interpretation in terms of odds and odds ratios (discussed below).

Overall, examining predicted probabilities for the cases in the sample provides an initial check of the model. To better understand and present the substantive findings, it is usually more effective to compute predictions at specific, substantively informative values. Our commands prvalue, prtab, and prgen are designed to make this simple.

4.6.2 Individual predicted probabilities with prvalue

A table of probabilities for ideal types of people (or countries, cows, or whatever you are studying) can quickly summarize the effects of key variables. In our example of labor force participation, we could compute predicted probabilities of labor force participation for women in these three types of families:

- young, low-income and low-education families with young children

- highly educated, middle-aged couples with no children at home

- an "average family" defined as having the mean on all variables.

This can be done with a series of calls to `prvalue` (see chapter 3 for a discussion of options for this command):[5]

```
. * young, low income, low education families with young children.
. prvalue, x(age=35 k5=2 wc=0 hc=0 inc=15) rest(mean)

logit: Predictions for lfp

Confidence intervals by delta method

                            95% Conf. Interval
    Pr(y=inLF|x):        0.1318   [ 0.0556,    0.2081]
    Pr(y=NotInLF|x):     0.8682   [ 0.7919,    0.9444]

              k5         k618        age         wc         hc         lwg
x=             2    1.3532537         35          0          0    1.0971148

             inc
x=            15
```

We have set the values of the independent variables to those that define our first type of family, with other variables held at their mean. The output shows the predicted probability of working, the confidence interval for that probability, and the specified values for the independent variables. At these values, we are 95% confident that the probability of being in the labor force is between .056 and .208. This process is repeated for the other ideal types.

```
. * highly educated families with no children at home.
. prvalue, x(age=50 k5=0 k618=0 wc=1 hc=1) rest(mean)

logit: Predictions for lfp

Confidence intervals by delta method

                            95% Conf. Interval
    Pr(y=inLF|x):        0.7166   [ 0.6333,    0.7999]
    Pr(y=NotInLF|x):     0.2834   [ 0.2001,    0.3667]

              k5         k618        age         wc         hc         lwg
x=             0            0         50          1          1    1.0971148

             inc
x=      20.128965
```

---

5. `mean` is the default setting for the `rest()` option, so `rest(mean)` does not need to be specified. We include it in many of our examples anyway, because its use emphasizes that the results are contingent on specified values for *all* of the independent variables.

```
. * an average person
. prvalue, rest(mean)

logit: Predictions for lfp

Confidence intervals by delta method
                                95% Conf. Interval
    Pr(y=inLF|x):       0.5778   [ 0.5392,    0.6164]
    Pr(y=NotInLF|x):    0.4222   [ 0.3836,    0.4608]
            k5         k618        age        wc         hc         lwg
x=    .2377158   1.3532537   42.537849   .2815405   .39176627   1.0971148
            inc
x=    20.128965
```

With predictions in hand, we can summarize the results and get a better general feel
for the factors affecting a wife's labor force participation.

Ideal type	Probability of LFP (95% CI)
Young, low-income, and low-education families with young children	.13 (.06, .21)
Highly educated, middle-aged couples with no children at home	.72 (.63, .80)
An "average" family	.58 (.54, .62)

4.6.3   Tables of predicted probabilities with prtab

Sometimes the focus might be on two or three categorical independent variables. Predictions for all combinations of the categories of these variables could be presented in a
table. For example,

No. of young children	Predicted probability		
	Did not attend college	Attended college	Difference
0	.61	.78	.17
1	.26	.44	.18
2	.08	.16	.08
3	.02	.04	.02

This table shows the strong effect on labor force participation of having young children
and how the effect differs according to the wife's education. One way to construct such
a table is by a series of calls to `prvalue` (we use the `brief` option to limit output):

```
. prvalue, x(k5=0 wc=0) rest(mean) brief
logit: Predictions for lfp
                                     95% Conf. Interval
    Pr(y=inLF|x):        0.6069    [ 0.5567,     0.6570]
    Pr(y=NotInLF|x):     0.3931    [ 0.3430,     0.4433]
. prvalue, x(k5=1 wc=0) rest(mean) brief
logit: Predictions for lfp
                                     95% Conf. Interval
    Pr(y=inLF|x):        0.2633    [ 0.1932,     0.3335]
    Pr(y=NotInLF|x):     0.7367    [ 0.6665,     0.8068]
    (and so on)
```

Even for a simple table, this approach is tedious and error prone. `prtab` automates the process by computing a table of predicted probabilities for all combinations of up to four categorical variables. For example,

```
. prtab k5 wc, rest(mean)
logit: Predicted probabilities of positive outcome for lfp
```

# kids < 6	Wife College: 1=yes 0=no	
	NoCol	College
0	0.6069	0.7758
1	0.2633	0.4449
2	0.0764	0.1565
3	0.0188	0.0412

```
              k5         k618        age         wc          hc         lwg
x=     .2377158   1.3532537   42.537849   .2815405   .39176627   1.0971148
             inc
x=     20.128965
```

The only disadvantage of using `prtab` is that it does not provide confidence intervals for the predictions.

4.6.4 Graphing predicted probabilities with prgen

When a variable of interest is continuous, you can either select values (e.g., quartiles) and construct a table or create a graph. For example, to examine the effects of `income` on labor force participation by `age`, we can use the estimated parameters to compute predicted probabilities as `income` changes for fixed values of `age`. This is shown in figure 4.4. The command `prgen` creates data that can be graphed in this way. The first step is to generate the predicted probabilities for those aged 30:

```
. prgen inc, from(0) to(100) generate(p30) x(age=30) rest(mean) n(11)
logit: Predicted values as inc varies from 0 to 100.
           k5        k618        age        wc        hc        lwg
x=   .2377158  1.3532537         30  .2815405  .39176627  1.0971148
          inc
x=   20.128965
. label var p30p1 "Age 30"
```

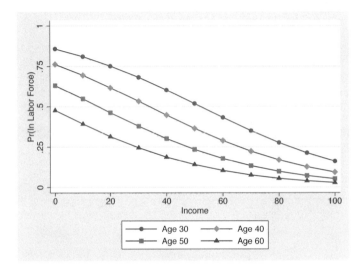

Figure 4.4: Graph of predicted probabilities created using prgen.

inc is the independent variable that we want to vary along the *x*-axis. The options that we use are

from(0) and to(100) specify the minimum and maximum values over which inc is to vary. The default is the variable's observed minimum and maximum values.

generate(p30) indicates the root name used in constructing new variables. prgen creates p30x that contains the values of inc that are used; p30p1 with the values of the probability of a 1 and p30p0 with values of the probability of a 0.

x(age=30) indicates that we want to hold the value of age at 30. By default, other variables will be held at their mean unless rest() is used to specify some other summary statistic.

n(11) indicates that 11 evenly spaced values of inc between 0 and 100 should be used. You should choose the value that corresponds to the number of symbols you want on your graph.

More calls of **prgen** are made holding **age** at different values:

```
. prgen inc, from(0) to(100) generate(p40) x(age=40) rest(mean) n(11)
  (output omitted )
. label var p40p1 "Age 40"
. prgen inc, from(0) to(100) generate(p50) x(age=50) rest(mean) n(11)
  (output omitted )
. label var p50p1 "Age 50"
. prgen inc, from(0) to(100) generate(p60) x(age=60) rest(mean) n(11)
  (output omitted )
. label var p60p1 "Age 60"
```

Listing the values for the first 11 observations in the dataset for some of the new variables **prgen** has created may help you understand better what this command does:

```
. list p30p1 p40p1 p50p1 p60p1 p60x in 1/11
```

	p30p1	p40p1	p50p1	p60p1	p60x
1.	.8575829	.7625393	.6313345	.4773258	0
2.	.8101358	.6947005	.5482202	.3928797	10
3.	.7514627	.6172101	.462326	.3143872	20
4.	.6817801	.5332655	.3786113	.2452419	30
5.	.6028849	.4473941	.3015535	.187153	40
6.	.5182508	.36455	.2342664	.1402662	50
7.	.4325564	.289023	.1781635	.1036283	60
8.	.3507161	.2236366	.1331599	.0757174	70
9.	.2768067	.1695158	.0981662	.0548639	80
10.	.2133547	.1263607	.071609	.0395082	90
11.	.1612055	.0929622	.0518235	.0283215	100

The predicted probabilities of labor force participation for those averages on all other variables at ages 30, 40, 50, and 60 are in the first four columns. The clear negative effect of age is shown by the increasingly small probabilities as we move across these columns in any row. The last column indicates the value of income for a given row, starting at 0 and ending at 100. We can see that the probabilities decrease as income increases.

The following **graph** command generates the plot:

```
. graph twoway connected p30p1 p40p1 p50p1 p60p1 p60x,             ///
> ytitle("Pr(In Labor Force)") ylabel(0(.25)1) xtitle("Income")
```

Because we have not used **graph** much yet, it is worth discussing some points that we find useful (also see section 2.16).

1. Recall that /// is a way of entering long lines in do-files.

2. **graph twoway** is the command for plotting a dependent variable on the y-axis against an independent variable along the x-axis. **graph twoway connected** specifies that the symbols used to mark the individual points be connected.

3. The variables to plot are p30p1 p40p1 p50p1 p60p1 p60x, where p60x, the last variable in the list, is the variable for the horizontal axis. All variables before the last variable are plotted on the vertical axis.

4. The options ytitle() and xtitle() specify the axis titles.

5. The ylabel() specifies which points on the *y*-axis to label.

4.6.5   Plotting confidence intervals

We can also use prgen to plot confidence intervals around our predictions by adding the ci option. Although you can use any of the options that control how confidence intervals are constructed (see page 123 for details), here we use the default options to keep things simple. We want to plot the probability of being in the labor force by age, adding the 95% confidence interval around the plot. The prgen command is

```
. prgen age, from(20) to(70) generate(prlfp) rest(mean) gap(2) ci
logit: Predicted values as age varies from 20 to 70.
            k5        k618        age         wc          hc          lwg
x=   .2377158   1.3532537   42.537849   .2815405   .39176627   1.0971148

            inc
x=   20.128965
. label var prlfpp1 "Predicted probability"
. label var prlfpp1ub "95% upper limit"
. label var prlfpp1lb "95% lower limit"
. label var prlfpx "Age"
```

We made several changes from the last time we used prgen. First, we added the ci option so that prgen would create variables with the estimates of the upper and lower bounds for the confidence interval of the predictions. These new variables are prlfpp1ub and prlfpp1lb. Second, rather than using the n() option to indicate the number of evenly spaced values of age to compute, we used the gap() option. With gap(), you simply indicate the spacing or gap between values. Here we want to compute values of age that increase by 2. Third, we have added variable labels for the variables created by prgen. These labels will be clearer in the graph than the default labels created by prgen. To plot the results, we use the following command:

```
. graph twoway                                       ///
>    (connected prlfpp1 prlfpx,                       ///
>        clcolor(black) clpat(solid) clwidth(medthick) ///
>        msymbol(i) mcolor(none))                      ///
>    (connected prlfpp1ub prlfpx,                      ///
>        msymbol(i) mcolor(none)                       ///
>        clcolor(black) clpat(dash) clwidth(thin))     ///
>    (connected prlfpp1lb prlfpx,                      ///
>        msymbol(i) mcolor(none)                       ///
>        clcolor(black) clpat(dash) clwidth(thin)),    ///
>    ytitle("Probability of Being in Labor Force")     ///
>    yscale(range(0 .35))                              ///
>    ylabel(, grid glwidth(medium) glpattern(solid))   ///
>    xscale(range(20 70)) xlabel(20(10)70)
```

This example of `graph twoway` shows how you can combine multiple plots into one graph. Each of the sections that begin with "(connected" and end with a ")" control the plotting of a different line. The resulting graph looks like this:

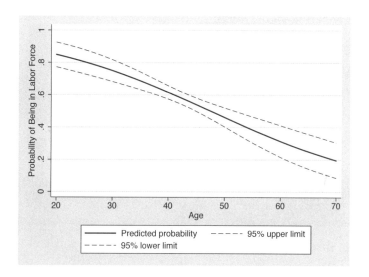

The graph shows that the confidence interval is smaller near the center of the data where age is 40 and increases as we move to younger or older ages.

Another way of showing the confidence intervals is to use shading. For this, we use the **rarea** type of plot. Specifying `rarea(`*yvar1 yvar2 xvar*`, color(`*color*`))` will shade the area on the *y*-axis between the values of *yvar1* and *yvar2*, with the *x*-axis specified with *xvar*. We define the **rarea** graph before the **connected** graph since Stata draws overlaid graphs in the order specified, and we want the line indicating the predicted probabilities to appear on top of the shading. The improved command is

```
. graph twoway                                          ///
>    (rarea prlfpp1lb prlfpp1ub prlfpx, color(gs14))    ///
>    (connected prlfpp1 prlfpx,                         ///
>       clcolor(black) clpat(solid) clwidth(medthick)   ///
>       msymbol(i) mcolor(none)),                        ///
>    ytitle("Probability of Being in Labor Force")      ///
>    yscale(range(0 .35))                                ///
>    ylabel(, grid glwidth(medium) glpattern(solid))    ///
>    xscale(range(20 70)) xlabel(20(10)70)               ///
>    legend(label (1 "95% confidence interval"))
```

(Continued on next page)

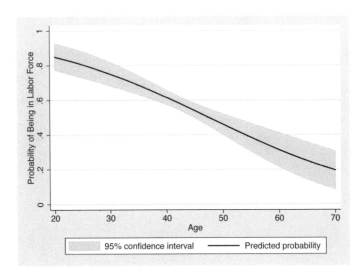

4.6.6   Changes in predicted probabilities

Although graphs are useful for showing how predicted probabilities are related to an independent variable, for even our simple example it is not practical to plot all possible combinations of the independent variables. And sometimes the plots show that a relationship is linear, making a graph is superfluous. In such circumstances, a useful summary measure is the change in the outcome as one variable changes, holding all other variables constant.

Marginal change

In economics, the *marginal effect* or *change* is commonly used:

$$\text{marginal change} = \frac{\partial \Pr(y = 1 \mid \mathbf{x})}{\partial x_k}$$

The marginal change is shown by the tangent to the probability curve in figure 4.5. The value of the marginal effect depends on the level of all variables in the model. It is often computed with all variables held at their mean or by computing the marginal change for each observation in the sample and then averaging across all values.

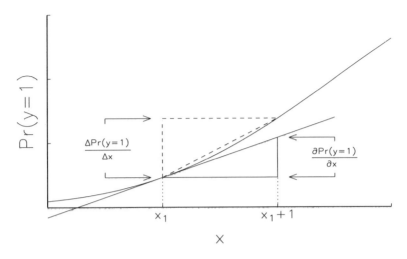

Figure 4.5: Marginal change compared with discrete change in the BRM.

Marginal change with prchange command

The command **prchange** computes the marginal at the values of the independent variables specified with x() or rest(). Running **prchange** with no options computes the marginal change (along with many other things discussed below) with all variables at their mean. Or, we can compute the marginal at specific values of the independent variables, such as when wc = 1 and age = 40. Here we request only the results for age:

```
. prchange age, x(wc=1 age=40) help
logit: Changes in Probabilities for lfp

        min->max      0->1      -+1/2     -+sd/2  MargEfct
age      -0.3940    -0.0017    -0.0121    -0.0971   -0.0121

           NotInLF     inLF
Pr(y|x)     0.2586    0.7414

              k5       k618       age       wc        hc       lwg       inc
   x=    .237716   1.35325        40        1   .391766   1.09711    20.129
sd(x)=   .523959   1.31987   8.07257  .450049   .488469   .587556   11.6348

 Pr(y|x): probability of observing each y for specified x values
Avg|Chg|: average of absolute value of the change across categories
Min->Max: change in predicted probability as x changes from its minimum to
          its maximum
    0->1: change in predicted probability as x changes from 0 to 1
  -+1/2: change in predicted probability as x changes from 1/2 unit below
          base value to 1/2 unit above
 -+sd/2: change in predicted probability as x changes from 1/2 standard
          dev below base to 1/2 standard dev above
MargEfct: the partial derivative of the predicted probability/rate with
          respect to a given independent variable
```

In plots that we do not show (but that we encourage you to create using prgen and graph), we found that the relationship between age and the probability of being in the labor force was essentially linear for those who attend college. Accordingly, we can take the marginal computed by prchange, multiply it by 10 to get the amount of change over 10 years, and report that for women who attend college, a 10-year increase in age decreases the probability of labor force participation by approximately .12, holding other variables at their mean.

When using the marginal, remember two points. First, the amount of change depends on the level of all variables. Second, as shown in figure 4.5, the marginal is the instantaneous rate of change. In general, it does not equal the actual change for a given finite change in the independent variable unless you are in a region of the probability curve that is approximately linear. Such linearity justifies the interpretation given above.

Marginal change with mfx command

The marginal change can also be computed using mfx, where the at() option is used to set values of the independent variables. Below we use mfx to estimate the marginal change for the same values that we used when calculating the marginal effect for age with prchange above:

```
. mfx, at(wc=1 age=40)
warning: no value assigned in at() for variables k5 k618 hc lwg inc;
    means used for k5 k618 hc lwg inc
Marginal effects after logit
      y  = Pr(lfp) (predict)
         = .74140317
```

variable	dy/dx	Std. Err.	z	P>|z|	[95% C.I.]	X
k5	-.2804763	.04221	-6.64	0.000	-.363212 -.197741	.237716
k618	-.0123798	.01305	-0.95	0.343	-.037959 .013199	1.35325
age	-.0120538	.00245	-4.92	0.000	-.016855 -.007252	40
wc*	.1802113	.04742	3.80	0.000	.087269 .273154	1
hc*	.0212952	.03988	0.53	0.593	-.056866 .099456	.391766
lwg	.1159345	.03229	3.59	0.000	.052643 .179226	1.09711
inc	-.0066042	.00163	-4.05	0.000	-.009802 -.003406	20.129

(*) dy/dx is for discrete change of dummy variable from 0 to 1

mfx is particularly useful if you need estimates of the standard errors of the marginal effects; however, mfx computes the estimates using numerical methods, and for some models the command can take a long time.

Discrete change

Given the nonlinearity of the model, we prefer the *discrete change* in the predicted probabilities for a given change in an independent variable. To define discrete change, we need two quantities:

$\Pr(y = 1 \mid \mathbf{x}, x_k)$ is the probability of an event given $\mathbf{x}$, noting in particular the value of x_k.

$\Pr(y = 1 \mid \mathbf{x}, x_k + \delta)$ is the probability of the event with only x_k increased by some quantity δ.

Then the *discrete change* for a change of δ in x_k equals

$$\frac{\Delta \Pr(y = 1 \mid \mathbf{x})}{\Delta x_k} = \Pr(y = 1 \mid \mathbf{x}, x_k + \delta) - \Pr(y = 1 \mid \mathbf{x}, x_k)$$

which can be interpreted that for a change in variable x_k from x_k to $x_k + \delta$, the predicted probability of an event changes by $\{\Delta \Pr(y = 1 \mid \mathbf{x})\} / \Delta x_k$, holding all other variables constant.

As shown in figure 4.5, in general, the two measures of change are not equal. That is,

$$\frac{\partial \Pr(y = 1 \mid \mathbf{x})}{\partial x_k} \neq \frac{\Delta \Pr(y = 1 \mid \mathbf{x})}{\Delta x_k}$$

The measures differ because the marginal change is the instantaneous rate of change, whereas the discrete change is the amount of change in the probability for a given finite change in one independent variable. The two measures are similar, however, when the change occurs over a region of the probability curve that is roughly linear.

The value of the discrete change depends on

1. The start level of the variable that is being changed. For example, do you want to examine the effect of age beginning at 30? At 40? At 50?
2. The amount of change in that variable. Are you interested in the effect of a change of 1 year in age? Of 5 years? Of 10 years?
3. The level of all other variables in the model. Do you want to hold all variables at their mean? Or, do you want to examine the effect for women? Or, do you want to compute changes separately for men and women?

Accordingly, a decision must be made regarding each of these factors. See chapter 3 for more discussion.

For our example, let's look at the discrete change with all variables held at their mean, which is computed by default by `prchange`, where the `help` option is used to get detailed descriptions of what the measures mean:

```
. prchange, help

logit: Changes in Probabilities for lfp
          min->max      0->1     -+1/2     -+sd/2   MargEfct
    k5    -0.6361    -0.3499   -0.3428   -0.1849   -0.3569
   k618   -0.1278    -0.0156   -0.0158   -0.0208   -0.0158
    age   -0.4372    -0.0030   -0.0153   -0.1232   -0.0153
     wc    0.1881     0.1881    0.1945    0.0884    0.1969
     hc    0.0272     0.0272    0.0273    0.0133    0.0273
    lwg    0.6624     0.1499    0.1465    0.0865    0.1475
    inc   -0.6415    -0.0068   -0.0084   -0.0975   -0.0084

           NotInLF     inLF
Pr(y|x)    0.4222    0.5778

              k5       k618       age        wc        hc       lwg       inc
     x=  .237716   1.35325   42.5378   .281541   .391766   1.09711    20.129
sd(x)=  .523959   1.31987   8.07257   .450049   .488469   .587556   11.6348

Pr(y|x): probability of observing each y for specified x values
Avg|Chg|: average of absolute value of the change across categories
Min->Max: change in predicted probability as x changes from its minimum to
          its maximum
   0->1: change in predicted probability as x changes from 0 to 1
  -+1/2: change in predicted probability as x changes from 1/2 unit below
          base value to 1/2 unit above
 -+sd/2: change in predicted probability as x changes from 1/2 standard
          dev below base to 1/2 standard dev above
MargEfct: the partial derivative of the predicted probability/rate with
          respect to a given independent variable
```

First, consider the results of changes from the minimum to the maximum. There is little to be learned by analyzing variables whose range of probabilities is small, such as hc, whereas age, k5, wc, lwg, and inc have *potentially* important effects. For these we can examine the value of the probabilities before and after the change by using the fromto option:

```
. prchange k5 age wc lwg inc, fromto

logit: Changes in Probabilities for lfp
          from:      to:      dif:      from:      to:      dif:      from:
          x=min    x=max   min->max      x=0      x=1      0->1     x-1/2
    k5    0.6596   0.0235   -0.6361    0.6596   0.3097   -0.3499    0.7398
    age   0.7506   0.3134   -0.4372    0.9520   0.9491   -0.0030    0.5854
     wc   0.5216   0.7097    0.1881    0.5216   0.7097    0.1881    0.4775
    lwg   0.1691   0.8316    0.6624    0.4135   0.5634    0.1499    0.5028
    inc   0.7326   0.0911   -0.6415    0.7325   0.7256   -0.0068    0.5820

            to:      dif:      from:      to:      dif:
          x+1/2     -+1/2    x-1/2sd   x+1/2sd    -+sd/2   MargEfct
    k5    0.3971   -0.3428    0.6675    0.4826   -0.1849   -0.3569
    age   0.5701   -0.0153    0.6382    0.5150   -0.1232   -0.0153
     wc   0.6720    0.1945    0.5330    0.6214    0.0884    0.1969
    lwg   0.6493    0.1465    0.5340    0.6204    0.0865    0.1475
    inc   0.5736   -0.0084    0.6258    0.5283   -0.0975   -0.0084

           NotInLF     inLF
Pr(y|x)    0.4222    0.5778

              k5       k618       age        wc        hc       lwg       inc
     x=  .237716   1.35325   42.5378   .281541   .391766   1.09711    20.129
sd(x)=  .523959   1.31987   8.07257   .450049   .488469   .587556   11.6348
```

We learn, for example, that varying `age` from its minimum of 30 to its maximum of 60 decreases the predicted probability by .44, from .75 to .31. Changing family income (`inc`) from its minimum to its maximum decreases the probability of a woman being in the labor force from .73 to .09. Interpreting other measures of change, the following interpretations can be made:

> *Using the unit change labeled* `-+1/2`: for a woman who is average on all characteristics, an additional young child decreases the probability of employment by .34.

> *Using the standard deviation change labeled* `-+sd/2`: a standard deviation change in age centered on the mean will decrease the probability of working by .12, holding other variables to their means.

> *Using a change from 0 to 1 labeled* `0->1`: if a woman attends college, her probability of being in the labor force is .19 greater than a woman who does not attend college, holding other variables at their mean.

What if you need to calculate discrete change for changes in the independent values that are not the default for `prchange` (e.g., a change of 10 years in age rather than 1 year)? This can be done in two ways:

Confidence intervals for discrete change using the prvalue command

`prchange` does not provide confidence intervals. Although this feature might be added in the future, for now you need to compute these intervals using a series of calls to `prvalue`. Let's start with a simple example and then show how the process can be automated. We can use `prchange` to compute the discrete change when `wc` changes from 0 to 1, holding other variables at their mean:

```
. prchange wc, brief
        min->max      0->1     -+1/2    -+sd/2  MargEfct
   wc    0.1881    0.1881    0.1945    0.0884    0.1969
```

Using `prvalue`, we `quietly` compute the predictions when `wc` is 0, save the results, and then compute predictions when `wc` is 1 using the `diff` option to compute discrete changes:

(Continued on next page)

```
. qui prvalue, x(wc=0) rest(mean) save

. prvalue, x(wc=1) diff

logit: Change in Predictions for lfp

Confidence intervals by delta method
                        Current    Saved     Change    95% CI for Change
       Pr(y=inLF|x):     0.7097    0.5216     0.1881   [ 0.0900,    0.2861]
       Pr(y=NotInLF|x):  0.2903    0.4784    -0.1881   [-0.2861,   -0.0900]

                    k5        k618        age          wc          hc         lwg
   Current=   .2377158   1.3532537   42.537849         1    .39176627   1.0971148
     Saved=   .2377158   1.3532537   42.537849         0    .39176627   1.0971148
      Diff=          0           0           0          1            0           0

                   inc
   Current=  20.128965
     Saved=  20.128965
      Diff=          0
```

The formal interpretation of these results requires that we imagine drawing repeated samples from the population and repeating all calculations for estimating the bounds of the confidence intervals for each sample (see page 88). About 95% of the computed confidence intervals would contain the true change in the predicted probability. In this sense, we are 95% confident that the true increase in the probability of a woman being in the labor force associated with her having been to college is between .09 and .29. More informally, we might say that we are 95% confident that the increase in the predicted probability of a woman being in the labor force associated with her having been to college is between .09 and .29. This confidence interval is the same as that computed by mfx, which uses numerical methods:

```
. mfx

Marginal effects after logit
      y  = Pr(lfp) (predict)
         =  .57779421
```

variable	dy/dx	Std. Err.	z	P>|z|	[95% C.I.]		X
k5	-.3568748	.04821	-7.40	0.000	-.451366	-.262383	.237716
k618	-.0157519	.01659	-0.95	0.342	-.048266	.016763	1.35325
age	-.0153371	.00311	-4.93	0.000	-.021434	-.00924	42.5378
wc*	.1880592	.05003	3.76	0.000	.09001	.286109	.281541
hc*	.0271985	.05004	0.54	0.587	-.070882	.125279	.391766
lwg	.1475137	.03674	4.01	0.000	.075496	.219532	1.09711
inc	-.0084031	.002	-4.19	0.000	-.012332	-.004474	20.129

(*) dy/dx is for discrete change of dummy variable from 0 to 1

Although mfx can compute confidence intervals for discrete changes for binary variables, it will not compute them for continuous variables and is slower than prvalue since mfx uses numerical methods to compute the confidence interval.

We can use a foreach loop (see [P] **foreach**) to automate the process of computing a series of discrete changes with confidence intervals. In the following example, we are computing a change from 0 to 1 for each of the variables k5, k618, and wc:

```
foreach v in k5 k618 wc {
      di _n "** Change from 0 to 1 in 'v'"
      qui prvalue, x('v'=0) rest(mean) save
          prvalue, x('v'=1) diff brief
}
```

Line 1 causes the block of code between the braces { and } to be repeated three times. During the first pass, the local macro v is equal to k5, during the second pass it is equal to k618, and during the third pass it equals wc. (The ' and ' around v tells Stata to insert the value of the local macro v.) For each pass, line 2 labels the output; 'v' inserts the name of the variable assigned in line 1, and line 3 quietly runs prvalue, assigning the variable represented by 'v' to equal 0 and saving the result. Similarly, for each pass line 4 assigns the variable 'v' to equal 1 and computes the difference between the new and saved results. Here is the output:

```
** Change from 0 to 1 in k5

logit: Change in Predictions for lfp
                     Current    Saved     Change     95% CI for Change
Pr(y=inLF|x):        0.3097    0.6596    -0.3499    [-0.4334,  -0.2663]
Pr(y=NotInLF|x):     0.6903    0.3404     0.3499    [ 0.2663,   0.4334]
** Change from 0 to 1 in k618

logit: Change in Predictions for lfp
                     Current    Saved     Change     95% CI for Change
Pr(y=inLF|x):        0.5833    0.5990    -0.0156    [-0.0475,   0.0163]
Pr(y=NotInLF|x):     0.4167    0.4010     0.0156    [-0.0163,   0.0475]
** Change from 0 to 1 in wc

logit: Change in Predictions for lfp
                     Current    Saved     Change     95% CI for Change
Pr(y=inLF|x):        0.7097    0.5216     0.1881    [ 0.0900,   0.2861]
Pr(y=NotInLF|x):     0.2903    0.4784    -0.1881    [-0.2861,  -0.0900]
```

The process is a bit more complicated when we want to compute the discrete, one-standard-deviation change centered around the mean. Again we use a foreach loop to repeat a block of code three times, once for age, again for lwg, and finally for inc. In the program below, line 2 runs the command summarize for the variable indicated by 'v'. In line 3, we create a start value equal to the mean minus one-half of a standard deviation. Line 4 computes the end value equal to the mean plus one-half of a standard deviation. The starting and ending values are then passed to prvalue in lines 6 and 7.

```
. foreach v in age lwg inc {
  2.      qui summarize 'v'
  3.      local start = r(mean) - (.5*r(sd))
  4.      local end = r(mean) + (.5*r(sd))
  5.      di _n "** Change from 'start' to 'end' in 'v'"
  6.      qui prvalue, x('v'='start') rest(mean) save
  7.          prvalue, x('v'='end') dif brief
  8. }
```

```
** Change from 38.50156159842585 to 46.57413561272952 in age
logit: Change in Predictions for lfp
                     Current     Saved    Change    95% CI for Change
    Pr(y=inLF|x):     0.5150    0.6382   -0.1232   [-0.1717,  -0.0747]
    Pr(y=NotInLF|x):  0.4850    0.3618    0.1232   [ 0.0747,   0.1717]
** Change from .8033366225286708 to 1.390893047643295 in lwg
logit: Change in Predictions for lfp
                     Current     Saved    Change    95% CI for Change
    Pr(y=inLF|x):     0.6204    0.5340    0.0865   [ 0.0445,   0.1285]
    Pr(y=NotInLF|x):  0.3796    0.4660   -0.0865   [-0.1285,  -0.0445]
** Change from 14.31156614524011 to 25.94636467863254 in inc
logit: Change in Predictions for lfp
                     Current     Saved    Change    95% CI for Change
    Pr(y=inLF|x):     0.5283    0.6258   -0.0975   [-0.1428,  -0.0522]
    Pr(y=NotInLF|x):  0.4717    0.3742    0.0975   [ 0.0522,   0.1428]
```

**Tip: Using estat summarize to get summary statistics for the estimation
sample.** Above we used `summarize` to compute values of the mean and standard deviation. If you want to view summary statistics restricted to the sample used to fit a model (that is, reflecting any `if` and `in` conditions you specified, as well as listwise deletion for observations with missing data), you can use `summarize` *varlist* `if e(sample)==1`. Or, more simply, `estat summarize` produces summary statistics restricted to the estimation sample for all variables in the model.

Nonstandard discrete changes with prvalue command

The command `prvalue` can be used to calculate the change in the probability for a discrete change of any magnitude in an independent variable. Say that we want to calculate the effect of a 10-year increase in `age` for a 30-year-old woman who is average on all other characteristics:

```
. prvalue, x(age=30) save brief
logit: Predictions for lfp
                                 95% Conf. Interval
    Pr(y=inLF|x):     0.7506    [ 0.6830,   0.8183]
    Pr(y=NotInLF|x):  0.2494    [ 0.1817,   0.3170]
. prvalue, x(age=40) diff brief
logit: Change in Predictions for lfp
                     Current     Saved    Change    95% CI for Change
    Pr(y=inLF|x):     0.6162    0.7506   -0.1345   [-0.1784,  -0.0906]
    Pr(y=NotInLF|x):  0.3838    0.2494    0.1345   [ 0.0906,   0.1784]
```

The `save` option preserves the results from the first call of `prvalue`. The second call adds the `diff` option to compute the differences between the two sets of predictions. We find that an increase in `age` from 30 to 40 years decreases a woman's probability of being in the labor force by .13.

Nonstandard discrete changes with prchange

We can also use `prchange` with the `delta()` and `uncentered` options. `delta(#)` specifies that the discrete change is to be computed for a change of # units instead of a one-unit change. `uncentered` specifies that the change should be computed starting at the base value (i.e., values set by the `x()` and `rest()` options), rather than being centered on the base. Here we want an uncentered change of 10 units, starting at `age=30`:

```
. prchange age, x(age=30) uncentered delta(10) rest(mean) brief
        min->max      0->1     +delta      +sd  MargEfct
age      -0.4372   -0.0030    -0.1345  -0.1062   -0.0118
```

The result under the heading `+delta` is the same as what we just calculated using `prvalue`.

4.7 Interpretation using odds ratios with listcoef

Effects for the logit model, but *not* probit, can be interpreted in terms of changes in the odds. Recall that for binary outcomes, we typically consider the odds of observing a positive outcome versus a negative one:

$$\Omega = \frac{\Pr(y=1)}{\Pr(y=0)} = \frac{\Pr(y=1)}{1 - \Pr(y=1)}$$

Recall also that the log of the odds is called the *logit* and that the logit model is *linear in the logit*, meaning that the log odds are a linear combination of the x's and βs. For example, consider a logit model with three independent variables:

$$\ln\left\{ \frac{\Pr(y=1 \mid \mathbf{x})}{1 - \Pr(y=1 \mid \mathbf{x})} \right\} = \ln \Omega(\mathbf{x}) = \beta_0 + \beta_1 x_1 + \beta_2 x_2 + \beta_3 x_3$$

We can interpret the coefficients as indicating that for a unit change in x_k, we expect the logit to change by β_k, holding all other variables constant.

This interpretation does *not* depend on the level of the other variables in the model. The problem is that a change of β_k in the log odds has little substantive meaning for most people (including us). By taking the exponential of both sides of this equation, we can also create a model that is multiplicative instead of linear but in which the outcome is the more intuitive measure, the odds:

$$\Omega\left(\mathbf{x}, x_2\right) = e^{\beta_0} e^{\beta_1 x_1} e^{\beta_2 x_2} e^{\beta_3 x_3}$$

where we take particular note of the value of x_2. If we let x_2 change by 1,

$$\Omega\left(\mathbf{x}, x_2 + 1\right) = e^{\beta_0} e^{\beta_1 x_1} e^{\beta_2 (x_2+1)} e^{\beta_3 x_3}$$

$$= e^{\beta_0} e^{\beta_0} e^{\beta_1 x_1} e^{\beta_2 x_2} e^{\beta_2} e^{\beta_3 x_3}$$

which leads to the *odds ratio:*

$$\frac{\Omega\left(\mathbf{x}, x_2 + 1\right)}{\Omega\left(\mathbf{x}, x_2\right)} = \frac{e^{\beta_0} e^{\beta_1 x_1} e^{\beta_2 x_2} e^{\beta_2} e^{\beta_3 x_3}}{e^{\beta_0} e^{\beta_1 x_1} e^{\beta_2 x_2} e^{\beta_3 x_3}} = e^{\beta_2}$$

Accordingly, we can interpret the exponential of the coefficient as follows:

> For a unit change in x_k, the *odds* are expected to change by a factor of $\exp(\beta_k)$, holding all other variables constant.

For $\exp(\beta_k) > 1$, you could say that the odds are "$\exp(\beta_k)$ times larger"; for $\exp(\beta_k) < 1$, you could say that the odds are "$\exp(\beta_k)$ times smaller". We can evaluate the effect of a standard deviation change in x_k instead of a unit change:

> For a standard deviation change in x_k, the odds are expected to change by a factor of $\exp(\beta_k \times s_k)$, holding all other variables constant.

The odds ratios for both a unit and a standard deviation change of the independent variables can be obtained with `listcoef`:

```
. listcoef, help

logit (N=753): Factor Change in Odds

  Odds of: inLF vs NotInLF
```

lfp	b	z	P>|z|	e^b	e^bStdX	SDofX
k5	-1.46291	-7.426	0.000	0.2316	0.4646	0.5240
k618	-0.06457	-0.950	0.342	0.9375	0.9183	1.3199
age	-0.06287	-4.918	0.000	0.9391	0.6020	8.0726
wc	0.80727	3.510	0.000	2.2418	1.4381	0.4500
hc	0.11173	0.542	0.588	1.1182	1.0561	0.4885
lwg	0.60469	4.009	0.000	1.8307	1.4266	0.5876
inc	-0.03445	-4.196	0.000	0.9661	0.6698	11.6348

```
        b = raw coefficient
        z = z-score for test of b=0
    P>|z| = p-value for z-test
      e^b = exp(b) = factor change in odds for unit increase in X
  e^bStdX = exp(b*SD of X) = change in odds for SD increase in X
    SDofX = standard deviation of X
```

Some examples of interpretations are as follows:

> For each additional young child, the odds of being employed decrease by a factor of .23, holding all other variables constant.

> For a standard deviation increase in the log of the wife's expected wages, the odds of being employed are 1.43 times greater, holding all other variables constant.

> Being 10 years older decreases the odds by a factor of .53 ($= e^{[-.063] \times 10}$), holding all other variables constant.

Other ways of computing odds ratios Odds ratios can also be computed with the
or option for `logit`. This approach does not, however, report the odds ratios for
a standard deviation change in the independent variables.

Multiplicative coefficients

When interpreting the odds ratios, remember that they are multiplicative. This means
that positive effects are greater than one and negative effects are between zero and
one. *Magnitudes of positive and negative effects should be compared by taking the
inverse of the negative effect (or vice versa).* For example, a positive factor change of
2 has the same magnitude as a negative factor change of $.5 = 1/2$. Thus a coefficient
of $.1 = 1/10$ indicates a stronger effect than a coefficient of 2. Another consequence
of the multiplicative scale is that to determine the effect on the odds of the event not
occurring, you simply take the inverse of the effect on the odds of the event occurring.
`listcoef` will automatically calculate this for you if you specify the `reverse` option:

```
. listcoef, reverse
logit (N=753): Factor Change in Odds
   Odds of: NotInLF vs inLF
```

lfp	b	z	P>|z|	e^b	e^bStdX	SDofX
k5	-1.46291	-7.426	0.000	4.3185	2.1522	0.5240
k618	-0.06457	-0.950	0.342	1.0667	1.0890	1.3199
age	-0.06287	-4.918	0.000	1.0649	1.6612	8.0726
wc	0.80727	3.510	0.000	0.4461	0.6954	0.4500
hc	0.11173	0.542	0.588	0.8943	0.9469	0.4885
lwg	0.60469	4.009	0.000	0.5462	0.7010	0.5876
inc	-0.03445	-4.196	0.000	1.0350	1.4930	11.6348

The header indicates that these are now the factor changes in the odds of `NotInLF`
versus `inLF`, whereas before we computed the factor change in the odds of `inLF` versus
`NotInLF`. We can interpret the result for `k5` as follows:

> For each additional child, the odds of not being employed are increased by
> a factor of 4.3 (=1/.23), holding other variables constant.

Effect of the base probability

The interpretation of the odds ratio assumes that the other variables have been held
constant, but it does not require that they be held at any specific values. Although the
odds ratio seems to resolve the problem of nonlinearity, remember: *a constant factor
change in the odds does not correspond to a constant change or constant factor change
in the probability.* For example, if the odds are 1/100, the corresponding probability

is $.01.$[6] If the odds double to $2/100$, the probability increases only by approximately
$.01$. Depending on your substantive purposes, this small change may be trivial or quite
important (such as when you identify a risk factor that makes it twice as likely that a
subject will contract a fatal disease). Meanwhile, if the odds are $1/1$ and double to $2/1$,
the probability increases by $.167$. Accordingly, the meaning of a given factor change in
the odds depends on the predicted probability, which in turn depends on the levels of
all variables in the model.

Percent change in the odds

Instead of a multiplicative or factor change in the outcome, some people prefer the
percent change,

$$100 \left\{ \exp \left(\beta_k \times \delta \right) - 1 \right\}$$

which is listed by `listcoef` with the `percent` option.

```
. listcoef, percent
logit (N=753): Percentage Change in Odds
    Odds of: inLF vs NotInLF
```

lfp	b	z	P>|z|	%	%StdX	SDofX
k5	-1.46291	-7.426	0.000	-76.8	-53.5	0.5240
k618	-0.06457	-0.950	0.342	-6.3	-8.2	1.3199
age	-0.06287	-4.918	0.000	-6.1	-39.8	8.0726
wc	0.80727	3.510	0.000	124.2	43.8	0.4500
hc	0.11173	0.542	0.588	11.8	5.6	0.4885
lwg	0.60469	4.009	0.000	83.1	42.7	0.5876
inc	-0.03445	-4.196	0.000	-3.4	-33.0	11.6348

With this option, the interpretations would be the following:

> For each additional young child, the odds of being employed decrease by
> 77%, holding all other variables constant.

> A standard deviation increase in the log of the wife's expected wages in-
> creases the odds of being employed by 43%, holding all other variables con-
> stant.

Percentage and factor change provide the same information; which you use for the binary
model is a matter of preference. Although we both tend to prefer percentage change,
methods for the graphical interpretation of the multinomial logit model (chapter 6)
work only with factor change coefficients.

---

6. The formula for computing probabilities from odds is $p = \Omega/1 + \Omega$.

Additional note If you report the odds ratios instead of the untransformed coefficients, the 95% confidence interval of the odds ratio is typically reported instead
of the standard error. The reason is that the odds ratio is a nonlinear transformation of the logit coefficient, so the confidence interval is asymmetric. For
example, if the logit coefficient is .75 with a standard error of .25, the 95% interval around the logit coefficient is approximately [.26, 1.24], but the confidence
interval around the odds ratio exp(.75)=2.12 is [exp(.26)=1.30, exp(1.24)=3.46].
Using the or option with the logit command reports odds ratios and includes
confidence intervals.

4.8   Other commands for binary outcomes

Logit and probit models are the most commonly used models for binary outcomes and
are the only ones that we consider in this book, but other models exist that can be
fitted in Stata. Among them, cloglog assumes a complementary log-log distribution
for the errors instead of a logistic or normal distribution. scobit fits a logit model that
relaxes the assumption that the marginal change in the probability is greatest when
$\Pr(y = 1) = .5$. hetprob allows the assumed variance of the errors in the probit model
to vary as a function of the independent variables. ivprobit fits a probit model where
one or more of the regressors are endogenously determined. biprobit simultaneously
fits two binary probits and can be used when errors are correlated with each other as
in the estimation of seemingly unrelated regression models for continuous dependent
variables.

5 Models for ordinal outcomes

Although the categories for an ordinal variable can be ordered, the distances between the categories are unknown. For example, in survey research, questions often provide the response categories of strongly agree, agree, disagree, and strongly disagree, but an analyst would probably not assume that the distance between strongly agreeing and agreeing is the same as the distance from agree to disagree. Educational attainments can be ordered as elementary education, high school diploma, college diploma, and graduate or professional degree. Ordinal variables also commonly result from limitations of data availability that require a coarse categorization of a variable that could, in principle, have been measured on an interval scale. For example, we might have a measure of income that is simply low, medium, or high.

Ordinal variables are often coded as consecutive integers from 1 to the number of categories. Perhaps because of this coding, it is tempting to analyze ordinal outcomes with the linear regression model. However, an ordinal dependent variable violates the assumptions of the LRM, which can lead to incorrect conclusions, as demonstrated strikingly by McKelvey and Zavoina (1975, 117) and Winship and Mare (1984, 521–523). Accordingly, with ordinal outcomes it is much better to use models that avoid the assumption that the distances between categories are equal. Although many different models have been designed for ordinal outcomes, in this chapter we focus on the logit and probit versions of the *ordinal regression model* (ORM), introduced by McKelvey and Zavoina (1975) in terms of an underlying latent variable and in biostatistics by McCullagh (1980), who referred to the logit version as the *proportional odds model*.

As with the binary regression model, the ORM is nonlinear, and the magnitude of the change in the outcome probability for a given change in one of the independent variables depends on the levels of all the independent variables. As with the BRM, the challenge is to summarize the effects of the independent variables to fully reflect key substantive processes without overwhelming and distracting detail. For ordinal outcomes, as well as for the models for nominal outcomes in chapter 6, the difficulty of this task is increased by having more than two outcomes to explain.

Before proceeding, we caution that researchers should think carefully before concluding that their outcome is indeed ordinal. Simply because the values of a variable *can* be ordered, do not assume that the variable *should* be analyzed as ordinal. A variable that can be ordered when considered for one purpose could be unordered or ordered differently when used for another purpose. Miller and Volker (1985) show how different assumptions about the ordering of occupations resulted in different conclusions. A variable might also reflect ordering on more than one dimension, such as attitude scales that reflect both the intensity and the direction of opinion. Moreover, surveys commonly include the category "don't know", which probably does not correspond to the middle category in a scale, even though analysts might be tempted to treat it this way. Overall, when the proper ordering is ambiguous, the models for nominal outcomes discussed in chapter 6 should be considered.

We begin by reviewing the statistical model, followed by an examination of testing, fit, and methods of interpretation. These discussions are intended as a review for those who are familiar with the models. For a complete discussion, see Long (1997). We end the chapter by considering several less common models for ordinal outcomes, which can be fitted using ado-files that others have developed. We also introduce the stereotype logit model, added in Stata 9 with the `slogit` command, but we postpone a full discussion until chapter 6. As always, you can obtain sample do-files and data files by downloading the `spost9_do` and `spost9_ado` packages (see chapter 1 for details).

5.1 The statistical model

The ORM can be developed in different ways, each of which leads to the same form of the model. These approaches to the model parallel those for the BRM. Indeed, the BRM can be viewed as a special case of the ordinal model in which the ordinal outcome has only two categories.

5.1.1 A latent-variable model

The ordinal regression model is commonly presented as a latent-variable model. Defining y^* as a latent variable ranging from $-\infty$ to ∞, the *structural model* is

$$y_i^* = \mathbf{x}_i\boldsymbol{\beta} + \varepsilon_i$$

Or, for the case of one independent variable,

$$y_i^* = \alpha + \beta x_i + \varepsilon_i$$

where i is the observation and ε is a random error, as discussed further below.

The *measurement model* for binary outcomes is expanded to divide y^* into J ordinal categories,

$$y_i = m \quad \text{if } \tau_{m-1} \le y_i^* < \tau_m \quad \text{for } m = 1 \text{ to } J$$

where the *cutpoints* τ_1 through τ_{J-1} are estimated. (Some authors refer to these as thresholds.) We assume $\tau_0 = -\infty$ and $\tau_J = \infty$ for reasons that will be clear shortly.

To illustrate the measurement model, consider the example used in this chapter. People are asked to respond to the following statement:

A working mother can establish just as warm and secure of a relationship with her child as a mother who does not work.

Possible responses are $1 =$ Strongly Disagree (SD), $2 =$ Disagree (D), $3 =$ Agree (A), and $4 =$ Strongly Agree (SA). The continuous latent variable can be thought of as the *propensity* to agree that working mothers can be warm and secure mothers. The observed response categories are tied to the latent variable by the measurement model:

$$y_i = \begin{cases} 1 \Rightarrow \text{SD} & \text{if } \tau_0 = -\infty \le y_i^* < \tau_1 \\ 2 \Rightarrow \text{D} & \text{if } \tau_1 \le y_i^* < \tau_2 \\ 3 \Rightarrow \text{A} & \text{if } \tau_2 \le y_i^* < \tau_3 \\ 4 \Rightarrow \text{SA} & \text{if } \tau_3 \le y_i^* < \tau_4 = \infty \end{cases}$$

Thus when the latent y^* crosses a cutpoint, the observed category changes. Anderson (1984) referred to ordinal variables created in this fashion as "grouped continuous" variables, and referred to what we call the ordinal regression model as the "grouped continuous model".

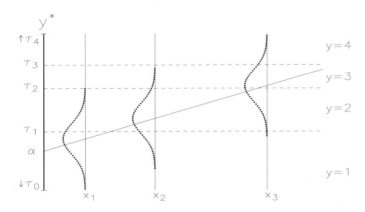

Figure 5.1: Relationship between observed y and latent y^* in ordinal regression model with one independent variable.

For one independent variable, the structural model is $y^* = \alpha + \beta x + \varepsilon$, which is plotted in figure 5.1 along with the cutpoints for the measurement model. This figure is similar to that for the binary regression model, except that there are now three horizontal lines representing the cutpoints τ_1, τ_2, and τ_3. The three cutpoints lead to four levels of y that are labeled on the right-hand side of the graph.

The probability of an observed outcome for a given value of x is the area under the curve between a pair of cutpoints. For example, the probability of observing $y = m$ for given values of the xs corresponds to the region of the distribution where y^* falls between τ_{m-1} and τ_m:

$$\Pr(y = m \mid \mathbf{x}) = \Pr(\tau_{m-1} \le y^* < \tau_m \mid \mathbf{x})$$

Substituting $\mathbf{x}\boldsymbol{\beta} + \varepsilon$ for y^* and using some algebra leads to the standard formula for the predicted probability in the ORM,

$$\Pr(y = m \mid \mathbf{x}) = F(\tau_m - \mathbf{x}\boldsymbol{\beta}) - F(\tau_{m-1} - \mathbf{x}\boldsymbol{\beta}) \tag{5.1}$$

where F is the cdf for ε. In ordinal probit, F is normal with $\text{Var}(\varepsilon) = 1$; in ordinal logit, F is logistic with $\text{Var}(\varepsilon) = \pi^2/3$. For $y = 1$, the second term on the right drops out because $F(-\infty - \mathbf{x}\boldsymbol{\beta}) = 0$, and for $y = J$, the first term equals $F(\infty - \mathbf{x}\boldsymbol{\beta}) = 1$.

Comparing these equations with those for the BRM shows that the ORM is identical to the binary regression model, with one exception. To show this, we fit chapter 4's binary model for labor-force participation using both logit and ologit (the command for ordinal logit):

```
. use http://www.stata-press.com/data/lf2/binlfp2, clear
(Data from 1976 PSID-T Mroz)
. logit lfp k5 k618 age wc hc lwg inc, nolog
 (output omitted)
. estimates store logit
. ologit lfp k5 k618 age wc hc lwg inc, nolog
 (output omitted)
. estimates store ologit
```

To compare the coefficients, we combine them using estimates table, which leads to the following table:[1]

```
. estimates table logit ologit, b(%9.3f) t label varwidth(30) equations(1:1)
```

Variable	logit	ologit
#1		
# kids < 6	-1.463	-1.463
	-7.43	-7.43
# kids 6-18	-0.065	-0.065
	-0.95	-0.95
Wife's age in years	-0.063	-0.063
	-4.92	-4.92
Wife College: 1=yes 0=no	0.807	0.807
	3.51	3.51
Husband College: 1=yes 0=no	0.112	0.112
	0.54	0.54
Log of wife's estimated wages	0.605	0.605
	4.01	4.01
Family income excluding wife's	-0.034	-0.034
	-4.20	-4.20
Constant	3.182	
	4.94	
cut1		
Constant		-3.182
		-4.94

legend: b/t

1. Because logit has a constant and ologit has a cutpoint, by default estimates table will not line up the coefficients from the two models. Rather, each of the independent variables will be listed twice. equations(1:1) tells estimates table to line up the coefficients. This is easiest to understand if you try our command without the equations(1:1) option.

The slope coefficients and their standard errors are identical, but for `logit` an intercept is reported (i.e., the coefficient associated with _cons), whereas for `ologit` the constant is replaced by the cutpoint labeled /`cut1`, which is equal but of opposite sign.

This difference is due to how the two models are identified. As the ORM has been presented, there are "too many" free parameters; that is, you cannot estimate $J - 1$ thresholds and the constant, too. For a unique set of ML estimates to exist, an identifying assumption needs to be made about either the intercept or one of the cutpoints. In Stata, the ORM is identified by assuming that the intercept is 0, and the values of all cutpoints are estimated. Some statistics packages for the ORM instead fix one of the cutpoints to 0 and estimate the intercept. In presenting the BRM, we immediately assumed that the value that divided y^* into observed 0s and 1s was 0. In effect, we identified the model by assuming a threshold of 0. Although different parameterizations can be confusing, keep in mind that the slope coefficients and predicted probabilities are the same under either parameterization (see Long 1997, 122–23 for more details).

5.1.2 A nonlinear probability model

The ordinal regression model can also be developed as a nonlinear probability model without appealing to the idea of a latent variable. Here we show how this can be done for the ordinal logit model. First, we define the odds that an outcome is less than or equal to m versus greater than m given $\mathbf{x}$:

$$\Omega_{\leq m|>m}\left(\mathbf{x}\right) \equiv \frac{\Pr\left(y \leq m \mid \mathbf{x}\right)}{\Pr\left(y > m \mid \mathbf{x}\right)} \quad \text{for } m = 1, J - 1$$

For example, we could compute the odds of disagreeing or strongly disagreeing (i.e., $m \leq 2$) versus agreeing or strongly agreeing $(m > 2)$. The log of the odds is assumed to equal

$$\ln \Omega_{\leq m|>m}\left(\mathbf{x}\right) = \tau_m - \mathbf{x}\boldsymbol{\beta} \tag{5.2}$$

For one independent variable and three categories (where we are fixing the intercept to equal 0),

$$\ln \frac{\Pr\left(y \leq 1 \mid \mathbf{x}\right)}{\Pr\left(y > 1 \mid \mathbf{x}\right)} = \tau_1 - \beta_1 x_1$$

$$\ln \frac{\Pr\left(y \leq 2 \mid \mathbf{x}\right)}{\Pr\left(y > 2 \mid \mathbf{x}\right)} = \tau_2 - \beta_1 x_1$$

Although it may seem confusing that the model subtracts βx rather than adding it, this is a consequence of computing the logit of $y \leq m$ versus $y > m$. While we agree that it would be simpler to stick with $\tau_m + \beta x$, this is not the way the model is normally presented.

5.2   Estimation using ologit and oprobit

The ordered logit and probit models can be fitted with the following commands and their basic options:

<u>olog</u>it *depvar* [*indepvars*] [*if*] [*in*] [*weight*] [, <u>r</u>obust <u>cl</u>uster(*varname*)
    <u>l</u>evel(#) <u>nolog</u>]

<u>opro</u>bit *depvar* [*indepvars*] [*if*] [*in*] [*weight*] [, <u>r</u>obust <u>cl</u>uster(*varname*)
    <u>l</u>evel(#) <u>nolog</u>]

In our experience, these models take more steps to converge than the models for either binary or nominal outcomes.

Variable lists

depvar is the dependent variable. The specific values assigned to the outcome categories are irrelevant, except that larger values are assumed to correspond to "higher" outcomes. For example, if you had three outcomes, you could use the values 1, 2, and 3, or −1.23, 2.3, and 999. Up to 50 outcomes are allowed in Intercooled Stata and Stata/SE; 20 outcomes are allowed in Small Stata.

indepvars is a list of independent variables. If *indepvars* is not included, Stata fits a model with only cutpoints.

Specifying the estimation sample

if and in qualifiers can be used to restrict the estimation sample. For example, if you want to fit an ordered logit model for only those in the 1989 sample, you could specify `ologit warm age ed prst male white if yr89==1`.

Listwise deletion Stata excludes cases in which there are missing values for any of the variables in the model. Accordingly, if two models are fitted using the same dataset but have different sets of independent variables, it is possible to have different samples. We recommend that you use `mark` and `markout` (discussed in chapter 3) to explicitly remove cases with missing data.

Weights

Both `ologit` and `oprobit` can be used with `fweights`, `pweights`, and `iweights`. See chapter 3 for more details.

Options

robust indicates that robust variance estimates are to be used. When cluster() is specified, robust standard errors are automatically used. See chapter 3 for more details.

cluster(*varname*) specifies that the observations are independent across the groups specified by unique values of *varname* but not necessarily within the groups. See chapter 3 for further details.

level(#) specifies the level of the confidence interval for estimated parameters. By default, Stata uses a 95% interval. You can also change the default level, say, to a 90% interval, with the command set level 90.

nolog suppresses the iteration history.

5.2.1 Example of attitudes toward working mothers

Our example is based on a question from the 1977 and 1989 General Social Survey. As we have already described, respondents were asked to evaluate the following statement: "A working mother can establish just as warm and secure of a relationship with her child as a mother who does not work". Responses were coded as: 1 = Strongly Disagree (SD), 2 = Disagree (D), 3 = Agree (A), and 4 = Strongly Agree (SA). A complete description of the data can be obtained by using describe, summarize, and tabulate:

```
. use http://www.stata-press.com/data/lf2/ordwarm2
(77 & 89 General Social Survey)
. describe warm yr89 male white age ed prst

              storage  display    value
variable name  type    format     label     variable label

warm           byte    %10.0g     SD2SA     Mom can have warm relations
                                              with child
yr89           byte    %10.0g     yrlbl     Survey year: 1=1989 0=1977
male           byte    %10.0g     sexlbl    Gender: 1=male 0=female
white          byte    %10.0g     race2lbl  Race: 1=white 0=not white
age            byte    %10.0g               Age in years
ed             byte    %10.0g               Years of education
prst           byte    %10.0g               Occupational prestige

. summarize warm yr89 male white age ed prst

    Variable |      Obs        Mean    Std. Dev.       Min        Max

        warm |     2293    2.607501    .9282156          1          4
        yr89 |     2293    .3986044    .4897178          0          1
        male |     2293    .4648932    .4988748          0          1
       white |     2293    .8765809    .3289894          0          1
         age |     2293    44.93546    16.77903         18         89

          ed |     2293    12.21805    3.160827          0         20
        prst |     2293    39.58526    14.49226         12         82
```

```
. tabulate warm
```

Mom can have warm relations with child	Freq.	Percent	Cum.
SD	297	12.95	12.95
D	723	31.53	44.48
A	856	37.33	81.81
SA	417	18.19	100.00
Total	2,293	100.00	

Using these data, we fitted the model

$$\Pr(\texttt{warm} = m \mid \mathbf{x}_i) = F(\tau_m - \mathbf{x}\boldsymbol{\beta}) - F(\tau_{m-1} - \mathbf{x}\boldsymbol{\beta})$$

where

$$\mathbf{x}\boldsymbol{\beta} = \beta_{\texttt{yr89}}\texttt{yr89} + \beta_{\texttt{male}}\texttt{male} + \beta_{\texttt{white}}\texttt{white} + \beta_{\texttt{age}}\texttt{age} + \beta_{\texttt{ed}}\texttt{ed} + \beta_{\texttt{prst}}\texttt{prst}$$

Here are the results from `ologit` and `oprobit`. We store each model with `estimates store` so that we can later make a table combining the results:

```
. ologit warm yr89 male white age ed prst, nolog
Ordered logistic regression                      Number of obs   =       2293
                                                 LR chi2(6)      =     301.72
                                                 Prob > chi2     =     0.0000
Log likelihood = -2844.9123                      Pseudo R2       =     0.0504
```

warm	Coef.	Std. Err.	z	P>|z|	[95% Conf. Interval]	
yr89	.5239025	.0798988	6.56	0.000	.3673037	.6805013
male	-.7332997	.0784827	-9.34	0.000	-.8871229	-.5794766
white	-.3911595	.1183808	-3.30	0.001	-.6231815	-.1591374
age	-.0216655	.0024683	-8.78	0.000	-.0265032	-.0168278
ed	.0671728	.015975	4.20	0.000	.0358624	.0984831
prst	.0060727	.0032929	1.84	0.065	-.0003813	.0125267
/cut1	-2.465362	.2389126			-2.933622	-1.997102
/cut2	-.630904	.2333155			-1.088194	-.173614
/cut3	1.261854	.2340179			.8031873	1.720521

```
. estimates store ologit
```

```
. oprobit warm yr89 male white age ed prst, nolog
Ordered probit regression                       Number of obs   =       2293
                                                LR chi2(6)      =     294.32
                                                Prob > chi2     =     0.0000
Log likelihood =  -2848.611                     Pseudo R2       =     0.0491
```

warm	Coef.	Std. Err.	z	P>|z|	[95% Conf.	Interval]
yr89	.3188147	.0468519	6.80	0.000	.2269867	.4106427
male	-.4170287	.0455459	-9.16	0.000	-.5062971	-.3277603
white	-.2265002	.0694773	-3.26	0.001	-.3626733	-.0903272
age	-.0122213	.0014427	-8.47	0.000	-.0150489	-.0093937
ed	.0387234	.0093241	4.15	0.000	.0204485	.0569982
prst	.003283	.001925	1.71	0.088	-.0004899	.0070559
/cut1	-1.428578	.1387742			-1.700571	-1.156586
/cut2	-.3605589	.1369219			-.6289209	-.092197
/cut3	.7681637	.1370564			.4995381	1.036789

```
. estimates store oprobit
```

The information in the header and the table of coefficients is in the same form as discussed in chapter 3. Since we stored the results for both models, we can compare the results using estimates table:

```
. estimates table ologit oprobit, b(%9.3f) t label varwidth(30)
```

Variable	ologit	oprobit
warm		
Survey year: 1=1989 0=1977	0.524	0.319
	6.56	6.80
Gender: 1=male 0=female	-0.733	-0.417
	-9.34	-9.16
Race: 1=white 0=not white	-0.391	-0.227
	-3.30	-3.26
Age in years	-0.022	-0.012
	-8.78	-8.47
Years of education	0.067	0.039
	4.20	4.15
Occupational prestige	0.006	0.003
	1.84	1.71
cut1		
Constant	-2.465	-1.429
	-10.32	-10.29
cut2		
Constant	-0.631	-0.361
	-2.70	-2.63
cut3		
Constant	1.262	0.768
	5.39	5.60

```
                                          legend: b/t
```

As with the BRM, the estimated coefficients differ from logit to probit by a factor of about 1.7, reflecting the differing scaling of the ordered logit and ordered probit models. Values of the z-tests are very similar because they are not affected by the scaling, but they are not identical because of slight differences in the shape of the assumed distribution of the errors.

5.2.2   Predicting perfectly

If the dependent variable does not vary within one of the categories of an independent variable, there will be a problem with estimation. To see what happens, let's transform the prestige variable prst into a dummy variable:

```
. gen dumprst = (prst<20 & warm==1)
. tab dumprst warm, miss
```

dumprst	Mom can have warm relations with child SD	D	A	SA	Total
0	257	723	856	417	2,253
1	40	0	0	0	40
Total	297	723	856	417	2,293

In all cases where dumprst is 1, respondents have values of SD for warm. That is, if you know dumprst is 1, you can predict perfectly that warm is 1 (i.e., SD). Although we purposely constructed dumprst so this would happen, perfect prediction can also occur in real data. If we fit the ORM using dumprst rather than prst, the perfectly predicted observations are dropped from the estimation sample.

```
. ologit warm yr89 male white age ed dumprst, nolog
```

Ordered logistic regression				Number of obs	=	2293
				LR chi2(6)	=	447.02
				Prob > chi2	=	0.0000
Log likelihood = -2772.2621				Pseudo R2	=	0.0746

warm	Coef.	Std. Err.	z	P>|z|	[95% Conf. Interval]	
yr89	.5268578	.0805997	6.54	0.000	.3688853	.6848303
male	-.7251825	.0792896	-9.15	0.000	-.8805872	-.5697778
white	-.4240687	.1197416	-3.54	0.000	-.658758	-.1893795
age	-.0210592	.0024462	-8.61	0.000	-.0258536	-.0162648
ed	.072143	.0133133	5.42	0.000	.0460494	.0982366
dumprst	-34.58373	1934739	-0.00	1.000	-3792053	3791983
/cut1	-2.776233	.243582			-3.253645	-2.298822
/cut2	-.8422903	.2363736			-1.305574	-.3790065
/cut3	1.06148	.236561			.5978287	1.525131

Note: 40 observations completely determined. Standard errors questionable.

The note at the bottom of the output above indicates the problem. In practice, the next step would be to delete the 40 cases in which dumprst equals 1 (you could use the command `drop if dumprst==1` to do this) and refit the model without `dumprst`. This corresponds to what is done automatically for binary models fitted by `logit` and `probit`.

5.3   Hypothesis testing with test and lrtest

Hypothesis tests of regression coefficients can be evaluated with the z-statistics in the estimation output, with `test` for Wald tests of simple and complex hypotheses, and with `lrtest` for the corresponding likelihood-ratio tests. We will briefly review each.

5.3.1   Testing individual coefficients

If the assumptions of the model hold, the ML estimators from `ologit` and `oprobit` are distributed asymptotically normally. The hypothesis $H_0: \beta_k = \beta^*$ can be tested with $z = \left(\widehat{\beta}_k - \beta^* \right) / \widehat{\sigma}_{\widehat{\beta}_k}$. Under the assumptions justifying ML, if H_0 is true, then z is distributed approximately normally with a mean of 0 and a variance of 1 for large samples. For example, consider the results for the variable `male` from the `ologit` output above:

```
. ologit warm male yr89 white age ed prst, nolog
 (output omitted )
```

warm	Coef.	Std. Err.	z	P>|z|	[95% Conf. Interval]
male	-.7332997	.0784827	-9.34	0.000	-.8871229 -.5794766

```
 (output omitted )
```

We conclude that sex significantly affects attitudes toward working mothers ($z = -9.34$, $p < 0.01$ for a two-tailed test).

Either a one-tailed or a two-tailed test can be used as discussed in chapter 4.

The z-test in the output of estimation commands is a Wald test, which can also be computed using `test`. For example, to test $H_0: \beta_{\text{male}} = 0$,

```
. test male
 ( 1)  [warm]male = 0

           chi2(  1) =    87.30
         Prob > chi2 =    0.0000
```

We conclude that sex significantly affects attitudes toward working mothers ($X^2 = 87.30$, $df = 1$, $p < 0.01$).

The value of a chi-squared test with 1 degree of freedom is identical to the square of the corresponding z-test, which can be demonstrated with the `display` command:

```
. display "z*z=" -9.343*-9.343
z*z=87.291649
```

Another way to verify this is to use the information that `test` leaves in memory. After `test`, we type

```
. return list

scalars:
              r(drop) =  0
              r(chi2) =  87.30028866314404
                r(df) =  1
                 r(p) =  9.32342950928e-21
```

The return `r(chi2)` contains the value of the chi-squared test statistics. We can take the square root of this quantity to confirm the relationship between the z-test and the chi-squared test.

```
. di "chi2=" r(chi2) "; sqrt(chi2)= " sqrt(r(chi2))
chi2=87.300289; sqrt(chi2)= 9.3434623
```

An LR test is computed by comparing the log likelihood from a full model with that of a restricted model. To test one coefficient, we begin by fitting the full model:

```
. ologit warm yr89 male white age ed prst, nolog
Ordered logistic regression                       Number of obs   =        2293
                                                  LR chi2(6)      =      301.72
                                                  Prob > chi2     =      0.0000
Log likelihood = -2844.9123                       Pseudo R2       =      0.0504
   (output omitted)
. estimates store fmodel
```

Then we fit the model, excluding `male`:

```
. ologit warm yr89 white age ed prst, nolog
Ordered logistic regression                       Number of obs   =        2293
                                                  LR chi2(5)      =      212.98
                                                  Prob > chi2     =      0.0000
Log likelihood =  -2889.278                       Pseudo R2       =      0.0355
   (output omitted)
. estimates store nmodel

. lrtest fmodel nmodel

Likelihood-ratio test                             LR chi2(1)  =       88.73
(Assumption: nmodel nested in fmodel)             Prob > chi2 =      0.0000
```

The resulting LR test can be interpreted to mean that the effect of being male is significant at the .01 level ($LRX^2 = 88.73$, $df = 1$, $p < .01$).

5.3.2   Testing multiple coefficients

We can also test a complex hypothesis that involves more than one coefficient. For example, our model has three demographic variables: `age`, `white`, and `male`. To test that all the demographic factors are simultaneously equal to zero, H_0: $\beta_{age} = \beta_{white} = \beta_{male} = 0$, we can use either a Wald or an LR test. For the Wald test, we fit the full model as before and then type

```
. test age white male
( 1)   [warm]age = 0
( 2)   [warm]white = 0
( 3)   [warm]male = 0

          chi2(  3) =   166.62
        Prob > chi2 =    0.0000
```

We conclude that the hypothesis that the demographic effects of age, race, and sex are simultaneously equal to zero can be rejected at the .01 level $(X^2 = 166.62, df = 3, p < .01)$.

test can also be used to test the equality of effects as shown in chapter 4.

To compute an LR test of multiple coefficients, we first fit the full model and save the results with lrtest, saving(0). Then to test H_0: $\beta_{age} = \beta_{white} = \beta_{male} = 0$, we fit the model that excludes these three variables and run lrtest:

```
. ologit warm yr89 male white age ed prst, nolog
  (output omitted )
. estimates store fmodel
. ologit warm yr89 ed prst, nolog
  (output omitted )
. estimates store nmodel
. lrtest fmodel nmodel
Likelihood-ratio test                              LR chi2(3)  =    171.58
(Assumption: nmodel nested in fmodel)              Prob > chi2 =    0.0000
```

We conclude that the hypothesis that the demographic effects of age, race, and sex are simultaneously equal to zero can be rejected at the .01 level $(X^2 = 171.58, df = 3, p < .01)$.

We find that the Wald and LR tests usually lead to the same decisions. When there are differences, they generally occur when the tests are near the cutoff for statistical significance. Because the LR test is invariant to reparameterization, we prefer the LR test.

5.4   Scalar measures of fit using fitstat

As we discuss more in chapter 3, scalar measures of fit can be useful in comparing competing models (see also Long 1997, 85–113). Several different measures can be computed after either ologit or oprobit with the SPost command fitstat:

```
. ologit warm yr89 male white age ed prst, nolog
  (output omitted)

. fitstat

Measures of Fit for ologit of warm
  Log-Lik Intercept Only:      -2995.770   Log-Lik Full Model:          -2844.912
  D(2284):                      5689.825   LR(6):                         301.716
                                           Prob > LR:                       0.000
  McFadden's R2:                   0.050   McFadden's Adj R2:               0.047
  ML (Cox-Snell) R2:               0.123   Cragg-Uhler(Nagelkerke) R2:      0.133
  McKelvey & Zavoina's R2:         0.127
  Variance of y*:                  3.768   Variance of error:               3.290
  Count R2:                        0.432   Adj Count R2:                    0.093
  AIC:                             2.489   AIC*n:                        5707.825
  BIC:                        -11982.891   BIC':                         -255.291
  BIC used by Stata:            5759.463   AIC used by Stata:            5707.825
```

Using simulations, both Hagle and Mitchell (1992) and Windmeijer (1995) find that, for ordinal outcomes, McKelvey and Zavoina's R^2 most closely approximates the R^2 obtained by fitting the linear regression model on the underlying latent variable.

5.5   Converting to a different parameterization*

Earlier, we noted that different software packages use different parameterizations to identify the model. Stata sets $\beta_0 = 0$ and estimates τ_1, whereas some programs fix $\tau_1 = 0$ and estimate β_0. Although all quantities of interest for the purpose of interpretation (e.g., predicted probabilities) are the same under both parameterizations, it is useful to see how Stata can fit the model under either parameterization. The key to understanding how this is done is the equation

$$\Pr\left(y = m \mid \mathbf{x}\right) = F\left([\tau_m - \delta] - [\beta_0 - \delta] - \mathbf{x}\boldsymbol{\beta}\right) - F\left([\tau_{m-1} - \delta] - [\beta_0 - \delta] - \mathbf{x}\boldsymbol{\beta}\right)$$

Without further constraints, it is possible to estimate only the differences $\tau_m - \delta$ and $\beta_0 - \delta$. Stata assumes $\delta = \beta_0$, which forces the estimate of β_0 to be 0, whereas some other programs assume $\delta = \tau_1$, which forces the estimate of τ_1 to be 0. For example,

Model parameter	Stata's estimate	Alternative parameterization
β_0	$\beta_0 - \beta_0 = 0$	$\beta_0 - \tau_1$
τ_1	$\tau_1 - \beta_0$	$\tau_1 - \tau_1 = 0$
τ_2	$\tau_2 - \beta_0$	$\tau_2 - \tau_1$
τ_3	$\tau_3 - \beta_0$	$\tau_3 - \tau_1$

Although you would only need to compute the alternative parameterization if you wanted to compare your results with those produced by another statistics package, seeing how this is done illustrates why the intercept and thresholds are arbitrary. To

estimate the alternative parameterization, we use `lincom` to estimate the difference
between Stata's estimates (see page 190) and the estimated value of the first cutpoint:[2]

```
. ologit warm yr89 male white age ed prst, nolog
  (output omitted)
. * intercept
. lincom 0 - _b[/cut1]
  ( 1) - [cut1]_cons = 0
```

warm	Coef.	Std. Err.	z	P>|z|	[95% Conf. Interval]	
(1)	2.465362	.2389126	10.32	0.000	1.997102	2.933622

Here we are computing the alternative parameterization of the intercept. `ologit` assumes that $\beta_0 = 0$, so we simply estimate $0 - \tau_1$; that is, 0-`_b[/cut1]`. The trick is
that the cutpoints are contained in the vector `_b[]`, using the labels `/cut1`, `/cut2`, and
`/cut3`. For the thresholds, we are estimating $\tau_2 - \tau_1$ and $\tau_3 - \tau_1$, which correspond to
`_b[/cut2]`-`_b[/cut1]` and `_b[/cut3]`-`_b[/cut1]`:

```
. * cutpoint 2
. lincom _b[/cut2] - _b[/cut1]
  ( 1) - [cut1]_cons + [cut2]_cons = 0
```

warm	Coef.	Std. Err.	z	P>|z|	[95% Conf. Interval]	
(1)	1.834458	.0630432	29.10	0.000	1.710895	1.95802

```
. * cutpoint 3
. lincom _b[/cut3] - _b[/cut1]
  ( 1) - [cut1]_cons + [cut3]_cons = 0
```

warm	Coef.	Std. Err.	z	P>|z|	[95% Conf. Interval]	
(1)	3.727216	.0826215	45.11	0.000	3.565281	3.889151

The estimate of $\tau_1 - \tau_1$ is, of course, 0.

5.6   The parallel regression assumption

Before discussing interpretation, it is important to understand an assumption that is
implicit in the ORM, known as both the *parallel regression assumption* and, for the
ordinal logit model, the *proportional odds assumption*. Using (5.1), the ORM can be
written as

---

2. Prior to version 9, Stata labeled the cutpoints as, for example, `_cut1` instead of the newer `/cut1`.
If you are using a version of Stata prior to version 9, simply replace the `/`s with `_`s in the commands in
this chapter.

$$\Pr(y = 1 \mid \mathbf{x}) = F(\tau_m - \mathbf{x}\boldsymbol{\beta})$$
$$\Pr(y = m \mid \mathbf{x}) = F(\tau_m - \mathbf{x}\boldsymbol{\beta}) - F(\tau_{m-1} - \mathbf{x}\boldsymbol{\beta}) \quad \text{for} m = 2 \text{ to } J - 1$$
$$\Pr(y = J \mid \mathbf{x}) = 1 - F(\tau_{m-1} - \mathbf{x}\boldsymbol{\beta})$$

These equations can be used to compute the cumulative probabilities, which have the simple form

$$\Pr(y \le m \mid \mathbf{x}) = F(\tau_m - \mathbf{x}\boldsymbol{\beta}) \quad \text{for } m = 1 \text{ to } J - 1$$

This equation shows that the ORM is equivalent to $J - 1$ binary regressions with the critical assumption that the slope coefficients are identical across each regression.

For example, with four outcomes and one independent variable, the equations are

$$\Pr(y \le 1 \mid \mathbf{x}) = F(\tau_1 - \beta x)$$
$$\Pr(y \le 2 \mid \mathbf{x}) = F(\tau_2 - \beta x)$$
$$\Pr(y \le 3 \mid \mathbf{x}) = F(\tau_3 - \beta x)$$

The intercept α is not in the equation since it has been assumed to equal 0 to identify the model. These equations lead to the following figure:[3]

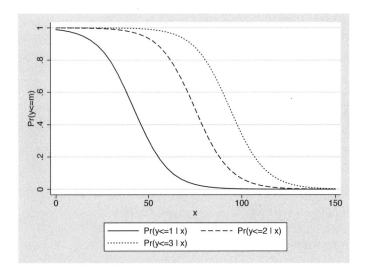

Each probability curve differs *only* in being shifted to the left or right. That is, they are parallel as a consequence of the assumption that the βs are equal for each equation.

---

3. This plot illustrates how graph can be used to construct graphs that are not based on real data. The commands for this graph are contained in st9ch5ordinal.do, which is part of the package spost9_do. See chapter 1 for details.

This figure suggests that the parallel regression assumptioncan be tested by comparing the estimate from the $J-1$ binary regressions,

$$\Pr(y \leq m \mid \mathbf{x}) = F(\tau_m - \mathbf{x}\boldsymbol{\beta}_m) \quad \text{for } m = 1, J-1$$

where the βs are allowed to differ across the equations. The parallel regression assumption implies that $\boldsymbol{\beta}_1 = \boldsymbol{\beta}_2 = \cdots = \boldsymbol{\beta}_{J-1}$. To the degree that the parallel regression assumption holds, the coefficients $\widehat{\boldsymbol{\beta}}_1, \widehat{\boldsymbol{\beta}}_2, \ldots, \widehat{\boldsymbol{\beta}}_{J-1}$ should be close. There are two commands in Stata that perform this test.

An approximate LR test

The command `omodel` (Wolfe and Gould 1998) is not part of official Stata but can be obtained by typing `net search omodel` and following the prompts. `omodel` computes an approximate LR test. Essentially, this method compares the log likelihood from `ologit` (or `oprobit`) with that obtained from pooling $J-1$ binary models fitted with `logit` (or `probit`), making an adjustment for the correlation between the binary outcomes defined by $y \leq m$. The syntax is

`omodel` [`logit`|`probit`] *depvar* [*indepvars*] [*if*] [*in*] [*weight*]

where the subcommand `logit` or `probit` indicates whether ordered logit or ordered probit is to be used. For example,

```
. omodel logit warm yr89 male white age ed prst
  (same output as for ologit warm yr89 male white age ed prst)
Approximate likelihood-ratio test of proportionality of odds
across response categories:
        chi2(12) =     48.91
      Prob > chi2 =    0.0000
```

Here the parallel regression assumption can be rejected at the .01 level.

A Wald test

The LR test is an omnibus test that the coefficients for all variables are simultaneously equal. Accordingly, you cannot determine whether the coefficients for some variables are identical across the binary equations while coefficients for other variables differ. To this end, a Wald test by Brant (1990) is useful since it tests the parallel regression assumption for each variable. The messy details of computing this test are found in Brant (1990) or Long (1997, 143–144). In Stata, the test is computed quickly with `brant`, which is part of SPost. After running `ologit` (`brant` does not work with `oprobit`), you run `brant` with the syntax:

`brant` [`, detail`]

The `detail` option provides a table of coefficients from each of the binary models. For
example,

```
. brant, detail

Estimated coefficients from j-1 binary regressions
                y>1           y>2           y>3
  yr89     .9647422    .56540626    .31907316
  male   -.30536425   -.69054232  -1.0837888
 white   -.55265759   -.31427081   -.39299842
   age    -.0164704   -.02533448   -.01859051
    ed    .10479624    .05285265    .05755466
  prst   -.00141118    .00953216    .00553043
 _cons   1.8584045    .73032873   -1.0245168

Brant Test of Parallel Regression Assumption
        Variable |     chi2   p>chi2    df

             All |    49.18    0.000    12

            yr89 |    13.01    0.001     2
            male |    22.24    0.000     2
           white |     1.27    0.531     2
             age |     7.38    0.025     2
              ed |     4.31    0.116     2
            prst |     4.33    0.115     2

A significant test statistic provides evidence that the parallel
regression assumption has been violated.
```

The chi-squared of 49.18 for the Brant test is close to the value of 48.91 from the LR
test. However, the Brant test shows that the largest violations are for `yr89` and `male`,
which indicates that there may be problems related to these variables.

Caveat regarding the parallel regression assumption We find that the parallel
regression assumption is frequently violated. When the assumption of parallel
regressions is rejected, alternative models that do not impose the constraint of
parallel regressions should be considered. Violation of the parallel regression
assumption is not a rationale for using ordinary least squares regression since
the assumptions implied by the application of the LRM to ordinal data are even
stronger. Alternative models that can be considered include models for nominal
outcomes discussed in chapter 6 or other models for ordinal outcomes discussed
in section 5.9.

5.7 Residuals and outliers using predict

Although no methods for detecting influential observations and outliers have been developed specifically for the ORM, Hosmer and Lemeshow (2000, 305) suggest applying the
methods for binary models to the $J-1$ cumulative probabilities that were discussed in

the last section. As noted by Hosmer and Lemeshow, the disadvantage of this approach is that you are evaluating only an approximation to the model you have fitted, because the coefficients of the binary models differ from those fitted in the ordinal model. But, if the parallel regression assumption is *not* rejected, you can be more confident in the results of your residual analysis.

To illustrate this approach, we start by generating three binary variables corresponding to `warm < 2`, `warm < 3`, and `warm < 4`:

```
. gen warmlt2 = (warm<2) if warm <.
. gen warmlt3 = (warm<3) if warm <.
. gen warmlt4 = (warm<4) if warm <.
```

For example, `warmlt3` is 1 if `warm` equals 1 or 2, else 0. Next we estimate binary logits for `warmlt2`, `warmlt3`, and `warmlt4` using the same independent variables as in our original `ologit` model. After estimating each logit, we generate standardized residuals using `predict` (for a detailed discussion of generating and inspecting these residuals, see chapter 4):

```
* warm < 2
. logit warmlt2 yr89 male white age ed prst
  (output omitted )
. predict rstd_lt2, rs
* warm < 3
. logit warmlt3 yr89 male white age ed prst
  (output omitted )
. predict rstd_lt3, rs
* warm < 4
. logit warmlt4 yr89 male white age ed prst
  (output omitted )
. predict rstd_lt4, rs
```

Next we create an index plot for each of the three binary equations. Using the results from the logit of `warmlt3` yields the following graph:

(Continued on next page)

```
. sort prst
. gen index = _n
. graph twoway scatter rstd_lt3 index, yline(0) ylabel(-4(2)4) ///
> xtitle("Observation Number")  xlabel(0(500)2293)              ///
> msymbol(Oh)
```

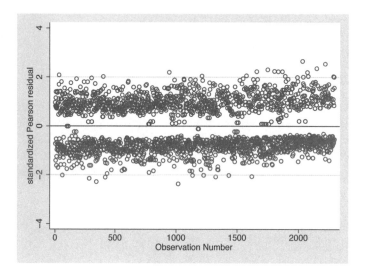

Given the size of the dataset, no residual stands out as being especially large. See section 4.4 for other ways in which you can examine outliers and influential cases.

5.8   Interpretation

If the idea of a latent variable makes substantive sense, simple interpretations are possible by rescaling y^* to compute standardized coefficients that can be used just like coefficients for the linear regression model. If the focus is on the categories of the ordinal variable (e.g., what affects the likelihood of strongly agreeing), the methods illustrated for the BRM can be extended to multiple outcomes. Because the ORM is nonlinear in the outcome probabilities, no approach can fully describe the relationship between a variable and the outcome probabilities. Consequently, you should consider each of these methods before deciding which approach is most effective in your application. For purposes of illustration, we continue to use the example of attitudes toward working mothers. Remember that the test of the parallel regression assumption suggests that this model is not appropriate for these data.

5.8.1 Marginal change in y*

In the ORM, $y^* = \mathbf{x}\boldsymbol{\beta} + \varepsilon$, and the marginal change in y^* with respect to x_k is

$$\frac{\partial y^*}{\partial x_k} = \beta_k$$

Because y^* is latent (and hence its metric is unknown), the marginal change cannot be interpreted without standardizing by the estimated standard deviation of y^*,

$$\widehat{\sigma}_{y^*}^2 = \widehat{\boldsymbol{\beta}}' \widehat{\mathrm{Var}}\,(\mathbf{x})\,\widehat{\boldsymbol{\beta}} + \mathrm{Var}\,(\varepsilon)$$

where $\widehat{\mathrm{Var}}\,(\mathbf{x})$ is the covariance matrix for the observed xs, $\widehat{\boldsymbol{\beta}}$ contains ML estimates, and $\mathrm{Var}(\varepsilon) = 1$ for ordered probit and $\pi^2/3$ for ordered logit. Then the y^*-*standardized* coefficient for x_k is

$$\beta_k^{Sy^*} = \frac{\beta_k}{\sigma_{y^*}}$$

which can be interpreted that for a unit increase in x_k, y^* is expected to increase by $\beta_k^{Sy^*}$ standard deviations, holding all other variables constant.

The *fully standardized coefficient* is

$$\beta_k^S = \frac{\sigma_k \beta_k}{\sigma_{y^*}} = \sigma_k \beta_k^{Sy^*}$$

which can be interpreted that for a standard deviation increase in x_k, y^* is expected to increase by β_k^S standard deviations, holding all other variables constant.

These coefficients can be computed with `listcoef` using the `std` option. For example, after fitting the ordered logit model,

```
. listcoef, std help
ologit (N=2293): Unstandardized and Standardized Estimates
  Observed SD: .9282156
    Latent SD: 1.9410634
```

warm	b	z	P>|z|	bStdX	bStdY	bStdXY	SDofX
yr89	0.52390	6.557	0.000	0.2566	0.2699	0.1322	0.4897
male	-0.73330	-9.343	0.000	-0.3658	-0.3778	-0.1885	0.4989
white	-0.39116	-3.304	0.001	-0.1287	-0.2015	-0.0663	0.3290
age	-0.02167	-8.778	0.000	-0.3635	-0.0112	-0.1873	16.7790
ed	0.06717	4.205	0.000	0.2123	0.0346	0.1094	3.1608
prst	0.00607	1.844	0.065	0.0880	0.0031	0.0453	14.4923

```
        b = raw coefficient
        z = z-score for test of b=0
   P>|z| = p-value for z-test
    bStdX = x-standardized coefficient
    bStdY = y-standardized coefficient
   bStdXY = fully standardized coefficient
    SDofX = standard deviation of X
```

If we think of the dependent variable as measuring support for mothers in the workplace, then the effect of the year of the interview can be interpreted as indicating that in 1989 support was .27 standard deviations higher than in 1977, holding all other variables constant.

To consider the effect of education, each standard deviation increase in education increases support by .11 standard deviations, holding all other variables constant.

5.8.2   Predicted probabilities

We usually prefer interpretations based somehow on predicted probabilities. These probabilities can be estimated with the formula

$$\widehat{\Pr}\left(y = m \mid \mathbf{x}\right) = F\left(\widehat{\tau}_m - \mathbf{x}\widehat{\boldsymbol{\beta}}\right) - F\left(\widehat{\tau}_{m-1} - \mathbf{x}\widehat{\boldsymbol{\beta}}\right)$$

with cumulative probabilities computed as

$$\widehat{\Pr}\left(y \le m \mid \mathbf{x}\right) = F\left(\tau_m - \mathbf{x}\widehat{\boldsymbol{\beta}}\right)$$

The values of $\mathbf{x}$ can be based on observations in the sample or can be hypothetical values of interest. The most basic command for computing probabilities is predict, but our SPost commands can be used to compute predicted probabilities in particularly useful ways.

5.8.3   Predicted probabilities with predict

After fitting a model with ologit or oprobit, a useful first step is to compute the in-sample predictions with the command

predict *newvar1* [*newvar2*[*newvar3...*]] [*if*] [*in*]

where you indicate one new variable name for each category of the dependent variable. For instance, in the following example predict specifies that the variables SDwarm, Dwarm, Awarm, and SAwarm should be created with predicted values for the four outcome categories:

```
. ologit warm yr89 male white age ed prst, nolog
  (output omitted)
. predict SDlogit Dlogit Alogit SAlogit
  (option p assumed; predicted probabilities)
```

The message (option p assumed; predicted probabilities) reflects that predict can compute many different quantities. Because we did not specify an option indicating which quantity to predict, option p for predicted probabilities was assumed.

   An easy way to see the range of the predictions is with dotplot, one of our favorite commands for quickly checking data:

```
. label var SDwarm "Pr(SD)"
. label var Dwarm "Pr(D)"
. label var Awarm "Pr(A)"
. label var SAwarm "Pr(SA)"
. dotplot SDwarm Dwarm Awarm SAwarm, ylabel(0(.25).75)
```

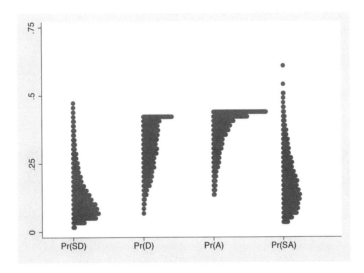

The predicted probabilities for the extreme categories tend to be less than .25, with most predictions for the middle categories falling between .25 and .5. In only a few cases is the probability of any outcome greater than .5.

Examining predicted probabilities within the sample provides a first, quick check of the model. To understand and present the substantive findings, however, it is usually more effective to compute predictions at specific, substantively informative values. Our commands `prvalue`, `prtab`, `prgen`, and `prchange` are designed to make this simple.

5.8.4 Individual predicted probabilities with prvalue

Predicted probabilities for individuals with a particular set of characteristics can be computed with `prvalue`. For example, we might want to examine the predicted probabilities for individuals with the following characteristics:

- Working-class men in 1977 who are near retirement.
- Young, highly educated women with prestigious jobs.
- An "average" individual in 1977.
- An "average" individual in 1989.

Each of these can be easily computed with `prvalue` (see chapter 3 for a discussion of options for this command). The predicted probabilities for older, working-class men are

```
. prvalue, x(yr89=0 male=1 prst=20 age=64 ed=16) rest(mean)
ologit: Predictions for warm
Confidence intervals by delta method
                            95% Conf. Interval
        Pr(y=SD|x):     0.2317   [ 0.1776,     0.2857]
        Pr(y=D|x):      0.4221   [ 0.3942,     0.4500]
        Pr(y=A|x):      0.2723   [ 0.2249,     0.3198]
        Pr(y=SA|x):     0.0739   [ 0.0523,     0.0954]
             yr89     male    white      age       ed     prst
    x=          0        1  .8765809       64       16       20
```

For young, highly educated women with prestigious jobs, they are

```
. prvalue, x(yr89=1 male=0 prst=80 age=30 ed=24) rest(mean) brief
ologit: Predictions for warm
                            95% Conf. Interval
        Pr(y=SD|x):     0.0164   [ 0.0106,     0.0222]
        Pr(y=D|x):      0.0781   [ 0.0554,     0.1008]
        Pr(y=A|x):      0.3147   [ 0.2636,     0.3658]
        Pr(y=SA|x):     0.5908   [ 0.5143,     0.6673]
```

and so on, for other sets of values.

We have set the values of the independent variables that define our hypothetical person using the `x()` and `rest()` options. The output from the first call of `prvalue` lists the values that have been set for all independent variables. This allows you to verify that `x()` and `rest()` did what you intended. For the second call, we added the `brief` option. This suppresses the output showing the levels of the independent variables. If you use this option, be certain that you have correctly specified the levels of all variables. Remember that the output of `prvalue` labels the categories according to the value labels assigned to the dependent variable. For example, `Pr(y=SD | x): 0.2317`. As it is easy to be confused about the outcome categories when using these models, it is prudent to assign clear value labels to your dependent variable (see chapter 2).

We can summarize the results in a table that lists the ideal types and provides a clear indication of which variables are important:

Ideal type	Probability for outcome category:			
	SD	D	A	SA
Working-class men in 1977 who are near retirement	.23	.42	.27	.07
Young, highly educated women in 1989 with prestigious jobs	.02	.08	.32	.59
An "average individual" in 1977	.13	.36	.37	.14
An "average individual" in 1989	.08	.28	.43	.21

5.8.5   Tables of predicted probabilities with prtab

In other cases, it can be useful to compute predicted probabilities for all combinations of a set of categorical independent variables. For example, the ideal types illustrate the importance of sex and the year when the question was asked. Using prtab, we can easily show the degree to which these variables affect opinions for those average on other characteristics.

```
. prtab yr89 male, novarlbl
ologit: Predicted probabilities for warm
Predicted probability of outcome 1 (SD)
```

		male
yr89	Women	Men
1977	0.0989	0.1859
1989	0.0610	0.1191

```
Predicted probability of outcome 2 (D)
```

		male
yr89	Women	Men
1977	0.3083	0.4026
1989	0.2282	0.3394

```
Predicted probability of outcome 3 (A)
```

		male
yr89	Women	Men
1977	0.4129	0.3162
1989	0.4406	0.3904

```
Predicted probability of outcome 4 (SA)
```

yr89	male Women	Men
1977	0.1799	0.0953
1989	0.2703	0.1510

```
          yr89       male      white       age         ed       prst
x=   .39860445  .46489315   .8765809  44.935456  12.218055  39.585259
```

(tables for other outcomes omitted)

Tip Sometimes the output of `prtab` is clearer without the variable labels. These can be suppressed with the `novarlbl` option.

The output from `prtab` can be rearranged into a table that clearly shows that men are more likely than women to strongly disagree or disagree with the proposition that working mothers can have relationships with their children that are as warm as those of mothers who do not work. The table also shows that between 1977 and 1989 there was a movement for both men and women toward more positive attitudes about working mothers.

1977	SD	D	A	SA
Men	.19	.40	.32	.10
Women	.10	.31	.41	.18
Difference	.09	.09	−.09	−.08

1989	SD	D	A	SA
Men	.12	.34	.39	.15
Women	.06	.23	.44	.27
Difference	.06	.11	−.05	−.12

Change from 1977 to 1989	SD	D	A	SA
Men	−.07	−.06	.07	.05
Women	−.04	−.08	.03	.09

`prtab` does not include confidence intervals. If we want confidence intervals for the probabilities in the table, we need to make separate calls to `prvalue`. For the first row in the above table, we could compute the confidence intervals as follows:

```
. prvalue, x(yr89=0 male=1)
ologit: Predictions for warm
Confidence intervals by delta method
                              95% Conf. Interval
        Pr(y=SD|x):     0.1859   [ 0.1631,     0.2088]
        Pr(y=D|x):      0.4026   [ 0.3777,     0.4275]
        Pr(y=A|x):      0.3162   [ 0.2925,     0.3400]
        Pr(y=SA|x):     0.0953   [ 0.0814,     0.1092]
             yr89     male    white       age       ed       prst
    x=          0        1  .8765809  44.935456  12.218055  39.585259
```

The confidence intervals for the other cells can be generated by the following commands:

```
. prvalue, x(yr89=0 male=0)
. prvalue, x(yr89=1 male=1)
. prvalue, x(yr89=1 male=0)
```

5.8.6   Graphing predicted probabilities with prgen

Graphing predicted probabilities for each outcome can also be useful for the ORM. Here we consider women in 1989 and show how predicted probabilities are affected by age. Of course, the plot could also be constructed for other sets of characteristics. The predicted probabilities as **age** ranges from 20 to 80 are generated by **prgen**:

```
. prgen age, from(20) to(80) generate(w89) x(male=0 yr89=1) ncases(13)
ologit: Predicted values as age varies from 20 to 80.
             yr89     male    white       age       ed       prst
    x=          1        0  .8765809  44.935456  12.218055  39.585259
```

You should be familiar with how x() operates, but it is useful to review the other options:

from(20) and to(80) specify the minimum and maximum values over which inc is to vary. The default is the variable's minimum and maximum values.

generate(w89) is the root name for the new variables.

ncases(13) indicates that 13 evenly spaced values of **age** between 20 and 80 should be generated.

Here w89x contains values of **age** ranging from 20 to 80. The p# variables contain the predicted probability for outcome # (e.g., w89p2 is the predicted probability of outcome 2). With ordinal outcomes, **prgen** also computes cumulative probabilities (i.e., summed) that are indicated by s (e.g., w89s2 is the sum of the predicted probability of outcomes 1 and 2).

```
. desc w89*

               storage  display    value
variable name  type     format     label      variable label

w89x           float    %9.0g                  Age in years
w89p1          float    %9.0g                  pr(SD)=Pr(1)
w89p2          float    %9.0g                  pr(D)=Pr(2)
w89p3          float    %9.0g                  pr(A)=Pr(3)
w89p4          float    %9.0g                  pr(SA)=Pr(4)
w89s1          float    %9.0g                  pr(y<=1)
w89s2          float    %9.0g                  pr(y<=2)
w89s3          float    %9.0g                  pr(y<=3)
w89s4          float    %9.0g                  pr(y<=4)
```

Although `prgen` assigns variable labels to the variables it creates, we can change these to improve the look of the plot that we are creating. Specifically,

```
. label var w89p1 "SD"
. label var w89p2 "D"
. label var w89p3 "A"
. label var w89p4 "SA"
. label var w89s1 "SD"
. label var w89s2 "SD or D"
. label var w89s3 "SD, D or A"
```

First, we plot the probabilities of individual outcomes using `graph`. Because the `graph` command is long, we use `///` to allow the commands to be longer than one line in our do-file.

```
. // step 1: graph predicted probabilities
. graph twoway connected w89p1 w89p2 w89p3 w89p4 w89x, ///
> title("Panel A: Predicted Probabilities")           ///
> xtitle("Age") xlabel(20(10)80) ylabel(0(.25).50)    ///
> yscale(noline) ylabel("") xline(44.93)              ///
> ytitle("") name(graph1, replace)
```

This `graph` command plots the four predicted probabilities against generated values for age contained in `w89x`. Standard options for `graph` are used to specify the axes and labels. The vertical line specified by `xline(44.93)` marks the average age in the sample. This line is used to illustrate the marginal effect discussed in section 5.8.7. Option `name(graph1, replace)` saves the graph in memory under the name `graph1` so that we can combine it with the next graph, which plots the cumulative probabilities:

```
. // step 2: graph cumulative probabilities
. graph twoway connected w89s1 w89s2 w89s3 w89x,    ///
> title("Panel B: Cumulative Probabilities")        ///
> xtitle("Age") xlabel(20(10)80) ylabel(0(.25)1)    ///
> yscale(noline) ylabel("") name(graph2, replace)   ///
> ytitle("")
```

Next we combine these two graphs (see chapter 2 for details on combining graphs):

```
. // step 3: combine graphs
. graph combine graph1 graph2, col(1) iscale(*.9) imargin(small) ///
> ysize(4.31) xsize(3.287)
```

This leads to figure 5.2. Panel A plots the predicted probabilities and shows that with age the probability of SA decreases rapidly, whereas the probability of D (and to a lesser degree, SD) increases. Panel B plots the cumulative probabilities. Both panels present the same information; which method you use is up to you.

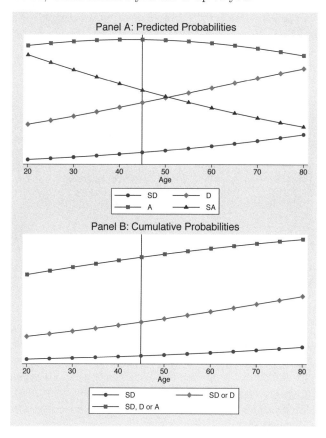

Figure 5.2: Plot of predicted probabilities for the ordered logit model.

5.8.7   Changes in predicted probabilities

When there are many variables in the model, it is impractical to plot them all. In such cases, measures of change in the outcome probabilities are useful for summarizing the effects of each variable. Before proceeding, however, we hasten to note that values of both discrete and marginal change depend on the levels of all variables in the model. We return to this point shortly.

Marginal change with prchange

The marginal change in the probability is computed as

$$\frac{\partial \Pr(y = m \mid \mathbf{x})}{\partial x_k} = \frac{\partial F(\tau_m - \mathbf{x}\boldsymbol{\beta})}{\partial x_k} - \frac{\partial F(\tau_{m-1} - \mathbf{x}\boldsymbol{\beta})}{\partial x_k}$$

which is the slope of the curve relating x_k to $\Pr(y{=}m \,|\, \mathbf{x})$, holding all other variables constant. Here we consider the marginal effect of age $\{\partial \Pr(y = m \mid \mathbf{x}) \,/\partial \texttt{age}\}$ for women in 1989 who are average on all other variables. This corresponds to the slope of the curves in panel A of figure 5.2 evaluated at the vertical line (recall that this line is drawn at the average age in the sample). The marginal is computed with prchange, where we specify that only the coefficients for age should be computed:

```
. prchange age, x(male=0 yr89=1) rest(mean)

ologit: Changes in Probabilities for warm

age
              Avg|Chg|          SD            D            A           SA
   Min->Max  .16441458    .10941909    .21941006   -.05462247   -.27420671
      -+1/2  .00222661    .00124099    .00321223    -.0001803   -.00427291
     -+sd/2   .0373125    .0208976     .05372739   -.00300205   -.07162295
   MargEfct  .00222662    .00124098    .00321226   -.00018032   -.00427292

                    SD            D            A           SA
   Pr(y|x)   .06099996    .22815652    .44057754    .27026597

                yr89        male       white         age          ed         prst
       x=         1           0     .876581     44.9355     12.2181      39.5853
   sd(x)=   .489718     .498875     .328989      16.779     3.16083      14.4923
```

The first thing to notice is the row labeled Pr(y|x), which is the predicted probabilities at the values set by x() and rest(). In panel A, these probabilities correspond to the intersection of the vertical line and the probability curves. The row MargEfct lists the slopes of the probability curves at the point of intersection with the vertical line in the figure. For example, the slope for SD (shown with circles) is .00124, whereas the slope for A (shown with squares) is negative and small. As with the BRM, the size of the slope indicates the instantaneous rate of change but does not correspond exactly to the amount of change in the probability for a change of one unit in the independent variable. However, when the probability curve is approximately linear, the marginal effect can be used to summarize the effect of a unit change in the variable on the probability of an outcome.

Marginal change with mfx

Marginal change can also be computed using mfx, where at() is used to set values of the independent variables. And it estimates the marginal effects for only one outcome category at a time, where the category is specified with the option predict(outcome(#)).

Using the same values for the independent variables as in the example above, we obtain the following results:

```
. mfx, at(male=0 yr89=1) predict(outcome(1))

warning: no value assigned in at() for variables white age ed prst;
   means used for white age ed prst

Marial effects after ologit
     y  = Pr(warm==1) (predict, outcome(1))
        =  .06099996
```

variable	dy/dx	Std. Err.	z	P>|z|	[95% C.I.]		X
yr89*	-.0378526	.00601	-6.30	0.000	-.049633	-.026072	1
male*	.0581355	.00731	7.95	0.000	.043803	.072468	0
white*	.0197511	.0055	3.59	0.000	.008972	.03053	.876581
age	.001241	.00016	7.69	0.000	.000925	.001557	44.9355
ed	-.0038476	.00097	-3.96	0.000	-.005754	-.001941	12.2181
prst	-.0003478	.00019	-1.83	0.068	-.000721	.000025	39.5853

```
(*) dy/dx is for discrete change of dummy variable from 0 to 1
```

The marginal for `age` is .001241, which matches the result obtained from `prchange`. The advantage of `mfx` is that it computes standard errors and confidence intervals.

Discrete change with prchange

As the marginal change can be misleading when the probability curve is changing rapidly or when an independent variable is a dummy variable, we prefer using discrete change (`mfx` computes discrete change for independent variables that are binary but not for other independent variables). The discrete change is the change in the predicted probability for a change in x_k from the start value x_S to the end value x_E (e.g., a change from $x_k = 0$ to $x_k = 1$). Formally,

$$\frac{\Delta \Pr(y = m \mid \mathbf{x})}{\Delta x_k} = \Pr(y = m \mid \mathbf{x}, x_k = x_E) - \Pr(y = m \mid \mathbf{x}, x_k = x_S)$$

where $\Pr(y = m \mid \mathbf{x}, x_k)$ is the probability that $y = m$ given $\mathbf{x}$, noting a specific value for x_k. The change is interpreted as indicating that when x_k changes from x_S to x_E, the predicted probability of outcome m changes by $\Delta \Pr(y = m \mid \mathbf{x}) / \Delta x_k$, holding all other variables at $\mathbf{x}$.

The value of the discrete change depends on (1) the value at which x_k starts, (2) the amount of change in x_k, and (3) the values of all other variables. Most often, each continuous variable except x_k is held at its mean. For dummy independent variables, the change could be computed for both values of the variable. For example, we could compute the discrete change for `age` separately for men and women.

Here the discrete-change coefficients for `male`, `age`, and `prst` for women in 1989, with other variables at their mean, are computed as follows:

```
. prchange male age prst, x(male=0 yr89=1) rest(mean)
ologit: Changes in Probabilities for warm
male
                 Avg|Chg|          SD           D            A           SA
       0->1     .08469636    .05813552    .11125721   -.05015317   -.11923955

age
                 Avg|Chg|          SD           D            A           SA
   Min->Max     .16441458    .10941909    .21941006   -.05462247   -.27420671
     -+1/2      .00222661    .00124099    .00321223    -.0001803   -.00427291
     -+sd/2     .0373125     .0208976     .05372739   -.00300205   -.07162295
   MargEfct     .00222662    .00124098    .00321226   -.00018032   -.00427292

prst
                 Avg|Chg|          SD           D            A           SA
   Min->Max     .04278038   -.02352008   -.06204067    .00013945    .08542132
     -+1/2      .00062411   -.00034784   -.00090037    .00005054    .00119767
     -+sd/2     .00904405   -.00504204   -.01304607    .00073212    .01735598
   MargEfct     .00062411   -.00034784   -.00090038    .00005054    .00119767

                       SD           D            A          SA
   Pr(y|x)     .06099996    .22815652    .44057754    .27026597

                 yr89      male     white       age        ed       prst
       x=          1         0    .876581   44.9355   12.2181   39.5853
   sd(x)=     .489718    .498875   .328989    16.779   3.16083   14.4923
```

For variables that are not binary, the discrete change can be interpreted for a unit change centered on the mean, for a standard deviation change centered on the mean, or as the variable changes from its minimum to its maximum value. The following are two examples:

> For a standard deviation increase in age, the probability of disagreeing increases by .05, holding other variables constant at their means.

> Moving from the minimum prestige to the maximum prestige changes the predicted probability of strongly agreeing by .08, holding all other variables constant at their means.

The J discrete-change coefficients for a variable can be summarized by computing the average of the *absolute values* of the changes across all the outcome categories:

$$\overline{\Delta} = \frac{1}{J} \sum_{j=1}^{J} \left| \frac{\Delta \Pr(y = j \mid \overline{\mathbf{x}})}{\Delta x_k} \right|$$

The absolute value must be used because the sum of the changes without taking the absolute value is necessarily zero. These are labeled as Avg|Chg|. For example, the effect of being a male is on average 0.08, which is larger than the average effect of a standard deviation change in either age or occupational prestige.

Confidence intervals for discrete changes

Although prchange computes discrete changes, it does not compute confidence intervals for these changes. If you want confidence intervals, you need to use prvalue with the save and diff options. In the first call to prvalue, you provide the starting values for the explanatory variables and specify save. In the second call, you provide the ending values and specify diff.

Consider the change in probabilities associated with a change from 0 to 1 in a dummy variable. With prchange, we calculated the difference in probabilities between men and women in 1989, holding the other variables at their means. We can compute this same change using prvalue, which will give us the confidence intervals as well.

```
. qui prvalue, x(male=0 yr89=1) rest(mean) save
. prvalue, x(male=1 yr89=1) rest(mean) diff
ologit: Change in Predictions for warm
Confidence intervals by delta method
```

	Current	Saved	Change	95% CI for Change
Pr(y=SD|x):	0.1191	0.0610	0.0581	[0.0438, 0.0725]
Pr(y=D|x):	0.3394	0.2282	0.1113	[0.0876, 0.1350]
Pr(y=A|x):	0.3904	0.4406	-0.0502	[-0.0679, -0.0324]
Pr(y=SA|x):	0.1510	0.2703	-0.1192	[-0.1448, -0.0937]

	yr89	male	white	age	ed	prst
Current=	1	1	.8765809	44.935456	12.218055	39.585259
Saved=	1	0	.8765809	44.935456	12.218055	39.585259
Diff=	0	1	0	0	0	0

If you compare the results listed in the Change column with the earlier results from prchange, you will see that they are the same. For example, the probability of strongly agreeing with the item about working mothers is .119 lower for men than for women. The estimated bounds of the 95% confidence interval for differences in probabilities between men and women in 1989 are −.145 and −.094, holding other variables at their means. Or, we might say, the results suggest that the difference between men and women in the predicted probability of strongly agreeing is .12 and the difference could be as small as .09 or as large as .14 with 95% confidence.

Now let's consider discrete change with a continuous variable. Earlier, prchange computed the changes associated with a standard deviation increase in age centered around the mean. The mean age is 44.9, with a standard deviation of 16.8. Accordingly, we want to compute the change in probabilities when age increases from 36.5 to 53.3. The start and end values for age are computed using the display command:

```
. di 44.935 - (.5*16.779)
36.5455
. di 44.935 + (.5*16.779)
53.3245
```

Using these values in a pair of `prvalue` commands, we find

```
. qui prvalue, x(male=0 yr89=1 age=36.5455) rest(mean) save
. prvalue, x(male=0 yr89=1 age=53.3245) rest(mean) diff
ologit: Change in Predictions for warm
Confidence intervals by delta method
                   Current      Saved      Change    95% CI for Change
    Pr(y=SD|x):     0.0723     0.0514      0.0209    [ 0.0155,    0.0262]
    Pr(y=D|x):      0.2556     0.2019      0.0537    [ 0.0414,    0.0661]
    Pr(y=A|x):      0.4362     0.4392     -0.0030    [-0.0118,    0.0058]
    Pr(y=SA|x):     0.2359     0.3076     -0.0716    [-0.0883,   -0.0549]

                yr89       male      white        age         ed        prst
    Current=       1          0   .8765809    53.3245  12.218055   39.585259
      Saved=       1          0   .8765809    36.5455  12.218055   39.585259
       Diff=       0          0          0     16.779          0           0
```

The amount of change matches the results from `prchange`, but now we have confidence intervals. For otherwise average women in 1989, a standard deviation increase in age, about 17 years, increases the probability of disagreeing by .054 with estimated bounds for the 95% confidence interval at .041 and .066.

Computing discrete change for a 10-year increase in age

In the example above, `age` was measured in years. Not surprisingly, the change in the predicted probability for a 1-year increase in `age` is trivially small. But, to characterize the effect of `age`, we could report the effect of a 10-year change in `age`.

Warning It is tempting to compute the discrete change for a 10-year change in `age` by simply multiplying the 1-year discrete change by 10. This will give you approximately the right answer *if* the probability curve is nearly linear over the range of change. But, when the curve is not linear, simply multiplying can give misleading results and even the wrong sign. To be safe, do not do it!

The `delta(#)` option for `prchange` computes the discrete change as an independent value changes from $\#/2$ units below the base value to $\#/2$ above. Here we use `delta(10)` and set the base value of `age` to its mean:

```
. prchange age, x(male=0 yr89=1) rest(mean) delta(10)
ologit: Changes in Probabilities for warm
(Note: d = 10)
age
              Avg|Chg|          SD           D           A          SA
Min->Max     .16441458    .10941909   .21941006  -.05462247  -.27420671
   -+d/2     .02225603    .01242571   .03208634  -.00179818  -.04271388
  -+sd/2      .0373125     .0208976   .05372739  -.00300205  -.07162295
MargEfct     .00222662    .00124098   .00321226  -.00018032  -.00427292

                    SD           D           A          SA
Pr(y|x)     .06099996   .22815652   .44057754   .27026597

              yr89       male      white       age        ed       prst
    x=           1          0    .876581   44.9355   12.2181    39.5853
sd(x)=     .489718    .498875    .328989    16.779   3.16083    14.4923
```

For females interviewed in 1989, the results in the -+d/2 row show the changes in the predicted probabilities associated with a 10-year increase in age centered on the mean.

5.8.8 Odds ratios using listcoef

For ologit, but not oprobit, we can interpret the results using odds ratios. Earlier, (5.2) defined the ordered logit model as

$$\Omega_{\leq m|>m}\left(\mathbf{x}\right) = \exp\left(\tau_m - \mathbf{x}\boldsymbol{\beta}\right)$$

For example, with four outcomes we would simultaneously estimate three equations:

$$\Omega_{\leq 1|>1}\left(\mathbf{x}\right) = \exp\left(\tau_1 - \mathbf{x}\boldsymbol{\beta}\right)$$
$$\Omega_{\leq 2|>2}\left(\mathbf{x}\right) = \exp\left(\tau_2 - \mathbf{x}\boldsymbol{\beta}\right)$$
$$\Omega_{\leq 3|>3}\left(\mathbf{x}\right) = \exp\left(\tau_3 - \mathbf{x}\boldsymbol{\beta}\right)$$

Using the same approach as shown for binary logit, the effect of a change in x_k of 1 equals

$$\frac{\Omega_{\leq m|>m}\left(\mathbf{x}, x_k + 1\right)}{\Omega_{\leq m|>m}\left(\mathbf{x}, x_k\right)} = e^{-\beta_k} = \frac{1}{e^{\beta_k}}$$

which can be interpreted as indicating that for a unit increase in x_k, the odds of an outcome being less than or equal to m is changed by the factor $\exp\left(-\beta_k\right)$, holding all other variables constant.

The value of the odds ratio does *not* depend on the value of m, which is why the parallel regression assumption is also known as the proportional odds assumption. We could interpret the odds ratio as follows:

> For a unit increase in x_k, the odds of a lower outcome compared with a higher outcome are changed by the factor $\exp\left(-\beta_k\right)$, holding all other variables constant.

or, for a change in x_k of δ,

$$\frac{\Omega_{\leq m|>m}(\mathbf{x}, x_k + \delta)}{\Omega_{\leq m|>m}(\mathbf{x}, x_k)} = \exp(-\delta \times \beta_k) = \frac{1}{\exp(\delta \times \beta_k)}$$

so that

> for an increase of δ in x_k, the odds of a lower outcome compared with a higher outcome change by the factor $\exp(-\delta \times \beta_k)$, holding all other variables constant.

In these results, we are discussing factor changes in the odds of lower outcomes compared with higher outcomes. This is done because the model is traditionally written as $\ln \Omega_{\leq m|>m}(\mathbf{x}) = \tau_m - \mathbf{x}\boldsymbol{\beta}$, which leads to the factor change coefficient of $\exp(-\beta_k)$. For purposes of interpretation, we could just as well consider the factor change in the odds of higher versus lower values; that is, changes in the odds $\Omega_{>m|\leq m}(\mathbf{x})$. This would equal $\exp(\beta_k)$.

The odds ratios for both a unit and a standard deviation change of the independent variables can be computed with `listcoef`, which lists the factor changes in the odds of higher versus lower outcomes. Here we request coefficients for only `male` and `age`:

```
. ologit warm yr89 male white age ed prst, nolog
(output omitted)
. listcoef male age, help
ologit (N=2293): Factor Change in Odds
    Odds of: >m vs <=m
```

warm	b	z	P>|z|	e^b	e^bStdX	SDofX
male	-0.73330	-9.343	0.000	0.4803	0.6936	0.4989
age	-0.02167	-8.778	0.000	0.9786	0.6952	16.7790

```
        b = raw coefficient
        z = z-score for test of b=0
    P>|z| = p-value for z-test
      e^b = exp(b) = factor change in odds for unit increase in X
 e^bStdX = exp(b*SD of X) = change in odds for SD increase in X
    SDofX = standard deviation of X
```

or to compute percent changes in the odds,

```
. listcoef male age, help percent
ologit (N=2293): Percentage Change in Odds
  Odds of: >m vs <=m
```

warm	b	z	P>|z|	%	%StdX	SDofX
male	-0.73330	-9.343	0.000	-52.0	-30.6	0.4989
age	-0.02167	-8.778	0.000	-2.1	-30.5	16.7790

```
    b = raw coefficient
    z = z-score for test of b=0
P>|z| = p-value for z-test
    % = percent change in odds for unit increase in X
%StdX = percent change in odds for SD increase in X
SDofX = standard deviation of X
```

These results can be interpreted as follows:

> The odds of having more positive attitudes toward working mothers are .48 times smaller for men than women, holding all other variables constant. Equivalently, the odds of having more positive values are 52% smaller for men than women, holding other variables constant.

> For a standard deviation increase in age, the odds of having more positive attitudes decrease by a factor of .69, holding all other variables constant.

When presenting odds ratios, we find that people find it easier to understand the results if you talk about *increases* in the odds rather than *decreases*. That is, it is clearer to say, "The odds increased by a factor of 2" than to say, "The odds decreased by a factor of .5". If you agree, then you can reverse the order when presenting odds. For example, we could say that

> The odds of having more *negative* attitudes toward working mothers are 2.08 times larger for men than women, holding all other variables constant.

This new factor change, 2.08, is just the inverse of the old value .48 (that is, 1/.48). listcoef computes the odds of a lower category versus a higher category if you specify the reverse option:

```
. listcoef male, reverse
ologit (N=2293): Factor Change in Odds
  Odds of: <=m vs >m
```

warm	b	z	P>|z|	e^b	e^bStdX	SDofX
male	-0.73330	-9.343	0.000	2.0819	1.4417	0.4989

The output now says `Odds of: <=m vs >m` instead of `Odds of: >m vs <=m`, as it did earlier.

When interpreting the odds ratios, remember two points that are discussed in detail in chapter 4. First, because odds ratios are multiplicative coefficients, *positive and negative effects should be compared by taking the inverse of the negative effect* (or vice versa). For example, a negative factor change of .5 has the same magnitude as a positive factor change of $2 = 1/.5$. Second, the interpretation assumes only that the other variables have been held constant, not held at any specific values (as was required for discrete change). But, *a constant factor change in the odds does not correspond to a constant change or constant factor change in the probability.*

5.9   Less common models for ordinal outcomes

Stata can also be used to fit several less commonly used models for ordinal outcomes. In concluding this chapter, we describe these models briefly and note their commands for estimation. Our SPost commands do not work with these models. For `ocratio`, this is mainly because these commands do not fully incorporate the new methods of returning information that were introduced with an earlier version of Stata.

5.9.1   The stereotype model

The *stereotype logistic model* (SLM), also referred to as the *stereotype ordered regression model* (SORM), was proposed by Anderson (1984) in response to the restrictive assumption of parallel regressions in the ordered regression model. This SLM, which can be fitted with in Stata using the `slogit` command, is a compromise between allowing the coefficients for each independent variable to vary by outcome category (as is the case with multinomial logit model considered in the next chapter) and restricting the coefficients to be identical across all outcomes as was the case with the OLM. The SLM is defined as

$$\ln \frac{\Pr(y = q \mid \mathbf{x})}{\Pr(y = r \mid \mathbf{x})} = (\theta_q - \theta_r) - (\phi_q - \phi_r)(\mathbf{x}\boldsymbol{\beta}) \qquad (5.3)$$

where $\boldsymbol{\beta}$ is a vector of coefficients associated with the independent variables, the θs are intercepts, and the ϕs are scale factors that mediate the effects of the x's. Because this model is so closely related to the multinomial logit model, and is sometimes used for nominal data, we will postpone discussion until the next chapter.

5.9.2   The generalized ordered logit model

The parallel regression assumption results from assuming the same coefficient vector $\boldsymbol{\beta}$ for all comparisons in the $J - 1$ equations

$$\ln \Omega_{\leq m|>m}(\mathbf{x}) = \tau_m - \mathbf{x}\boldsymbol{\beta}$$

where $\Omega_{\leq m|>m}(\mathbf{x}) = \{\Pr(y \leq m \mid \mathbf{x})\} / \{\Pr(y > m \mid \mathbf{x})\}$. The generalized ordered logit model (GOLM) allows $\boldsymbol{\beta}$ to differ for each of the $J - 1$ comparisons. That is,

$$\ln \Omega_{\leq m|>m}(\mathbf{x}) = \tau_m - \mathbf{x}\boldsymbol{\beta}_m \quad \text{for } j = 1 \text{ to } J - 1$$

where predicted probabilities are computed as

$$\Pr(y = 1 \mid \mathbf{x}) = \frac{\exp(\tau_1 - \mathbf{x}\boldsymbol{\beta}_1)}{1 + \exp(\tau_1 - \mathbf{x}\boldsymbol{\beta}_1)}$$

$$\Pr(y = j \mid \mathbf{x}) = \frac{\exp(\tau_j - \mathbf{x}\boldsymbol{\beta}_j)}{1 + \exp(\tau_j - \mathbf{x}\boldsymbol{\beta}_j)} - \frac{\exp(\tau_{j-1} - \mathbf{x}\boldsymbol{\beta}_{j-1})}{1 + \exp(\tau_{j-1} - \mathbf{x}\boldsymbol{\beta}_{j-1})} \quad \text{for } j = 2 \text{ to } J - 1$$

$$\Pr(y = J \mid \mathbf{x}) = 1 - \frac{\exp(\tau_{J-1} - \mathbf{x}\boldsymbol{\beta}_{J-1})}{1 + \exp(\tau_{J-1} - \mathbf{x}\boldsymbol{\beta}_{J-1})}$$

To ensure that the $\Pr(y = j \mid \mathbf{x})$ is between 0 and 1, the condition

$$(\tau_j - \mathbf{x}\boldsymbol{\beta}_j) \geq (\tau_{j-1} - \mathbf{x}\boldsymbol{\beta}_{j-1})$$

must hold. Once predicted probabilities are computed, all the approaches used to interpret the ORM results can be readily applied. This model has been discussed by Clogg and Shihadeh (1994, 146–147), Fahrmeir and Tutz (1994, 91), and McCullagh and Nelder (1989, 155). It can be fitted in Stata with the add-on command `gologit` (Fu 1998). This command has not been recently updated and thus is no longer supported in SPost. More recently, Williams (2005) has written `gologit2`, which extends the original command to fit two special cases of the general models: the proportional odds model and the partial proportional odds model (see Lall et al. 2002; Peterson and Harrell 1990). These models are less restrictive than the ordinal logit model fitted by `ologit` but more parsimonious than the multinomial logit model fitted by `mlogit`. We plan to add support for `gologit2` to SPost.

5.9.3   The continuation ratio model

The *continuation ratio model* was proposed by Fienberg (1980, 110) and was designed for ordinal outcomes in which the categories represent the progression of events or stages in some process through which an individual can advance. For example, the outcome could be faculty rank, where the stages are assistant professor, associate professor, and full professor. A key characteristic of the process is that an individual must pass through each stage. For example, to become an associate professor you must be an assistant professor; to be a full professor, an associate professor. Although there are versions of this model based on other binary models (e.g., probit), here we consider the logit version.

If $\Pr(y = m \mid \mathbf{x})$ is the probability of being in stage m given $\mathbf{x}$ and $\Pr(y > m \mid \mathbf{x})$ is the probability of being in a stage later than m, the continuation ratio model for the log odds is

$$\ln\left\{\frac{\Pr\left(y=m\mid\mathbf{x}\right)}{\Pr\left(y>m\mid\mathbf{x}\right)}\right\}=\tau_m-\mathbf{x}\boldsymbol{\beta}\quad\text{for }m=1\text{ to }J-1$$

where the βs are constrained to be equal across outcome categories, whereas the constant term τ_m differs by stage. As with other logit models, we can also express the model in terms of the odds:

$$\frac{\Pr\left(y=m\mid\mathbf{x}\right)}{\Pr\left(y>m\mid\mathbf{x}\right)}=\exp\left(\tau_m-\mathbf{x}\boldsymbol{\beta}\right)$$

Accordingly, $\exp\left(-\beta_k\right)$ can be interpreted as the effect of a unit increase in x_k on the odds of being in m compared with being in a higher category given that an individual is in category m or higher, holding all other variables constant. From this equation, the predicted probabilities can be computed as

$$\Pr\left(y=m\mid\mathbf{x}\right)=\frac{\exp\left(\tau_m-\mathbf{x}\boldsymbol{\beta}\right)}{\prod_{j=1}^{m}\left\{1+\exp\left(\tau_j-\mathbf{x}\boldsymbol{\beta}\right)\right\}}\quad\text{for }m=1\text{ to }J-1$$

$$\Pr\left(y=J\mid\mathbf{x}\right)=1-\sum_{j=1}^{J-1}\Pr\left(y=j\mid\mathbf{x}\right)$$

These predicted probabilities can be used for interpreting the model. In Stata, this model can be fitted using `ocratio` by Wolfe (1998); type `net search ocratio` and follow the prompts to download.

6 Models for nominal outcomes with case-specific data

An outcome is nominal when the categories are assumed to be unordered. For example, marital status can be grouped nominally into the categories of divorced, never married, married, or widowed. Occupations might be organized as professional, white collar, blue collar, craft, and menial, which is the example we use in this chapter. Other examples include reasons for leaving the parents' home, the organizational context of scientific work (e.g., industry, government, and academia), and the choice of language in a multilingual society. Further, in some cases a researcher might prefer to treat an outcome as nominal, even though it is ordered or partially ordered. For example, if the response categories are strongly agree, agree, disagree, strongly disagree, and don't know, the category "don't know" invalidates models for ordinal outcomes. Or, you might decide to use a nominal regression model when the assumption of parallel regressions is rejected. In general, if you have concerns about the ordinality of the dependent variable, the potential loss of efficiency in using models for nominal outcomes is outweighed by avoiding potential bias.

This chapter focuses on three closely related models for nominal (and sometimes ordinal) outcomes with case-specific data. The multinomial logit model (MNLM) is the most frequently used nominal regression model. In this model, you are essentially estimating a separate binary logit for each pair of outcome categories. Next we consider the multinomial probit model with uncorrelated errors, which is the normal counterpart to the MNLM. We then discuss the stereotype logistic regression model (SLM). Although this model is often used for ordinal outcomes, it is closely related to the MNLM. All these models assume that the data are case specific, meaning that each independent variable has one value for each individual. Examples of such variables are an individual's race or education. In the next chapter, we consider models that include alternative-specific data.

Models for nominal outcomes, both in this chapter and the next, require us to be more exacting about some basic terminology. Until now we have used "individual", "observation", and "case" interchangeably to refer to observational units, where each observational unit corresponds to a single row or record in the dataset. In the next two chapters, we will use only the term "case" for this purpose. Most of the time, we use the word "alternative" to refer to a possible outcome. Sometimes we refer to an alternative as an outcome category or a comparison group in order to be consistent with the usual terminology for a model or the output generated by Stata. The term "choice" refers to the alternative that is actually observed, which can be thought of as the "most preferred" alternative. For example, if the dependent variable is the party voted for in the last presidential election, the alternatives might be Republican, Democrat, and Independent. If the person corresponding to a given case voted for the alternative of Democrat, we would say that the choice for this case is Democrat. But you should not infer from the term "choice" that the models we describe can be used only for data where the outcome occurs through a process of choice. For example, if we were modeling the type of injuries that people (i.e., cases) entering the emergency room of a hospital have, we would use the term "choice" even though the injury sustained is unlikely to be a choice. We will continue with this terminology in chapter 7, but with one complication. Chapter 7 deals with alternative-specific variables that vary not only by case but also by the alternative. For example, if a commuter is selecting one of three modes of travel, an alternative-specific predictor might be her travel time using each alternative. Each case has three rows of data, one for each of the alternatives, since this is the easiest way to organize the data. We discuss this more fully in the next chapter.

We begin by discussing the MNLM, where the biggest challenge is that the model includes many parameters and it is easy to be overwhelmed by the complexity of the results. This complexity is compounded by the nonlinearity of the model, which leads to the same difficulties of interpretation found for models in prior chapters. Although fitting the model is straightforward, interpretation involves many challenges that are the focus of this chapter. We begin by reviewing the statistical model, followed by a discussion of testing, fit, and finally methods of interpretation. These discussions are intended as a review for those who are familiar with the models. For a complete discussion, see Long (1997). As always, you can obtain sample do-files and data files by downloading the `spost9_do` and `spost9_ado` packages (see chapter 1 for details).

6.1   The multinomial logit model

The MNLM can be thought of as simultaneously estimating binary logits for all comparisons among the alternatives. For example, let `occ3` be a nominal outcome with the categories M for manual jobs, W for white-collar jobs, and P for professional jobs. Assume that there is one independent variable, `ed`, measuring years of education. We can examine the effect of `ed` on `occ3` by estimating three binary logits,

$$\ln\left\{\frac{\Pr(P\mid \mathbf{x})}{\Pr(M\mid \mathbf{x})}\right\} = \beta_{0,P\mid M} + \beta_{1,P\mid M}\,\texttt{ed}$$

$$\ln\left\{\frac{\Pr(W\mid \mathbf{x})}{\Pr(M\mid \mathbf{x})}\right\} = \beta_{0,W\mid M} + \beta_{1,W\mid M}\,\texttt{ed}$$

$$\ln\left\{\frac{\Pr(P\mid \mathbf{x})}{\Pr(W\mid \mathbf{x})}\right\} = \beta_{0,P\mid W} + \beta_{1,P\mid W}\,\texttt{ed}$$

where the subscripts to the βs indicate which comparison is being made (e.g., $\beta_{1,P\mid M}$ is the coefficient for the first independent variable for the comparison of P and M).

The three binary logits include redundant information. Because $\ln a/b = \ln a - \ln b$, the following equality must hold:

$$\ln\left\{\frac{\Pr\left(P\mid \mathbf{x}\right)}{\Pr\left(M\mid \mathbf{x}\right)}\right\} - \ln\left\{\frac{\Pr\left(W\mid \mathbf{x}\right)}{\Pr\left(M\mid \mathbf{x}\right)}\right\} = \ln\left\{\frac{\Pr\left(P\mid \mathbf{x}\right)}{\Pr\left(W\mid \mathbf{x}\right)}\right\}$$

This implies that

$$\beta_{0,P|M} - \beta_{0,W|M} = \beta_{0,P|W} \tag{6.1}$$
$$\beta_{1,P|M} - \beta_{1,W|M} = \beta_{1,P|W}$$

In general, with J alternatives, only $J-1$ binary logits need to be estimated. Estimates for the remaining coefficients can be computed using equalities of the sort shown in (6.1).

The problem with fitting the MNLM by estimating a series of binary logits is that each binary logit is based on a different sample. For example, in the logit comparing P with M, those in W are dropped. To see this, we can look at the output from a series of binary logits. First, we estimate a binary logit comparing manual and professional workers:

```
. use http://www.stata-press.com/data/lf2/nomintro2, clear
(1982 General Social Survey)
. tab prof_man, miss

  prof_man │     Freq.    Percent      Cum.
───────────┼──────────────────────────────────
    Manual │       184      54.60     54.60
      Prof │       112      33.23     87.83
         . │        41      12.17    100.00
───────────┼──────────────────────────────────
     Total │       337     100.00

. logit prof_man ed, nolog
```

```
Logistic regression                        Number of obs   =        296
                                           LR chi2(1)      =     139.78
                                           Prob > chi2     =     0.0000
Log likelihood = -126.43879                Pseudo R2       =     0.3560
```

prof_man	Coef.	Std. Err.	z	P>|z|	[95% Conf. Interval]
ed	.7184599	.0858735	8.37	0.000	.550151	.8867688
_cons	-10.19854	1.177457	-8.66	0.000	-12.50632	-7.89077

Forty-one cases are missing for `prof_man` and have been deleted. These correspond to respondents who have white-collar occupations. Likewise, the next two binary logits also exclude cases corresponding to the excluded category:

```
. tab wc_man, miss
```

wc_man	Freq.	Percent	Cum.
Manual	184	54.60	54.60
WhiteCol	41	12.17	66.77
.	112	33.23	100.00
Total	337	100.00	

```
. logit wc_man ed, nolog
```

Logistic regression

Number of obs	=	225
LR chi2(1)	=	16.00
Prob > chi2	=	0.0001
Pseudo R2	=	0.0749

Log likelihood = -98.818194

wc_man	Coef.	Std. Err.	z	P>|z|	[95% Conf. Interval]
ed	.3418255	.0934517	3.66	0.000	.1586636 .5249875
_cons	-5.758148	1.216291	-4.73	0.000	-8.142035 -3.374262

```
. tab prof_wc, miss
```

prof_wc	Freq.	Percent	Cum.
WhiteCol	41	12.17	12.17
Prof	112	33.23	45.40
.	184	54.60	100.00
Total	337	100.00	

```
. logit prof_wc ed, nolog
```

Logistic regression

Number of obs	=	153
LR chi2(1)	=	23.34
Prob > chi2	=	0.0000
Pseudo R2	=	0.1312

Log likelihood = -77.257045

prof_wc	Coef.	Std. Err.	z	P>|z|	[95% Conf. Interval]
ed	.3735466	.0874469	4.27	0.000	.2021538 .5449395
_cons	-4.332833	1.227293	-3.53	0.000	-6.738283 -1.927382

The results from the binary logits can be compared with the output from `mlogit`, the command that fits the MNLM:

```
. tab occ3, miss
```

occ3	Freq.	Percent	Cum.
Manual	184	54.60	54.60
WhiteCol	41	12.17	66.77
Prof	112	33.23	100.00
Total	337	100.00	

```
. mlogit occ3 ed, nolog
Multinomial logistic regression                    Number of obs   =        337
                                                   LR chi2(2)      =     145.89
                                                   Prob > chi2     =     0.0000
Log likelihood = -248.14786                        Pseudo R2       =     0.2272
```

occ3	Coef.	Std. Err.	z	P>|z|	[95% Conf. Interval]	
WhiteCol						
ed	.3000735	.0841358	3.57	0.000	.1351703	.4649767
_cons	-5.232602	1.096086	-4.77	0.000	-7.380892	-3.084312
Prof						
ed	.7195673	.0805117	8.94	0.000	.5617671	.8773674
_cons	-10.21121	1.106913	-9.22	0.000	-12.38072	-8.041698

(occ3==Manual is the base outcome)

The output from `mlogit` is divided into two panels. The top panel is labeled `WhiteCol`, which is the value label for the second category of the dependent variable; the second panel is labeled `Prof`, which corresponds to the third outcome category. The key to understanding the two panels is the last line of output: `occ3==Manual is the base outcome`. This means that the panel `WhiteCol` presents coefficients from the comparison of W to M. The second panel, labeled `Prof`, holds the comparison of P to M. Accordingly, the top panel should be compared with the coefficients from the binary logit for W and M (outcome variable `wc_man`) listed above. For example, the coefficient for the comparison of W to M from `mlogit` is $\hat{\beta}_{1,W|M} = .3000735$ with $z = 3.57$, whereas the `logit` estimate is $\hat{\beta}_{1,W|M} = .3418255$ with $z = 3.66$. Overall, the estimates from the binary model are close to those from the MNLM but not exactly the same.

Although theoretically $\beta_{1,P|M} - \beta_{1,W|M} = \beta_{1,P|W}$, the estimates from the *binary* logits are $\hat{\beta}_{1,P|M} - \hat{\beta}_{1,W|M} = .7184599 - .3418255 = .3766344$, which does not equal the binary logit estimate $\hat{\beta}_{1,P|W} = .3735466$. A series of binary logits using `logit` does *not* impose the constraints among coefficients that are implicit in the definition of the model. When fitting the model with `mlogit`, the constraints are imposed. Indeed, the output from `mlogit` presents only two of the three comparisons from our example, namely, W versus M and P versus M. The remaining comparison, W versus P, is the difference between the two sets of estimated coefficients. Details on using `listcoef` to automatically compute the remaining comparisons are given below.

6.1.1 Formal statement of the model

Formally, the MNLM can be written as

$$\ln \Omega_{m|b}(\mathbf{x}) = \ln \frac{\Pr(y = m \mid \mathbf{x})}{\Pr(y = b \mid \mathbf{x})} = \mathbf{x}\boldsymbol{\beta}_{m|b} \quad \text{for } m = 1 \text{ to } J$$

where b is the base category, which is also referred to as the comparison group. As $\ln \Omega_{b|b}(\mathbf{x}) = \ln 1 = 0$, it must hold that $\boldsymbol{\beta}_{b|b} = 0$. That is, the log odds of an outcome compared with itself are always 0, and thus the effects of any independent variables must also be 0. These J equations can be solved to compute the predicted probabilities:

$$\Pr(y = m \mid \mathbf{x}) = \frac{\exp\left(\mathbf{x}\boldsymbol{\beta}_{m|b}\right)}{\sum_{j=1}^{J} \exp\left(\mathbf{x}\boldsymbol{\beta}_{j|b}\right)}$$

Although the predicted probability will be the same regardless of the base outcome, b, changing the base outcome can be confusing since the resulting output from `mlogit` *appears* to be quite different. Suppose that you have three outcomes and fit the model with alternative 1 as the base category. Your probability equations would be

$$\Pr(y = m \mid \mathbf{x}) = \frac{\exp\left(\mathbf{x}\boldsymbol{\beta}_{m|1}\right)}{\sum_{j=1}^{J} \exp\left(\mathbf{x}\boldsymbol{\beta}_{j|1}\right)}$$

and you would obtain estimates $\widehat{\boldsymbol{\beta}}_{2|1}$ and $\widehat{\boldsymbol{\beta}}_{3|1}$, where $\boldsymbol{\beta}_{1|1} = 0$. If someone else set up the model with base category 2, their equations would be

$$\Pr(y = m \mid \mathbf{x}) = \frac{\exp\left(\mathbf{x}\boldsymbol{\beta}_{m|2}\right)}{\sum_{j=1}^{J} \exp\left(\mathbf{x}\boldsymbol{\beta}_{j|2}\right)}$$

and they would obtain $\widehat{\boldsymbol{\beta}}_{1|2}$ and $\widehat{\boldsymbol{\beta}}_{3|2}$, where $\boldsymbol{\beta}_{2|2} = 0$. Although the estimated parameters are different, they are only different *parameterizations* that provide the same predicted probabilities. The confusion arises only if you are not clear about which parameterization you are using. Unfortunately, some software packages—but *not* Stata—make it hard to tell which set of parameters is being estimated. We return to this issue when we discuss how Stata's `mlogit` parameterizes the model in the next section.

6.2   Estimation using mlogit

The multinomial logit model is fitted with the following command and its basic options:

<u>mlo</u>git *depvar* [*indepvars*] [*if*] [*in*] [*weight*] [, <u>noc</u>onstant

    <u>b</u>aseoutcome(*#*) <u>c</u>onstraints(*clist*) <u>r</u>obust <u>cl</u>uster(*varname*) <u>l</u>evel(*#*)

    <u>rrr</u> <u>nolog</u>]

In our experience, the model converges quickly, even when there are many outcome categories and independent variables.

Variable lists

    depvar is the dependent variable. The actual values taken on by the dependent variable are irrelevant. For example, if you had three outcomes, you could use the values

1, 2, and 3 or −1, 0, and 999. Up to 50 outcomes are allowed in Stata/SE and Intercooled Stata, and 20 outcomes are allowed in Small Stata.

indepvars is a list of independent variables. If *indepvars* is not included, Stata fits a model with only constants.

Specifying the estimation sample

if and in qualifiers can be used to restrict the estimation sample. For example, if you want to fit the model with only white respondents, use the command `mlogit occ ed exper if white==1`.

Listwise deletion Stata excludes cases in which there are missing values for any of the variables. Accordingly, if two models are fitted using the same dataset but have different sets of independent variables, it is possible to have different samples. We recommend that you use `mark` and `markout` (discussed in chapter 3) to explicitly remove cases with missing data.

Weights

`mlogit` can be used with `fweights`, `pweights`, and `iweights`. In chapter 3, we provide a brief discussion of the different types of weights and how weights are specified in Stata's syntax.

Options

`noconstant` excludes the constant terms from the model.

`baseoutcome(#)` specifies the value of *depvar* that is the base category (i.e., reference group) for the coefficients that are listed. This determines how the model is parameterized. If the `baseoutcome()` option is not specified, the most frequent outcome in the estimation sample is chosen as the base. The base category is always reported immediately below the estimates; for example, `Outcome occ3==Manual is the base outcome`.

`constraints(clist)` specifies the linear constraints to be applied during estimation. The default is to perform unconstrained estimation. Constraints are defined with the `constraint` command. This option is illustrated in section 6.3.3 when we discuss an LR test for combining outcome categories.

`robust` indicates that robust variance estimates are to be used. When `cluster()` is specified, robust standard errors are automatically used. See chapter 3 for more details.

`cluster(varname)` specifies that the observations be independent across the groups specified by unique values of *varname* but not necessarily independent within the groups. See chapter 3 for more details.

`level(#)` specifies the level of the confidence interval for estimated parameters. By default, Stata uses 95% intervals. You can also change the default level to, say, a 90% interval, with the command `set level 90`.

`rrr` reports the estimated coefficients transformed to relative risk ratios, defined as $\exp(b)$ rather than b, along with standard errors and confidence intervals for these ratios.

`nolog` suppresses the iteration history.

6.2.1 Example of occupational attainment

The 1982 General Social Survey asked respondents their occupation, which we recoded into five broad categories: menial jobs (M), blue collar jobs (B), craft jobs (C), white collar jobs (W), and professional jobs (P). Three independent variables are considered: `white` indicating the race of the respondent, `ed` measuring years of education, and `exper` measuring years of work experience.

```
. summarize white ed exper
    Variable |       Obs        Mean    Std. Dev.       Min        Max
-------------+--------------------------------------------------------
       white |       337    .9169139    .2764227          0          1
          ed |       337    13.09496    2.946427          3         20
       exper |       337    20.50148    13.95936          2         66
```

The distribution among outcome categories is

```
. tab occ
  Occupation |      Freq.     Percent        Cum.
-------------+-----------------------------------
      Menial |         31        9.20        9.20
     BlueCol |         69       20.47       29.67
       Craft |         84       24.93       54.60
    WhiteCol |         41       12.17       66.77
        Prof |        112       33.23      100.00
-------------+-----------------------------------
       Total |        337      100.00
```

Using these variables, the following MNLM was fitted:

$$\ln\Omega_{M|P}(\mathbf{x}_i) = \beta_{0,M|P} + \beta_{1,M|P}\texttt{white} + \beta_{2,M|P}\texttt{ed} + \beta_{3,M|P}\texttt{exper}$$
$$\ln\Omega_{B|P}(\mathbf{x}_i) = \beta_{0,B|P} + \beta_{1,B|P}\texttt{white} + \beta_{2,B|P}\texttt{ed} + \beta_{3,B|P}\texttt{exper}$$
$$\ln\Omega_{C|P}(\mathbf{x}_i) = \beta_{0,C|P} + \beta_{1,C|P}\texttt{white} + \beta_{2,C|P}\texttt{ed} + \beta_{3,C|P}\texttt{exper}$$
$$\ln\Omega_{W|P}(\mathbf{x}_i) = \beta_{0,W|P} + \beta_{1,W|P}\texttt{white} + \beta_{2,W|P}\texttt{ed} + \beta_{3,W|P}\texttt{exper}$$

where we specify the fifth outcome P as the base category:

```
. mlogit occ white ed exper, baseoutcome(5) nolog
Multinomial logistic regression              Number of obs   =        337
                                             LR chi2(12)     =     166.09
                                             Prob > chi2     =     0.0000
Log likelihood = -426.80048                  Pseudo R2       =     0.1629
```

occ	Coef.	Std. Err.	z	P>|z|	[95% Conf. Interval]	
Menial						
white	-1.774306	.7550543	-2.35	0.019	-3.254186	-.2944273
ed	-.7788519	.1146293	-6.79	0.000	-1.003521	-.5541826
exper	-.0356509	.018037	-1.98	0.048	-.0710028	-.000299
_cons	11.51833	1.849356	6.23	0.000	7.893659	15.143
BlueCol						
white	-.5378027	.7996033	-0.67	0.501	-2.104996	1.029391
ed	-.8782767	.1005446	-8.74	0.000	-1.07534	-.6812128
exper	-.0309296	.0144086	-2.15	0.032	-.05917	-.0026893
_cons	12.25956	1.668144	7.35	0.000	8.990061	15.52907
Craft						
white	-1.301963	.647416	-2.01	0.044	-2.570875	-.0330509
ed	-.6850365	.0892996	-7.67	0.000	-.8600605	-.5100126
exper	-.0079671	.0127055	-0.63	0.531	-.0328693	.0169351
_cons	10.42698	1.517943	6.87	0.000	7.451864	13.40209
WhiteCol						
white	-.2029212	.8693072	-0.23	0.815	-1.906732	1.50089
ed	-.4256943	.0922192	-4.62	0.000	-.6064407	-.2449479
exper	-.001055	.0143582	-0.07	0.941	-.0291967	.0270866
_cons	5.279722	1.684006	3.14	0.002	1.979132	8.580313

```
(occ==Prof is the base outcome)
```

Methods of testing coefficients and interpretation of the estimates will be considered after we discuss the effects of using different base categories.

6.2.2 Using different base categories

By default, mlogit sets the base category to the alternative with the most observations. Or, as illustrated in the last example, you can select the base category with baseoutcome(). mlogit then reports coefficients for the effect of each independent variable on each category relative to the base category. However, you should also examine the effects on other pairs of outcome categories. For example, you might be interested in how race affects the allocation of workers between Craft and BlueCol (e.g., $\beta_{1,B|C}$), which was not estimated in the output listed above. Although this coefficient can be estimated by rerunning mlogit with a different base category (e.g., mlogit occ white ed exper, baseoutcome(3)), it is easier to use listcoef, which presents estimates for *all* combinations of outcome categories. Because listcoef can generate much output, we show two options that limit which coefficients are listed. First, you can include a list of variables, and only coefficients for those variables will be listed. For example,

```
. listcoef white, help

mlogit (N=337): Factor Change in the Odds of occ

Variable: white (sd=.27642268)
```

Odds comparing Alternative 1 to Alternative 2	b	z	P>|z|	e^b	e^bStdX
Menial -BlueCol	-1.23650	-1.707	0.088	0.2904	0.7105
Menial -Craft	-0.47234	-0.782	0.434	0.6235	0.8776
Menial -WhiteCol	-1.57139	-1.741	0.082	0.2078	0.6477
Menial -Prof	-1.77431	-2.350	0.019	0.1696	0.6123
BlueCol -Menial	1.23650	1.707	0.088	3.4436	1.4075
BlueCol -Craft	0.76416	1.208	0.227	2.1472	1.2352
BlueCol -WhiteCol	-0.33488	-0.359	0.720	0.7154	0.9116
BlueCol -Prof	-0.53780	-0.673	0.501	0.5840	0.8619
Craft -Menial	0.47234	0.782	0.434	1.6037	1.1395
Craft -BlueCol	-0.76416	-1.208	0.227	0.4657	0.8096
Craft -WhiteCol	-1.09904	-1.343	0.179	0.3332	0.7380
Craft -Prof	-1.30196	-2.011	0.044	0.2720	0.6978
WhiteCol -Menial	1.57139	1.741	0.082	4.8133	1.5440
WhiteCol -BlueCol	0.33488	0.359	0.720	1.3978	1.0970
WhiteCol -Craft	1.09904	1.343	0.179	3.0013	1.3550
WhiteCol -Prof	-0.20292	-0.233	0.815	0.8163	0.9455
Prof -Menial	1.77431	2.350	0.019	5.8962	1.6331
Prof -BlueCol	0.53780	0.673	0.501	1.7122	1.1603
Prof -Craft	1.30196	2.011	0.044	3.6765	1.4332
Prof -WhiteCol	0.20292	0.233	0.815	1.2250	1.0577

```
        b = raw coefficient
        z = z-score for test of b=0
    P>|z| = p-value for z-test
      e^b = exp(b) = factor change in odds for unit increase in X
 e^bStdX = exp(b*SD of X) = change in odds for SD increase in X
```

Or, you can limit the output to those coefficients that are significant at a given level using the `pvalue(#)` option, which specifies that only coefficients significant at the # significance level or smaller will be printed. For example,

(Continued on next page)

```
. listcoef, pvalue(.05)
```
mlogit (N=337): Factor Change in the Odds of occ when P>|z| < 0.05
Variable: white (sd=.27642268)

Odds comparing Alternative 1 to Alternative 2	b	z	P>|z|	e^b	e^bStdX
Menial −Prof	−1.77431	−2.350	0.019	0.1696	0.6123
Craft −Prof	−1.30196	−2.011	0.044	0.2720	0.6978
Prof −Menial	1.77431	2.350	0.019	5.8962	1.6331
Prof −Craft	1.30196	2.011	0.044	3.6765	1.4332

Variable: ed (sd=2.9464271)

Odds comparing Alternative 1 to Alternative 2	b	z	P>|z|	e^b	e^bStdX
Menial −WhiteCol	−0.35316	−3.011	0.003	0.7025	0.3533
Menial −Prof	−0.77885	−6.795	0.000	0.4589	0.1008
BlueCol −Craft	−0.19324	−2.494	0.013	0.8243	0.5659
BlueCol −WhiteCol	−0.45258	−4.425	0.000	0.6360	0.2636
BlueCol −Prof	−0.87828	−8.735	0.000	0.4155	0.0752
Craft −BlueCol	0.19324	2.494	0.013	1.2132	1.7671
Craft −WhiteCol	−0.25934	−2.773	0.006	0.7716	0.4657
Craft −Prof	−0.68504	−7.671	0.000	0.5041	0.1329
WhiteCol−Menial	0.35316	3.011	0.003	1.4236	2.8308
WhiteCol−BlueCol	0.45258	4.425	0.000	1.5724	3.7943
WhiteCol−Craft	0.25934	2.773	0.006	1.2961	2.1471
WhiteCol−Prof	−0.42569	−4.616	0.000	0.6533	0.2853
Prof −Menial	0.77885	6.795	0.000	2.1790	9.9228
Prof −BlueCol	0.87828	8.735	0.000	2.4067	13.3002
Prof −Craft	0.68504	7.671	0.000	1.9838	7.5264
Prof −WhiteCol	0.42569	4.616	0.000	1.5307	3.5053

Variable: exper (sd=13.959364)

Odds comparing Alternative 1 to Alternative 2	b	z	P>|z|	e^b	e^bStdX
Menial −Prof	−0.03565	−1.977	0.048	0.9650	0.6079
BlueCol −Prof	−0.03093	−2.147	0.032	0.9695	0.6494
Prof −Menial	0.03565	1.977	0.048	1.0363	1.6449
Prof −BlueCol	0.03093	2.147	0.032	1.0314	1.5400

If you do not need to see the comparisons between all pairs of alternatives, you can limit the output with the gt or lt options of listcoef. By default, listcoef lists comparisons in both directions. For example, it will show you the effect on the odds of alternative 1 versus alternative 2 and the effect on the odds of 2 versus 1. The gt option limits comparisons to those in which the first alternative is greater than the second; lt shows comparisons when the first alternative is less than the second. For example,

```
. listcoef ed, pvalue(.05) gt nolabel
mlogit (N=337): Factor Change in the Odds of occ when P>|z| < 0.05
Variable: ed (sd=2.9464271)
```

Odds comparing Alternative 1 to Alternative 2		b	z	P>|z|	e^b	e^bStdX
3	-2	0.19324	2.494	0.013	1.2132	1.7671
4	-1	0.35316	3.011	0.003	1.4236	2.8308
4	-2	0.45258	4.425	0.000	1.5724	3.7943
4	-3	0.25934	2.773	0.006	1.2961	2.1471
5	-1	0.77885	6.795	0.000	2.1790	9.9228
5	-2	0.87828	8.735	0.000	2.4067	13.3002
5	-3	0.68504	7.671	0.000	1.9838	7.5264
5	-4	0.42569	4.616	0.000	1.5307	3.5053

We used the `nolabel` option to show the category values of the two alternatives rather than their value labels, and the `pvalue(.05)` option limits the coefficients that are printed to those that are significant at the .05 level.

6.2.3 Predicting perfectly

`mlogit` handles perfect prediction somewhat differently than the estimations commands for binary and ordinal models that we have discussed. `logit` and `probit` automatically remove the observations that imply perfect prediction and compute estimates accordingly. `ologit` and `oprobit` keep these observations in the model, fit the z for the problem variable as 0, and provide an incorrect LR chi-squared but also warn that a given number of observations are completely determined. You should delete these observations and refit the model. `mlogit` is just like `ologit` and `oprobit`, except that *you do not receive a warning message.* You will see, however, that all coefficients associated with the variable causing the problem have $z = 0$ (and $p > |z| = 1$). You should refit the model, excluding the problem variable and deleting the observations that imply the perfect predictions. Using the `tabulate` command to generate a cross-tabulation of the problem variable and the dependent variable should reveal the combination that results in perfect prediction.

6.3 Hypothesis testing of coefficients

In the MNLM, you can test individual coefficients with the reported z-statistics, with a Wald test using `test`, or with an LR test using `lrtest`. As the methods of testing one coefficient that were discussed in chapters 4 and 5 still apply fully, they are not considered further here. However, in the MNLM there are new reasons for testing groups of coefficients. First, testing that a variable has no effect requires a test that $J - 1$ coefficients are simultaneously equal to zero. Second, testing whether the independent variables as a group differentiate between two alternatives requires a test of K coefficients. This section focuses on these two kinds of tests.

Caution regarding specification searches Given the difficulties of interpretation
that are associated with the MNLM, it is tempting to search for a more parsimonious model by excluding variables or combining outcome categories based on a
sequence of tests. Such a search requires great care. First, these tests involve
multiple coefficients. Although the overall test might indicate that *as a group* the
coefficients are not significantly different from zero, an *individual* coefficient can
still be substantively and statistically significant. Accordingly, you should examine the individual coefficients involved in each test before deciding to revise your
model. Second, as with all searches that use repeated, sequential tests, there is a
danger of overfitting the data. When models are constructed based on prior testing
using the same data, significance levels should be used only as rough guidelines.

6.3.1 mlogtest for tests of the MNLM

Although the tests in this section can be computed using `test` or `lrtest`, in practice
this is tedious. The `mlogtest` command (Freese and Long 2000) makes the computation
of these tests easy. The syntax is

mlogtest $[$ *varlist* $]$ $[$, <u>all</u> <u>lr</u> <u>w</u>ald <u>c</u>ombine lrcomb

$\quad$ <u>set</u>(*varlist* $[\backslash$ *varlist* $[\backslash...]]$) <u>iia</u> <u>h</u>ausman <u>smh</u>siao <u>det</u>ail <u>base</u> $]$

varlist indicates that the variables for which tests of significance should be computed.
If no *varlist* is given, tests are run for all independent variables.

Options

lr requests a likelihood-ratio (LR) test for each variable in *varlist*. If *varlist* is not
$\quad$ specified, tests for all variables are computed.

wald requests a Wald test for each variable in *varlist*. If *varlist* is not specified, tests
$\quad$ for all variables are computed.

combine requests Wald tests of whether dependent categories can be combined.

lrcomb requests LR tests of whether dependent categories can be combined. These tests
$\quad$ use constrained estimation and overwrite constraint #999 if it is already defined.

set(*varlist* $[\backslash$ *varlist* $[\backslash...]]$) specifies that a set of variables is to be considered together
$\quad$ for the LR test or Wald test. \ is used to specify multiple sets of variables. For
$\quad$ example, mlogtest, lr set(age age2 \ iscat1 iscat2) computes one LR test for
$\quad$ the hypothesis that the effects of age and age2 are jointly 0 and a second LR test
$\quad$ that the effects of iscat1 and iscat2 are jointly 0.

Other options for mlogtest are discussed later in the chapter.

6.3.2   Testing the effects of the independent variables

With J dependent categories, there are $J - 1$ nonredundant coefficients associated with each independent variable x_k. For example, in our logit on occupation, there are four coefficients associated with ed: $\beta_{2,M|P}$, $\beta_{2,B|P}$, $\beta_{2,C|P}$, and $\beta_{2,W|P}$. The hypothesis that x_k does not affect the dependent variable can be written as

$$H_0\text{: } \beta_{k,1|b} = \cdots = \beta_{k,J|b} = 0$$

where b is the base category. Because $\beta_{k,b|b}$ is necessarily 0, the hypothesis imposes constraints on $J - 1$ parameters. This hypothesis can be tested with either a Wald or an LR test.

A likelihood-ratio test

The LR test involves (1) fitting the full model, including all the variables, resulting in the likelihood-ratio statistic LR_F^2; (2) fitting the restricted model that excludes variable x_k, resulting in LR_R^2; and (3) computing the difference $LR_{RvsF}^2 = LR_F^2 - LR_R^2$, which is distributed as chi-squared with $J-1$ degrees of freedom if the null hypothesis is true. This can be done using lrtest:

```
. use http://www.stata-press.com/data/lf2/nomocc2, clear
(1982 General Social Survey)
. mlogit occ white ed exper, baseoutcome(5) nolog
  (output omitted )
. estimates store fmodel
. mlogit occ ed exper, baseoutcome(5) nolog
  (output omitted )
. estimates store nmodel_white
. lrtest fmodel nmodel_white
Likelihood-ratio test                        LR chi2(4)  =      8.10
(Assumption: nmodel_white nested in fmodel)  Prob > chi2 =    0.0881
. mlogit occ white exper, baseoutcome(5) nolog
  (and so on )
```

Although using lrtest is straightforward, the command mlogtest, lr is even simpler because it automatically computes the tests for all variables by making repeated calls to lrtest:

```
. mlogit occ white ed exper, baseoutcome(5) nolog
  (output omitted )
. mlogtest, lr
**** Likelihood-ratio tests for independent variables (N=337)
Ho: All coefficients associated with given variable(s) are 0.
```

occ	chi2	df	P>chi2
white	8.095	4	0.088
ed	156.937	4	0.000
exper	8.561	4	0.073

The results of the LR test, regardless of how they are computed, can be interpreted as follows:

> The effect of race on occupation is significant at the .10 level but not at the .05 level ($X^2 = 8.10$, $df = 4$, $p = .09$). The effect of education is significant at the .01 level ($X^2 = 156.94$, $df = 4$, $p < .01$).

Or, it can be stated more formally:

> The hypothesis that all the coefficients associated with education are simultaneously equal to 0 can be rejected at the .01 level ($X^2 = 156.94$, $df = 4$, $p < .01$).

A Wald test

Although the LR test is generally considered superior, its computational costs can be prohibitive if the model is complex or the sample is very large. K Wald tests can also be computed using test without fitting additional models. For example,

```
. mlogit occ white ed exper, baseoutcome(5) nolog
  (output omitted)
. test white

 ( 1)   [Menial]white = 0
 ( 2)   [BlueCol]white = 0
 ( 3)   [Craft]white = 0
 ( 4)   [WhiteCol]white = 0

           chi2(  4) =      8.15
         Prob > chi2 =    0.0863

. test ed

 ( 1)   [Menial]ed = 0
 ( 2)   [BlueCol]ed = 0
 ( 3)   [Craft]ed = 0
 ( 4)   [WhiteCol]ed = 0

           chi2(  4) =     84.97
         Prob > chi2 =    0.0000

. test exper

 ( 1)   [Menial]exper = 0
 ( 2)   [BlueCol]exper = 0
 ( 3)   [Craft]exper = 0
 ( 4)   [WhiteCol]exper = 0

           chi2(  4) =      7.99
         Prob > chi2 =    0.0918
```

The output from test makes explicit which coefficients are being tested. Here we see the way in which Stata labels parameters in models with multiple equations. For example, [Menial]white is the coefficient for the effect of white in the equation comparing the outcome Menial with the base category Prof; [BlueCol]white is the coefficient for the effect of white in the equation comparing the outcome BlueCol with the base category Prof.

As with the LR test, `mlogtest`, `wald` automates this process:

```
. mlogtest, wald
**** Wald tests for independent variables (N=337)
Ho: All coefficients associated with given variable(s) are 0.
          occ |     chi2   df   P>chi2
       -------+---------------------------
        white |    8.149    4    0.086
           ed |   84.968    4    0.000
        exper |    7.995    4    0.092
       -------+---------------------------
```

These tests can be interpreted in the same way as shown for the LR test above.

Testing multiple independent variables

The logic of the Wald or LR tests can be extended to test that the effects of two or more independent variables are simultaneously zero. For example, the hypothesis to test that x_k and x_ℓ have no effect is

$$H_0: \beta_{k,1|b} = \cdots = \beta_{k,J|b} = \beta_{\ell,1|b} = \cdots = \beta_{\ell,J|b} = 0$$

The `set(`*varlist*`[\` *varlist*`[\...]])` option in `mlogtest` specifies which variables are to be simultaneously tested. For example, to test the hypothesis that the effects of `ed` and `exper` are simultaneously equal to 0, we could use `lrtest` as follows:

```
. mlogit occ white ed exper, baseoutcome(5) nolog
  (output omitted)
. estimates store fmodel
. mlogit occ white, baseoutcome(5) nolog
  (output omitted)
. estimates store nmodel
. lrtest fmodel nmodel
Likelihood-ratio test                           LR chi2(8)  =    160.77
(Assumption: nmodel nested in fmodel)           Prob > chi2 =    0.0000
```

or, using `mlogtest`,

```
. mlogit occ white ed exper, baseoutcome(5) nolog
  (output omitted)
. mlogtest, lr set(ed exper)
**** Likelihood-ratio tests for independent variables (N=337)
  Ho: All coefficients associated with given variable(s) are 0.
```

occ	chi2	df	P>chi2
white	8.095	4	0.088
ed	156.937	4	0.000
exper	8.561	4	0.073
set_1: ed exper	160.773	8	0.000

6.3.3   Tests for combining alternatives

If none of the independent variables significantly affect the odds of alternative m versus alternative n, we say that m and n are *indistinguishable* with respect to the variables in the model (Anderson 1984). Alternatives m and n's being indistinguishable corresponds to the hypothesis that

$$H_0\colon \beta_{1,m|n} = \cdots \beta_{K,m|n} = 0$$

which can be tested with either a Wald or an LR test. In our experience, the two tests provide similar results. If alternatives are indistinguishable with respect to the variables in the model, then you can obtain more efficient estimates by combining them. To test whether alternatives are indistinguishable, you can use `mlogtest`.

A Wald test for combining alternatives

The command `mlogtest, combine` computes Wald tests of the null hypothesis that two alternatives can be combined for all pairs of alternatives. For example,

(Continued on next page)

```
. mlogit occ white ed exper, baseoutcome(5) nolog
  (output omitted)
. mlogtest, combine
**** Wald tests for combining alternatives (N=337)
  Ho: All coefficients except intercepts associated with a given pair
      of alternatives are 0 (i.e., alternatives can be combined).
```

Alternatives tested	chi2	df	P>chi2
Menial- BlueCol	3.994	3	0.262
Menial- Craft	3.203	3	0.361
Menial-WhiteCol	11.951	3	0.008
Menial- Prof	48.190	3	0.000
BlueCol- Craft	8.441	3	0.038
BlueCol-WhiteCol	20.055	3	0.000
BlueCol- Prof	76.393	3	0.000
Craft-WhiteCol	8.892	3	0.031
Craft- Prof	60.583	3	0.000
WhiteCol- Prof	22.203	3	0.000

For example, we can reject the hypothesis that categories `Menial` and `Prof` are indistinguishable, whereas we cannot reject that `Menial` and `BlueCol` are indistinguishable.

Using test [category]*

The `mlogtest` command computes the tests for combining categories with the `test` command. For example, to test that `Menial` is indistinguishable from the base category `Prof`, type

```
. test [Menial]
 ( 1)   [Menial]white = 0
 ( 2)   [Menial]ed = 0
 ( 3)   [Menial]exper = 0
          chi2(  3) =     48.19
        Prob > chi2 =      0.0000
```

which matches the results from `mlogtest` in row `Menial-Prof`. [*outcome*] in `test` is used to indicate which equation is being referenced in multiple equation commands. `mlogit` is a multiple equation command because it is in effect estimating $J - 1$ binary logit equations.

The test is more complicated when neither outcome is the base category. For example, to test that m and n are indistinguishable when the base category b is neither m nor n, the hypothesis you want to test is

$$H_0: \left(\beta_{1,m|b} - \beta_{1,n|b}\right) = \cdots = \left(\beta_{K,m|b} - \beta_{K,n|b}\right) = 0$$

That is, you want to test the difference between two sets of coefficients. This can be done with `test [outcome1=outcome2]`. For example, to test if `Menial` and `Craft` can be combined, type

```
. test [Menial=Craft]

 ( 1)  [Menial]white - [Craft]white = 0
 ( 2)  [Menial]ed - [Craft]ed = 0
 ( 3)  [Menial]exper - [Craft]exper = 0

           chi2(  3) =     3.20
         Prob > chi2 =    0.3614
```

Again the results are identical to those from `mlogtest`.

An LR test for combining alternatives

An LR test of combining m and n can be computed by first fitting the full model with no constraints, with the resulting LR statistic LR_F^2. Then we fit a restricted model M_R in which outcome m is used as the base category and all the coefficients except the constant in the equation for outcome n are constrained to 0, with the resulting test statistic LR_R^2. The test statistic is the difference $LR_{RvsF}^2 = LR_F^2 - LR_R^2$, which is distributed as chi-squared with K degrees of freedom. The command `mlogtest,` `lrcomb` computes $J \times (J-1)$ tests for all pairs of outcome categories. For example,

```
. mlogit occ white ed exper, baseoutcome(5) nolog
  (output omitted )

. mlogtest, lrcomb

**** LR tests for combining alternatives (N=337)

Ho: All coefficients except intercepts associated with a given pair
    of alternatives are 0 (i.e., alternatives can be collapsed).
```

Alternatives tested	chi2	df	P>chi2
Menial- BlueCol	4.095	3	0.251
Menial- Craft	3.376	3	0.337
Menial-WhiteCol	13.223	3	0.004
Menial- Prof	64.607	3	0.000
BlueCol- Craft	9.176	3	0.027
BlueCol-WhiteCol	22.803	3	0.000
BlueCol- Prof	125.699	3	0.000
Craft-WhiteCol	9.992	3	0.019
Craft- Prof	95.889	3	0.000
WhiteCol- Prof	26.736	3	0.000

Using constraint with lrtest[*]

The command `mlogtest, lrcomb` computes the test by using the powerful `constraint` command. To show this, we use the test comparing `Menial` and `BlueCol` reported by `mlogtest, lrcomb` above. First, we fit the full model and save the results of `lrtest`:

```
. mlogit occ white ed exper, nolog
  (output omitted )

. estimates store fmodel
```

Second, we define a constraint using the command

```
. constraint define 999 [Menial]
```

This defines constraint 999, where the number is arbitrary. The expression [Menial] indicates that all the coefficients except the constant from the Menial equation should be constrained to 0. Third, we refit the model with this constraint. The base category must be BlueCol, so that the coefficients indicated by [Menial] are comparisons of BlueCol and Menial:

```
. mlogit occ exper ed white, base(2) constraint(999) nolog
Multinomial logistic regression              Number of obs   =      337
                                             LR chi2(9)      =   161.99
                                             Prob > chi2     =   0.0000
Log likelihood = -428.84791                  Pseudo R2       =   0.1589
 ( 1)  [Menial]exper = 0
 ( 2)  [Menial]ed = 0
 ( 3)  [Menial]white = 0
```

occ	Coef.	Std. Err.	z	P>|z|	[95% Conf. Interval]	
Menial						
exper	(dropped)					
ed	(dropped)					
white	(dropped)					
_cons	-.8001193	.2162194	-3.70	0.000	-1.223901	-.3763371
Craft						
exper	.0242824	.0113959	2.13	0.033	.0019469	.0466179
ed	.1599345	.0693853	2.31	0.021	.0239418	.2959273
white	-.2381783	.4978563	-0.48	0.632	-1.213959	.7376021
_cons	-1.969087	1.054935	-1.87	0.062	-4.036721	.098547
WhiteCol						
exper	.0312007	.0143598	2.17	0.030	.0030561	.0593454
ed	.4195709	.0958978	4.38	0.000	.2316147	.607527
white	.8829927	.843371	1.05	0.295	-.7699841	2.535969
_cons	-7.140306	1.623401	-4.40	0.000	-10.32211	-3.958498
Prof						
exper	.032303	.0133779	2.41	0.016	.0060827	.0585233
ed	.8445092	.093709	9.01	0.000	.6608429	1.028176
white	1.097459	.6877939	1.60	0.111	-.2505923	2.44551
_cons	-12.42143	1.569897	-7.91	0.000	-15.49837	-9.344489

```
(occ==BlueCol is the base outcome)
```

mlogit requires the option constraint(999) to indicate that estimation should impose this constraint. The output clearly indicates which constraints have been imposed. Finally, we use lrtest to compute the test:

```
. estimates store nmodel
. lrtest fmodel nmodel
Likelihood-ratio test                        LR chi2(3)   =     4.09
(Assumption: nmodel nested in fmodel)        Prob > chi2 =     0.2514
```

6.4   Independence of irrelevant alternatives

Both the MNLM and the conditional logit model (discussed below) make the assumption known as the *independence of irrelevant alternatives* (IIA). Here we describe the assumption in terms of the MNLM. In this model,

$$\frac{\Pr\left(y = m \mid \mathbf{x}\right)}{\Pr\left(y = n \mid \mathbf{x}\right)} = \exp\left\{\mathbf{x}\left(\boldsymbol{\beta}_{m|b} - \boldsymbol{\beta}_{n|b}\right)\right\}$$

where the odds do not depend on other alternatives that are available. In this sense, these alternatives are "irrelevant". What this means is that adding or deleting alternatives does not affect the odds among the remaining alternatives. This point is often made with the red bus–blue bus example. Suppose that you have the choice of a red bus or a car to get to work and that the odds of taking a red bus compared with those of taking a car are 1:1. IIA implies that the odds will remain 1:1 between these two alternatives, even if a new *blue* bus company comes to town that is identical to the red bus company, except for the color of the bus. Thus the probability of driving a car can be made arbitrarily small by adding enough different colors of buses! More reasonably, we might expect that the odds of a red bus compared with those of a car would be reduced to 1:2 since half of those riding the red bus would be expected to ride the blue bus.

Tests of IIA involve comparing the estimated coefficients from the full model to those from a restricted model that excludes at least one of the alternatives. If the test statistic is significant, the assumption of IIA is rejected indicating that the MNLM is inappropriate. In this section, we consider the two most common tests of IIA: the Hausman–McFadden (HM) test (1984) and the Small–Hsiao (SH) test (1985). For details on other tests, see Fry and Harris (1996, 1998). In a model with J alternatives, there are $J - 1$ ways of computing each test. If you remove the first alternative and refit the model, you get the first restricted model. If you remove the second alternative, the second, and so on, for a total of $J - 1$ restricted models, each of these restricted models will lead to a different test statistic, as we demonstrate below.

Both the HM and the SH tests are computed by `mlogtest`, and for both tests we compute $J - 1$ variations. As many users of `mlogtest` have told us, the HM and SH tests often provide conflicting information on whether IIA has been violated (i.e., some of the tests reject the null hypothesis, whereas others do not). To explore this further, Cheng and Long (2005) ran Monte Carlo experiments to examine the properties of these tests. Their results show that the HM test has poor size properties even with sample sizes of more than 1,000. For some data structures, the SH test has reasonable size properties for samples of 500 or more. But, with other data structures the size properties are extremely poor and do not get better as the sample size increases. Overall, they conclude that these tests are not useful for assessing violations of the IIA property. It appears that the best advice regarding IIA goes back to an early statement by McFadden (1973), who wrote that the multinomial and conditional logit models should be used only in cases where the alternatives "can plausibly be assumed to be distinct and weighted independently in the eyes of each decision maker". Similarly, Amemiya

(1981, 1,517) suggests that the MNLM works well when the alternatives are dissimilar. Care in specifying the model to involve distinct alternatives that are not substitutes for one another seems to be reasonable, albeit unfortunately ambiguous, advice. Nonetheless, we continue to include these tests in `mlogtest`, but we do not encourage their use. As we will show here, these tests can produce contradictory results.

Hausman test of IIA

The Hausman test of IIA involves the following steps:

1. Fit the full model with all J alternatives included, with estimates in $\widehat{\boldsymbol{\beta}}_F$.

2. Fit a restricted model by eliminating one or more alternatives, with estimates in $\widehat{\boldsymbol{\beta}}_R$.

3. Let $\widehat{\boldsymbol{\beta}}_F^*$ be a subset of $\widehat{\boldsymbol{\beta}}_F$ after eliminating coefficients not fitted in the restricted model. The test statistic is

$$H = \left(\widehat{\boldsymbol{\beta}}_R - \widehat{\boldsymbol{\beta}}_F^*\right)' \left\{\widehat{\mathrm{Var}}\left(\widehat{\boldsymbol{\beta}}_R\right) - \widehat{\mathrm{Var}}\left(\widehat{\boldsymbol{\beta}}_F^*\right)\right\}^{-1} \left(\widehat{\boldsymbol{\beta}}_R - \widehat{\boldsymbol{\beta}}_F^*\right)$$

where H is asymptotically distributed as chi-squared with degrees of freedom equal to the rows in $\widehat{\boldsymbol{\beta}}_R$ if IIA is true. Significant values of H indicate that the IIA assumption has been violated.

The Hausman test of IIA can be computed with `mlogtest`. Here the results are

```
. mlogit occ white ed exper, baseoutcome(5) nolog
  (output omitted)
. mlogtest, hausman base
**** Hausman tests of IIA assumption (N=337)
Ho: Odds(Outcome-J vs Outcome-K) are independent of other alternatives.
Omitted  |    chi2    df    P>chi2   evidence
---------+----------------------------------------
Menial   |   7.324    12    0.835    for Ho
BlueCol  |   0.320    12    1.000    for Ho
Craft    | -14.436    12    1.000    for Ho
WhiteCol |  -5.541    11    1.000    for Ho
Prof     |  -0.119    12    1.000    for Ho
```

Five tests of IIA are reported. The first four correspond to excluding one of the four nonbase categories. The fifth test, in row `Prof`, is computed by refitting the model using the largest remaining outcome as the base category.[1] Although none of the tests reject the H_0 that IIA holds, the results differ considerably, depending on the outcome considered. Further, three of the test statistics are negative, which we find to be very

---

1. Even though `mlogtest` fits other models to compute various tests, when the command ends it restores the estimates from your original model. Accordingly, other commands that require results from your original `mlogit`, such as `predict` and `prvalue`, will still work correctly.

common. Hausman and McFadden (1984, 1226) note this possibility and conclude that a negative result is evidence that IIA has *not* been violated. A further sense of the variability of the results can be seen by rerunning `mlogit` with a different base category and then running `mlogtest, hausman base`.

Small–Hsiao test of IIA

To compute Small and Hsiao's test, the sample is divided randomly into two subsamples of about equal size. The unrestricted MNLM is fitted on both subsamples, where $\widehat{\boldsymbol{\beta}}_u^{S_1}$ contains estimates from the unrestricted model on the first subsample and $\widehat{\boldsymbol{\beta}}_u^{S_2}$ is its counterpart for the second subsample. A weighted average of the coefficients is computed as

$$\widehat{\boldsymbol{\beta}}_u^{S_1 S_2} = \left(\frac{1}{\sqrt{2}} \right) \widehat{\boldsymbol{\beta}}_u^{S_1} + \left\{ 1 - \left(\frac{1}{\sqrt{2}} \right) \right\} \widehat{\boldsymbol{\beta}}_u^{S_2}$$

Next a restricted sample is created from the second subsample by eliminating all cases with a chosen value of the dependent variable. The MNLM is fitted using the restricted sample, yielding the estimates $\widehat{\beta}_r^{S_2}$ and the likelihood $L(\widehat{\beta}_r^{S_2})$. The Small–Hsiao statistic is

$$SH = -2 \left\{ L(\widehat{\boldsymbol{\beta}}_u^{S_1 S_2}) - L(\widehat{\boldsymbol{\beta}}_r^{S_2}) \right\}$$

which is asymptotically distributed as a chi-squared with the degrees of freedom equal to the number of coefficients that are fitted both in the full model and the restricted model.

To compute the Small–Hsiao test, you use the command `mlogtest, smhsiao` (our program uses code from `smhsiao` by Nick Winter, available at the SSC-IDEAS archive). For example,

```
. mlogtest, smhsiao
**** Small-Hsiao tests of IIA assumption (N=337)
Ho: Odds(Outcome-J vs Outcome-K) are independent of other alternatives.
 Omitted | lnL(full)  lnL(omit)   chi2   df   P>chi2   evidence
---------+---------------------------------------------------------
  Menial | -182.140   -169.907  24.466   12   0.018    against Ho
 BlueCol | -148.711   -140.054  17.315   12   0.138    for Ho
   Craft | -131.801   -119.286  25.030   12   0.015    against Ho
WhiteCol | -161.436   -148.550  25.772   12   0.012    against Ho
```

In three variations of the SH test, we reject the null, whereas the HM test accepted the null in all cases.

Because the Small–Hsiao test requires randomly dividing the data into subsamples, the results will differ with successive calls of the command, as the sample will be divided differently. To obtain test results that can be replicated, you must explicitly set the seed used by the random-number generator. For example,

```
. set seed 8675309

. mlogtest, smhsiao

**** Small-Hsiao tests of IIA assumption (N=337)
 Ho: Odds(Outcome-J vs Outcome-K) are independent of other alternatives.
  Omitted │ lnL(full)  lnL(omit)   chi2   df   P>chi2   evidence
 ─────────┼──────────────────────────────────────────────────────
   Menial │ -169.785   -161.523  16.523   12   0.168    for Ho
  BlueCol │ -131.900   -125.871  12.058   12   0.441    for Ho
    Craft │ -136.934   -129.905  14.058   12   0.297    for Ho
 WhiteCol │ -155.364   -150.239  10.250   12   0.594    for Ho
```

Using a new seed, we accept the null in each case, illustrating a common problem when using the SH test—you can get quite different results depending on how the sample is randomly divided.

Advanced: setting the random seed The random numbers that divide the sample for the Small–Hsiao test are based on Stata's `uniform()` function, which uses a pseudorandom number generator. This generator creates a sequence of numbers based on a seed number. Although these numbers appear to be random, the same sequence will be generated each time you start with the same seed number. In this sense (and some others), these numbers are pseudorandom rather than random. If you specify the seed with `set seed #`, you ensure that you can replicate your results later. See the *Data Management Reference Manual* for more details.

6.5   Measures of fit

As with the binary and ordinal models, scalar measures of fit for the MNLM model can be computed with the SPost command `fitstat`. The same caveats against overstating the importance of these scalar measures apply here as to the other models we consider (see also chapter 3). To examine the fit of individual observations, you can estimate the series of binary logits implied by the multinomial logit model and use the established methods of examining the fit of observations to binary logit estimates. This is the same approach that was recommended in chapter 5 for ordinal models.

6.6   Interpretation

Although the MNLM is a mathematically simple extension of the binary model, interpretation is made difficult by the many possible comparisons. Even in our simple example with five outcomes, we have many possible comparisons: $M|P$, $B|P$, $C|P$, $W|P$, $M|W$, $B|W$, $C|W$, $M|C$, $B|C$, and $M|B$. It is tedious to write all the comparisons, let alone to interpret each of them for each of the independent variables. Thus the key to interpretation is to avoid being overwhelmed by the many comparisons. Most of the methods we propose are similar to those for ordinal outcomes, and accordingly, these are treated briefly. However, methods of plotting discrete changes and factor changes are new, so these are considered in greater detail.

6.6.1   Predicted probabilities

Predicted probabilities can be computed with the formula

$$\widehat{\Pr}\left(y=m\mid \mathbf{x}\right)=\frac{\exp\left(\mathbf{x}\widehat{\boldsymbol{\beta}}_{m|J}\right)}{\sum_{j=1}^{J}\exp\left(\mathbf{x}\widehat{\boldsymbol{\beta}}_{j|J}\right)}$$

where $\mathbf{x}$ can contain values from individuals in the sample or hypothetical values. The most basic command for computing probabilities is `predict`, but we also illustrate a series of SPost commands that compute predicted probabilities in useful ways.

6.6.2   Predicted probabilities with predict

After fitting the model with `mlogit`, the predicted probabilities within the sample can be calculated with the command

predict *newvar1* $\left[newvar2\ldots[newvarJ]\right]$ $\left[if\right]$ $\left[in\right]$

where you must provide one new variable name for each of the J categories of the dependent variable, ordered from the lowest to highest numerical values. For example,

```
. mlogit occ white ed exper, baseoutcome(5) nolog
  (output omitted)
. predict ProbM ProbB ProbC ProbW ProbP
(option p assumed; predicted probabilities)
```

The variables created by `predict` are

```
. desc Prob*

              storage  display     value
variable name  type    format      label       variable label

ProbM          float   %9.0g                   Pr(occ==1)
ProbB          float   %9.0g                   Pr(occ==2)
ProbC          float   %9.0g                   Pr(occ==3)
ProbW          float   %9.0g                   Pr(occ==4)
ProbP          float   %9.0g                   Pr(occ==5)

. summarize Prob*

    Variable |     Obs        Mean    Std. Dev.        Min        Max

       ProbM |     337    .0919881     .059396    .0010737   .3281906
       ProbB |     337    .2047478    .1450568    .0012066   .6974148
       ProbC |     337    .2492582    .1161309    .0079713    .551609
       ProbW |     337    .1216617    .0452844    .0083857   .2300058
       ProbP |     337    .3323442    .2870992    .0001935   .9597512
```

Using predict to compare mlogit and ologit

An interesting way to illustrate how predictions can be plotted is to compare predictions
from ordered logit and multinomial logit when the models are applied to the same
data. Recall from chapter 5 that the range of the predicted probabilities for middle
categories abruptly ended, whereas predictions for the end categories had a more gradual
distribution. To illustrate this point, the example in chapter 5 is estimated using `ologit`
and `mlogit`, with predicted probabilities computed for each case:

```
. use http://www.stata-press.com/data/lf2/ordwarm2,clear
(77 & 89 General Social Survey)
. ologit warm yr89 male white age ed prst, nolog
  (output omitted)
. predict SDologit Dologit Aologit SAologit
(option p assumed; predicted probabilities)
. label var Dologit "ologit-D"
. mlogit warm yr89 male white age ed prst, nolog
  (output omitted)
. predict SDmlogit Dmlogit Amlogit SAmlogit
(option p assumed; predicted probabilities)
. label var Dmlogit "mlogit-D"
```

We can plot the predicted probabilities of disagreeing in the two models with the command `dotplot Dologit Dmlogit, ylabel(0(.25).75)`, which leads to

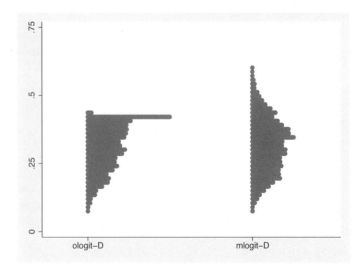

Although the two sets of predictions have a correlation of .92 (computed by the command
`correlate Dologit Dmlogit`), the abrupt truncation of the distribution for the ordered
logit model strikes us as substantively unrealistic.

6.6.3 Predicted probabilities and discrete change with prvalue

Predicted probabilities for individuals with specified characteristics can be computed with prvalue. For example, we might compute the probabilities of each occupational outcome to compare nonwhites and whites who are average on education and experience:

```
. use http://www.stata-press.com/data/lf2/nomocc2, clear
(1982 General Social Survey)
. mlogit occ white ed exper, baseoutcome(5) nolog
  (output omitted )
. quietly prvalue, x(white=0) rest(mean) save
. prvalue, x(white=1) rest(mean) diff

mlogit: Change in Predictions for occ

Confidence intervals by delta method
                    Current      Saved     Change    95% CI for Change
  Pr(y=Menial|x):    0.0860     0.2168    -0.1309    [-0.3056,   0.0439]
  Pr(y=BlueCol|x):   0.1862     0.1363     0.0498    [-0.0897,   0.1893]
  Pr(y=Craft|x):     0.2790     0.4387    -0.1597    [-0.3686,   0.0491]
  Pr(y=WhiteCol|x):  0.1674     0.0877     0.0797    [-0.0477,   0.2071]
  Pr(y=Prof|x):      0.2814     0.1204     0.1611    [ 0.0277,   0.2944]

               white         ed        exper
  Current=         1   13.094955   20.501484
    Saved=         0   13.094955   20.501484
     Diff=         1           0            0
```

This example also shows how to use prvalue to compute differences between two sets of probabilities. Our first call of prvalue is done quietly, but we save the results. The second call uses the diff option, and the output compares the results for the first and second set of values computed. By using prvalue with the save and diff options, we obtain confidence intervals for the discrete changes. The predicted difference between blacks and whites in the probability of having professional jobs is the only case in which the 95% confidence interval does not include zero.

6.6.4 Tables of predicted probabilities with prtab

If you want predicted probabilities for all combinations of a set of categorical independent variables, prtab is useful. For example, we might want to know how white and nonwhite respondents differ in their probability of having a menial job by years of education:

(Continued on next page)

```
. label def lwhite 0 NonWhite 1 White
. label val white lwhite
. prtab ed white, novarlbl outcome(1)
mlogit: Predicted probabilities of outcome 1 (Menial) for occ
```

| | white | |
ed	NonWhite	White
3	0.2847	0.1216
6	0.2987	0.1384
7	0.2988	0.1417
8	0.2963	0.1431
9	0.2906	0.1417
10	0.2814	0.1366
11	0.2675	0.1265
12	0.2476	0.1104
13	0.2199	0.0883
14	0.1832	0.0632
15	0.1393	0.0401
16	0.0944	0.0228
17	0.0569	0.0120
18	0.0310	0.0060
19	0.0158	0.0029
20	0.0077	0.0014

```
        white         ed       exper
x=   .91691395   13.094955   20.501484
```

Tip: outcome() option Here we use the outcome() option to restrict the output to
one outcome category. Without this option, prtab will produce a separate table
for each outcome category.

The table produced by prtab shows the substantial differences between whites and
nonwhites in the probabilities of having menial jobs and how these probabilities are
affected by years of education. However, given the number of categories for ed, plotting
these predicted probabilities with prgen is probably a more useful way to examine the
results.

6.6.5   Graphing predicted probabilities with prgen

Predicted probabilities can be plotted using the same methods considered for the ordinal
regression model. After fitting the model, we use prgen to compute the predicted
probabilities for whites with average working experience as education increases from 6
years to 20 years:

```
. prgen ed, x(white=1) from(6) to(20) generate(wht) ncases(15)
mlogit: Predicted values as ed varies from 6 to 20.
          white         ed      exper
x=            1  13.094955  20.501484
```

Here is what the options specify:

x(white=1) sets white to 1. Because the rest() option is not included, all other
   variables are set to their means by default.

from(6) and to(20) set the minimum and maximum values over which ed is to vary.
   The default is to use the variable's minimum and maximum values.

ncases(15) indicates that 15 evenly spaced values of ed between 6 and 20 are to be
   generated. We chose 15 for the number of values from 6 to 20, inclusive.

gen(wht) specifies the root name for the new variables generated by prgen. For example, the variable whtx contains values of ed, the p-variables (e.g., whtp2) contain the
   predicted probabilities for each outcome, and the s-variables contain the summed
   probabilities:

```
. desc wht*

                 storage  display    value
variable name    type     format     label      variable label
```
variable name	storage type	display format	value label	variable label
whtx	float	%9.0g		Years of education
whtp1	float	%9.0g		pr(Menial)=Pr(1)
whtp2	float	%9.0g		pr(BlueCol)=Pr(2)
whtp3	float	%9.0g		pr(Craft)=Pr(3)
whtp4	float	%9.0g		pr(WhiteCol)=Pr(4)
whtp5	float	%9.0g		pr(Prof)=Pr(5)
whts1	float	%9.0g		pr(y<=1)
whts2	float	%9.0g		pr(y<=2)
whts3	float	%9.0g		pr(y<=3)
whts4	float	%9.0g		pr(y<=4)
whts5	float	%9.0g		pr(y<=5)

The same thing can be done to compute predicted probabilities for nonwhites:

```
. prgen ed, x(white=0) from(6) to(20) generate(nwht) ncases(15)
mlogit: Predicted values as ed varies from 6 to 20.
          white         ed      exper
x=            0  13.094955  20.501484
```

Plotting probabilities for one outcome and two groups

The variables nwhtp1 and whtp1 contain the predicted probabilities of having menial
jobs for nonwhites and whites. Plotting these provides clearer information than the
results of prtab given above:

```
. label var whtp1 "Whites"

. label var nwhtp1 "Nonwhites"

. graph twoway connected whtp1 nwhtp1 nwhtx,
        xtitle("Years of Education")
        ytitle("Pr(Menial Job)")
        ylabel(0(.25).50) xlabel(6 8 12 16 20)
```

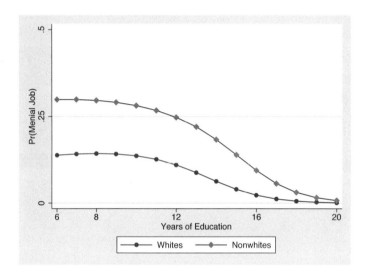

Graphing probabilities for all outcomes for one group

Even though nominal outcomes are not ordered, plotting the summed probabilities can
be a useful way to show predicted probabilities for all outcome categories. To show this,
we construct a graph to show how education affects the probability of each occupation for
whites (a similar graph could be plotted for nonwhites). This is done using the *roots#*
variables created by prgen, which provide the probability of being in an outcome less
than or equal to some value. For example, the label for whts3 is pr(y<=3), which
indicates that all nominal categories coded as 3 or less are added together. To plot
these probabilities, the first thing we do is change the variable labels to the name of the
highest category in the sum, which makes the graph clearer (as you will see below):

```
. label var whts1 "Menial"

. label var whts2 "Blue Collar"

. label var whts3 "Craft"

. label var whts4 "White Collar"
```

To create the summed plot, we use the following command:

```
. graph twoway connected whts1 whts2 whts3 whts4 whtx, ///
> xtitle("Whites: Years of Education")                 ///
> ytitle("Summed Probability")                         ///
> xlabel(6(2)20)                                        ///
> ylabel(0(.25)1)
```

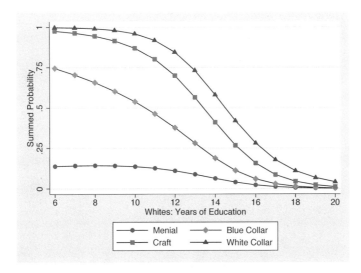

The graph plots the four summed probabilities against whtx, where standard options for graph are used. This graph is not ideal, but before revising it, let's make sure we understand what is being plotted. The lowest line with circles, labeled "Menial" in the key, plots the probability of having a menial job for a given year of education. This is the same information as plotted in our prior graph for whites. The next line with small diamonds, labeled "Blue Collar" in the key, plots the sum of the probability of having a menial job or a blue-collar job. Thus the area between the line with circles and the line with diamonds is the probability of having a blue-collar job, and so on.

Because what we really want to illustrate are the regions between the curves, this graph is not as effective as we would like. In the graph command below, we use the rarea plot type to shade the regions between the curves. The syntax for an rarea plot[2] is

graph twoway rarea *y1var* *y2var* *xvar* [*if*] [*in*] [, *rarea_options*]

where *y1var* defines the lower boundary and *y2var* defines the upper boundary of the region for each *x*-value given in the variable *xvar*.

Continuing with our example, as the probabilities are bounded between zero and one, we begin by creating variables that hold these extreme values.

---

2. Type help twoway rarea for more information.

```
. gen zero = 0
. gen one  = 1
```

Now we are ready to draw the full graph.

```
. graph twoway (rarea zero whts1 whtx, bc(gs1))              ///
>              (rarea whts1 whts2 whtx, bc(gs4))             ///
>              (rarea whts2 whts3 whtx, bc(gs8))             ///
>              (rarea whts3 whts4 whtx, bc(gs11))            ///
>              (rarea whts4 one whtx, bc(gs14)),             ///
>              ytitle("Summed Probability")                 ///
>              legend( order( 1 2 3 4 5)                     ///
>              label( 1 "Menial")                            ///
>              label( 2 "Blue Collar") label( 3 "Craft")     ///
>              label(4 "White Collar") label(5 "Professional")) ///
>              xtitle("Whites: Years of Education")          ///
>              xlabel(6 8 12 16 20) ylabel(0(.25)1)          ///
>              plotregion(margin(zero)))
```

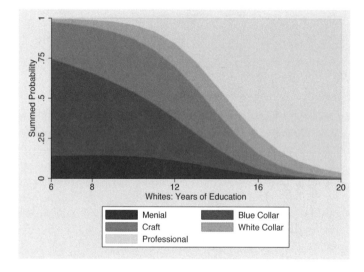

Figure 6.1: Whites: years of education.

The changes in the shaded regions in figure 6.1 clearly illustrate how the probability of selecting any one occupation changes as education increases.

6.6.6 Changes in predicted probabilities

Marginal and discrete change can be used in the same way as in models for ordinal outcomes. As before, both can be computed using `prchange`.

Marginal change is defined as

$$
\frac{\partial \Pr\left(y = m \mid \mathbf{x}\right)}{\partial x_k} = \Pr\left(y = m \mid \mathbf{x}\right) \left\{ \beta_{k,m|J} - \sum_{j=1}^{J} \beta_{k,j|J} \Pr(y = j \mid \mathbf{x}) \right\}
$$

As this equation combines all the $\beta_{k,j|J}$s, the value of the marginal change depends on the levels of all variables in the model. Further, as the value of x_k changes, the sign of the marginal can change. For example, at one point the marginal effect of education on having a craft occupation could be positive, whereas at another point the marginal effect could be negative.

Discrete change is defined as

$$
\frac{\Delta \Pr\left(y = m \mid \mathbf{x}\right)}{\Delta x_k} = \Pr\left(y = m \mid \mathbf{x}, x_k = x_E\right) - \Pr\left(y = m \mid \mathbf{x}, x_k = x_S\right)
$$

where the magnitude of the change depends on the levels of all variables and the size of the change that is being made. The J discrete-change coefficients for a variable (one for each outcome category) can be summarized by computing the average of the *absolute values* of the changes across all the outcome categories,

$$
\overline{\Delta} = \frac{1}{J} \sum_{j=1}^{J} \left| \frac{\Delta \Pr\left(y = j \mid \overline{\mathbf{x}}\right)}{\Delta x_k} \right|
$$

where the absolute value is taken because the sum of the changes without taking the absolute value is necessarily zero.

Computing marginal and discrete change with prchange

Discrete and marginal changes are computed with `prchange` (the full syntax for which is provided in chapter 3). For example,

(*Continued on next page*)

```
. mlogit occ white ed exper
  (output omitted)

. prchange

mlogit: Changes in Probabilities for occ

white
                Avg|Chg|      Menial       BlueCol         Craft      WhiteCol
      0->1     .11623582   -.13085523     .04981799    -.15973434     .07971004

                              Prof
      0->1                 .1610615

ed
                Avg|Chg|      Menial       BlueCol         Craft      WhiteCol
Min->Max      .39242268   -.13017954    -.70077323    -.15010394     .02425591
   -+1/2      .05855425   -.02559762    -.06831616    -.05247185     .01250795
   -+sd/2      .1640657   -.07129153    -.19310513    -.14576758     .03064777
MargEfct      .05894859   -.02579097    -.06870635    -.05287415     .01282041

                              Prof
Min->Max      .95680079
   -+1/2      .13387768
   -+sd/2      .37951647
MargEfct      .13455107

exper
                Avg|Chg|      Menial       BlueCol         Craft      WhiteCol
Min->Max      .12193559   -.11536534    -.18947365     .03115708     .09478889
   -+1/2      .00233425   -.00226997    -.00356567     .00105992      .0016944
   -+sd/2      .03253578   -.03167491    -.04966453     .01479983     .02360725
MargEfct      .00233427   -.00226997    -.00356571     .00105992     .00169442

                              Prof
Min->Max      .17889298
   -+1/2      .00308132
   -+sd/2      .04293236
MargEfct      .00308134

                 Menial       BlueCol         Craft      WhiteCol          Prof
Pr(y|x)       .09426806    .18419114     .29411051     .16112968     .26630062

                 white          ed        exper
        x=      .916914      13.095      20.5015
     sd(x)=     .276423     2.94643      13.9594
```

The first thing to notice is the output labeled `Pr(y|x)`, which is the predicted probabilities at the values set by `x()` and `rest()`. Marginal change is listed in the rows `MargEfct`. For variables that are not binary, discrete change is reported over the range of the variable (reported as `Min->Max`), for changes of one unit centered on the base values (reported as `-+1/2`), and for changes of one standard deviation centered on the base values (reported as `-+sd/2`). If the `uncentered` option is used, the changes begin at the value specified by `x()` or `rest()` and increase one unit or one standard deviation from there. For binary variables, the discrete change from 0 to 1 is the only appropriate quantity and is the only quantity that is presented. Looking at the results for `white` above, we can see that for someone who is average in education and experience, the predicted probability of having a professional job is .16 higher for whites than nonwhites. The average change is listed in the column `Avg|Chg|`. For example, for `white`, $\overline{\Delta} = 0.12$, the average absolute change in the probability of various occupational categories for being white as opposed to nonwhite is .12.

Marginal change with mfx

The marginal change can also be computed using `mfx`, where the `at()` option is used to set values of the independent variables. Like `prchange`, the `mfx` command sets all values of the independent variables to their means by default. Also we must estimate the marginal effects for one outcome at a time, using the `predict(outcome(#))` option to specify the outcome for which we want marginal effects:

```
. mfx, predict(outcome(1))
Marginal effects after mlogit
      y  = Pr(occ==1) (predict, outcome(1))
         =  .09426806
```

variable	dy/dx	Std. Err.	z	P>|z|	[95% C.I.]		X
white*	-.1308552	.08914	-1.47	0.142	-.305562	.043852	.916914
ed	-.025791	.00688	-3.75	0.000	-.039269	-.012312	13.095
exper	-.00227	.00126	-1.80	0.071	-.004737	.000197	20.5015

(*) dy/dx is for discrete change of dummy variable from 0 to 1

These results are for the Menial category (occ==1). Estimates for `exper` and `ed` match the results in the `MargEfct` rows of the `prchange` output above. Meanwhile, for the binary variable `white`, the discrete change from 0 to 1 is presented, which also matches the corresponding result from `prchange`. An advantage of `mfx` is that standard errors for the effects are also provided; a disadvantage is that `mfx` can take a long time to produce results after `mlogit`, especially if the number of observations and independent variables is large.

6.6.7 Plotting discrete changes with prchange and mlogview

One difficulty with nominal outcomes is the many coefficients that need to be considered: one for each variable times the number of outcome categories minus one. To help you sort out all this information, discrete-change coefficients can be plotted using our program `mlogview`. After fitting the model with `mlogit` and computing discrete changes with `prchange`, executing `mlogview` opens the following dialog box:

(Continued on next page)

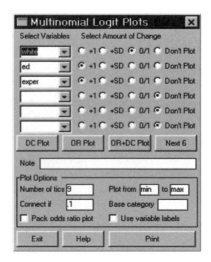

Dialog boxes are easier to use than to explain. So, as we describe various features, the best advice is to generate the dialog box shown above and experiment.

Selecting variables If you click and hold a 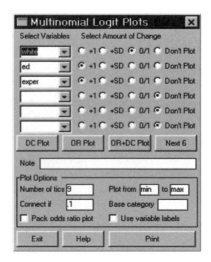 button, you can select a variable to be plotted. The same variable can be plotted more than once, for example, showing the effects of different amounts of change.

Selecting the amount of change The radio buttons allow you to select the type of discrete-change coefficient to plot for each selected variable: +1 selects coefficients for a change of one unit; +SD selects coefficients for a change of one standard deviation; 0/1 selects changes from 0 to 1; and Don't Plot is self-explanatory.

Making a plot Even though there are more options to explain, you should try plotting your selections by clicking on DC Plot, which produces a graph. The command `mlogview` works by generating the syntax for the command `mlogplot`, which actually draws the plot. In the Results window, you will see the `mlogplot` command that was used to generate your graph (full details on `mlogplot` are given in section 6.6.9). If there is an error in the options you select, the error message will appear in the Results window.

On the assumption that everything has worked, we generate the following graph:

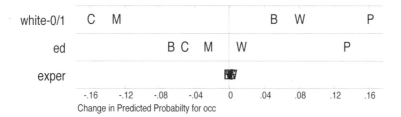

The graph immediately shows how a unit increase in each variable affects the probability of each outcome. Although it appears that the effects of being white are the largest, changes of one unit in education and (especially) experience are often too small to be as informative. It would make more sense to look at the effects of a standard deviation change in these variables. To do this, we return to the dialog box and click on the radio button +SD. Before we see what this does, let's consider several other options that can be used.

Adding labels The box Note allows you to enter text that will be placed at the top of the graph. Clicking the box for Use variable labels replaces the names of the variables on the left axis with the variable labels associated with each variable. When you do this, you may find that the labels are too long. If so, you can use the `label variable` command to change them.

Tick marks The values for the tick marks are determined by specifying the minimum and maximum values to plot and the number of tick marks. For example, we could specify a plot from $-.2$ to .4 with seven tick marks. This will lead to labels every .1 units.

Using some of the features discussed above, our dialog box would look like this:

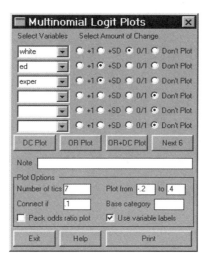

Clicking on DC Plot produces the following graph:

White Worker-0/1	C M		B W P	
Yrs of Education-std	B C M		W	P
Yrs of Experience-std		BM CWP		

```
        -.2          -.1          0          .1        .2        .3        .4
             Change in Predicted Probabilty for occ
```

You can see that the effects of education are largest and that those of experience are smallest. Or, each coefficient can be interpreted individually, such as the following:

> The effects of a standard deviation change in education are largest, with an increase of more than .35 in the probability of having a professional occupation.

> The effects of race are also substantial, with average blacks being less likely to enter blue-collar, white-collar, or professional jobs than average whites.

> Expected changes due to a standard deviation change in experience are much smaller and show that experience increases the probabilities of more highly skilled occupations.

In using these graphs, remember that different values for discrete change are obtained at different levels of the variables, which are specified with the x() and rest() options for prchange.

Value labels with mlogview The value labels for the different categories of the dependent variables must begin with different letters because the plots generated with mlogview use the first letter of the value label.

6.6.8   Odds ratios using listcoef and mlogview

Discrete change does little to illuminate the dynamics among the outcomes. For example, a decrease in education increases the probability of both blue-collar and craft jobs, but how does it affect the odds of a person choosing a craft job relative to a blue-collar job? To deal with these issues, odds ratios (also referred to as factor change coefficients) can be used. Holding other variables constant, the factor change in the odds of outcome m versus outcome n as x_k increases by δ equals

$$\frac{\Omega_{m|n}\left(\mathbf{x}, x_k + \delta\right)}{\Omega_{m|n}\left(\mathbf{x}, x_k\right)} = e^{\beta_{k,m|n}\delta}$$

If the amount of change is $\delta = 1$, the odds ratio can be interpreted as follows:

> For a unit change in x_k, the odds of m versus n are expected to change by a factor of $\exp(\beta_{k,m|n})$, holding all other variables constant.

If the amount of change is $\delta = s_{x_k}$, then the odds ratio can be interpreted as follows:

> For a standard deviation change in x_k, the odds of m versus n are expected to change by a factor of $\exp(\beta_{k,m|n} \times s_k)$, holding all other variables constant.

Listing odds ratios with listcoef

The difficulty in interpreting odds ratios for the MNLM is that, to understand the effect of a variable, you need to examine the coefficients for comparisons among all pairs of outcomes. The standard output from `mlogit` includes only $J - 1$ comparisons with the base category. Although you could estimate coefficients for all possible comparisons by rerunning `mlogit` with different base categories (e.g., `mlogit occ white ed exper, baseoutcome(3)`), using `listcoef` is much simpler. For example, to examine the effects of race, type

```
. listcoef white, help

mlogit (N=337): Factor Change in the Odds of occ

Variable: white (sd=.27642268)
```

Odds comparing Alternative 1 to Alternative 2	b	z	P>|z|	e^b	e^bStdX
Menial -BlueCol	-1.23650	-1.707	0.088	0.2904	0.7105
Menial -Craft	-0.47234	-0.782	0.434	0.6235	0.8776
Menial -WhiteCol	-1.57139	-1.741	0.082	0.2078	0.6477
Menial -Prof	-1.77431	-2.350	0.019	0.1696	0.6123
BlueCol -Menial	1.23650	1.707	0.088	3.4436	1.4075
BlueCol -Craft	0.76416	1.208	0.227	2.1472	1.2352
BlueCol -WhiteCol	-0.33488	-0.359	0.720	0.7154	0.9116
BlueCol -Prof	-0.53780	-0.673	0.501	0.5840	0.8619
Craft -Menial	0.47234	0.782	0.434	1.6037	1.1395
Craft -BlueCol	-0.76416	-1.208	0.227	0.4657	0.8096
Craft -WhiteCol	-1.09904	-1.343	0.179	0.3332	0.7380
Craft -Prof	-1.30196	-2.011	0.044	0.2720	0.6978
WhiteCol-Menial	1.57139	1.741	0.082	4.8133	1.5440
WhiteCol-BlueCol	0.33488	0.359	0.720	1.3978	1.0970
WhiteCol-Craft	1.09904	1.343	0.179	3.0013	1.3550
WhiteCol-Prof	-0.20292	-0.233	0.815	0.8163	0.9455
Prof -Menial	1.77431	2.350	0.019	5.8962	1.6331
Prof -BlueCol	0.53780	0.673	0.501	1.7122	1.1603
Prof -Craft	1.30196	2.011	0.044	3.6765	1.4332
Prof -WhiteCol	0.20292	0.233	0.815	1.2250	1.0577

```
        b = raw coefficient
        z = z-score for test of b=0
   P>|z| = p-value for z-test
     e^b = exp(b) = factor change in odds for unit increase in X
e^bStdX = exp(b*SD of X) = change in odds for SD increase in X
```

The odds ratios of interest are in the column labeled e^b. For example, the odds ratio for the effect of race on having a professional versus a menial job is 5.90, which can be interpreted as follows:

> The odds of having a professional occupation relative to a menial occupation are 5.90 times greater for whites than for blacks, holding education and experience constant.

Remember: the gt, lt, and pvalue options control which comparisons are printed by listcoef. See pages 233–234 for more details.

Plotting odds ratios

However, examining all the coefficients for even a single variable with only five dependent categories is complicated. An *odds-ratio plot* makes it easy to quickly see patterns in results for even a complex MNLM (see Long 1997, chapter 6 for full details). To explain how to interpret an odds ratio plot, we begin with some hypothetical output from a MNLM with three outcomes and three independent variables:

		Logit coefficient for:		
Comparison		x_1	x_2	x_3
$B \mid A$	$\beta_{B\mid A}$	−0.693	0.693	0.347
	$\exp(\beta_{B\mid A})$	0.500	2.000	1.414
	p	0.04	0.01	0.42
$C \mid A$	$\beta_{C\mid A}$	0.347	−0.347	0.693
	$\exp(\beta_{C\mid A})$	1.414	0.707	2.000
	p	0.21	0.04	0.37
$C \mid B$	$\beta_{C\mid B}$	1.040	−1.040	0.346
	$\exp(\beta_{C\mid B})$	2.828	0.354	1.414
	p	0.02	0.03	0.21

These coefficients were constructed to have some fixed relationships among categories and variables:

- The effects of x_1 and x_2 on $B \mid A$ (which you can read as B versus A) are equal but of opposite size. The effect of x_3 is half as large.
- The effects of x_1 and x_2 on $C \mid A$ are half as large (and in opposite directions) as the effects on $B \mid A$, whereas the effect of x_3 is in the same direction but twice as large.

In the odds-ratio plot, the independent variables are each represented on a separate row, and the horizontal axis indicates the relative magnitude of the β coefficients associated with each outcome. Here is the plot, where the letters correspond to the outcome categories:

```
Factor Change Scale Relative to Category A
    .5          .63         .79         1          1.26        1.59         2

x1     B                          A            C

x2                    C           A                                    B

x3                                A                       B            C

    −.69        −.46        −.23        0          .23         .46         .69
    Logit Coefficient Scale Relative to Category A
```

The plot reveals much information, which we now summarize.

Sign of coefficients

If a letter is to the right of another letter, increases in the independent variable make the outcome to the right more likely. Thus relative to outcome A, an increase in x_1 makes it more likely that we will observe outcome C and less likely that we will observe outcome B. This corresponds to the positive sign of the $\beta_{1,C|A}$ coefficient and the negative sign of the $\beta_{1,B|A}$ coefficient. The signs of these coefficients are reversed for x_2, and accordingly, the odds-ratio plot for x_2 is a mirror image of that for x_1.

Magnitude of effects

The distance between a pair of letters indicates the magnitude of the effect. For both x_1 and x_2, the distance between A and B is twice the distance between A and C, which reflects that $\beta_{B|A}$ is twice as large as $\beta_{C|A}$ for both variables. For x_3, the distance between A and B is half the distance between A and C, reflecting that $\beta_{3,C|A}$ is twice as large as $\beta_{3,B|A}$.

The additive relationship

The additive relationships among coefficients shown in (6.1) are also fully reflected in this graph. For any of the independent variables, $\beta_{C|A} = \beta_{B|A} + \beta_{C|B}$. Accordingly, the distance from A to C is the sum of the distances from A to B and B to C.

The base category

The additive scale on the bottom axis measures the value of the $\beta_{k,m|n}$s. The multiplicative scale on the top axis measures the $\exp\left(\beta_{k,m|n}\right)$s. The As are stacked on top of one another because the plot uses A as its base category for graphing the coefficients. The choice of base category is arbitrary. We could have used alternative B instead. If we had, the rows of the graph would be shifted to the left or right so that the Bs lined up. Doing this leads to the following graph:

Factor Change Scale Relative to Category B

	.35	.5	.71	1	1.41	2	2.83
x1				B		A	C
x2	C		A	B			
x3			A	B	C		
	−1.04	−.69	−.35	0	.35	.69	1.04

Logit Coefficient Scale Relative to Category B

Creating odds-ratio plots

These graphs can be created using `mlogview` after running `mlogit`. Using our example and after changing a few options, we obtain this dialog box:

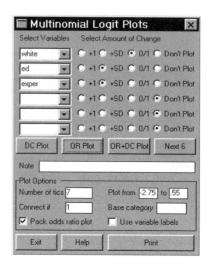

Clicking on OR Plot gives

Factor Change Scale Relative to Category Prof

	.06	.11	.19	.33	.58	1	1.73
white–0/1			M	C	B	W P	
ed–std		B M C		W		P	
exper–std					MB	CWP	
	−2.75	−2.2	−1.65	−1.1	−.55	0	.55

Logit Coefficient Scale Relative to Category Prof

Several things are immediately apparent. The effect of experience is the smallest, although increases in experience make it more likely that one will be in a craft, white-collar, or professional occupation relative to a menial or blue-collar one. We also see that education has the largest effect; as expected, increases in education increase the odds of having a professional job relative to any other type.

Adding significance levels

The current graph does not reflect statistical significance. This is added by drawing a line between categories for which there is *not* a significant coefficient. The *lack* of statistical significance is shown by a connecting line, suggesting that those two outcomes are "tied together". You can add the significance level to the plot with the Connect if box on the dialog box. For example, if we enter .1 in this box and uncheck the "pack odds ratio plot" box, we obtain

(Continued on next page)

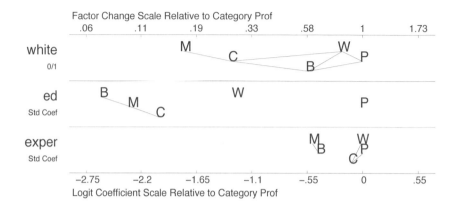

To make the connecting lines clear, vertical spacing is added to the graph. *This vertical spacing has no meaning and is used only to make the lines clearer.* The graph shows that race orders occupations from menial to craft to blue collar to white collar to professional, but the connecting lines show that none of the adjacent categories are significantly differentiated by race. Being white increases the odds of being a craft worker relative to having a menial job, but the effect is not significant. However, being white significantly increases the odds of being a blue-collar worker, a white-collar worker, or a professional, relative to having a menial job. The effects of `ed` and `exper` can be interpreted similarly.

Adding discrete change

In chapter 4, we emphasized that *whereas the factor change in the odds is constant across the levels of all variables, the discrete change gets larger or smaller at different values of the variables.* For example, if the odds increase by a factor of 10 but the current odds are 1 in 10,000, the substantive impact is small. But if the current odds were 1 in 5, the impact is large. Information on the discrete change in probability can be incorporated in the odds-ratio graph by making the size of the letter proportional to the discrete change in the odds (specifically, the area of the letter is proportional to the size of the discrete change). This can easily be added to our graph. First, after estimating the MNLM, run `prchange` at the levels of the variables that you want. Then enter `mlogview` to open the dialog box. Set any of the options, and then click the OR+DC Plot button:

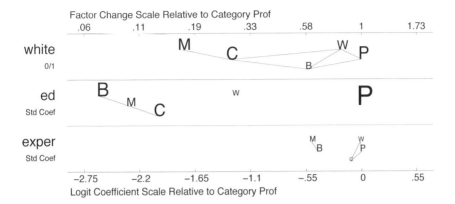

With a little practice, you can quickly create and interpret these graphs.

6.6.9   Using mlogplot*

The dialog box `mlogview` does not actually draw the plots but only sends the options you select to `mlogplot`, which creates the graph. Once you click a plot button in `mlogview`, the necessary `mlogplot` command, including options, appears in the Results window. This is done because `mlogview` invokes a dialog box and so cannot be used effectively in a do-file. But once you create a plot using the dialog box you can copy the generated `mlogplot` command from the Results window and paste it into a do-file. This should be clear by looking at the following screenshot:

(Continued on next page)

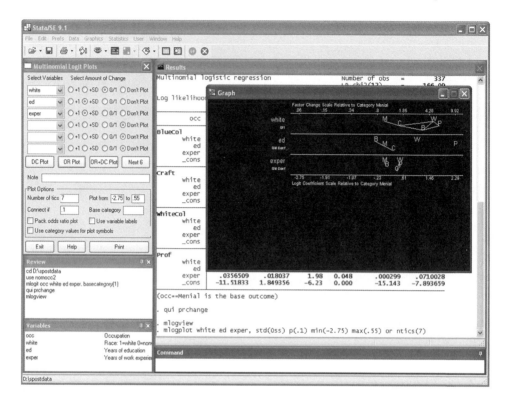

The dialog box with selected options appears in the upper left of the screen. After we clicked on the OR Plot button, the graph in the upper right appeared along with the following command in the Results window:

```
. mlogplot white ed exper, std(Oss) p(.1) min(-2.75) max(.55) or ntics(7)
```

If you enter this command from the Command window or run it from a do-file, the same graph will be generated. The full syntax for `mlogplot` is described in appendix A.

6.6.10   Plotting estimates from matrices with mlogplot[*]

You can also use `mlogplot` to construct odds-ratio plots (but not discrete-change plots) using coefficients that are to be contained in matrices. For example, you can plot coefficients from published papers or generate examples like those we used above. To do this, you must construct matrices containing the information to be plotted and add the option `matrix` to the command. The easiest way to see how this is done is with an example, followed by details on each matrix. The commands

```
. matrix mnlbeta = (-.693, .693, .347   .347, -.347, .693 )
. matrix mnlsd   = (1, 2, 4)
. global mnlname = "x1 x2 x3"
```

```
. global mnlcatnm = "B C A"
. global mnldepnm "depvar"
. mlogplot, matrix std(uuu) vars(x1 x2 x3) packed
```

create the following plot:

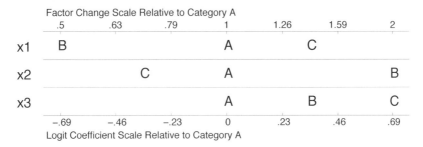

Options for using matrices with mlogplot

matrix indicates that the coefficients to be plotted are contained in matrices.

vars(*varlist*) contains the names of the variables to be plotted. This list must contain names from mnlname, which will be described next, but does not need to be in the same order as in mnlname. The list can contain the same name more than once and can select a subset of the names from mnlname.

Global macros and matrices used by mlogplot

mnlname is a string containing the names of the variables corresponding to the columns of the matrix mnlbeta. For example, global mnlname = "x1 x2 x3".

mnlbeta is a matrix with the βs, where element (i, j) is the coefficient $\beta_{j,i|b}$. That is, rows i are for different contrasts; columns j are for variables. For example, matrix mnlbeta = (-.693, .693, .347 \ .347, -.347, .693). As constant terms are *not* plotted, they are not included in mnlbeta.

mnlsd is a vector with the standard deviations for the variables listed in mnlname. For example, matrix mnlsd = (1, 2, 4). If you do not want to view standardized coefficients, this matrix can be made all 1s.

mnlcatnm is a string with labels for the outcome categories with each label separated by a space. For example, global mnlcatnm = "B C A". The first label corresponds to the first row of mnlbeta, the second to the second, and so on. *The label for the base category is last.*

Example

Suppose that you want to compare the logit coefficients estimated from two groups, such as whites and nonwhites from the example used in this chapter. We begin by estimating the logit coefficients for whites:

```
. use http://www.stata-press.com/data/lf2/nomocc2, clear
(1982 General Social Survey)

. mlogit occ ed exper if white==1, base(5) nolog
```

Multinomial logistic regression

			Number of obs	=	309
			LR chi2(8)	=	154.60
			Prob > chi2	=	0.0000
Log likelihood = -388.21313			Pseudo R2	=	0.1660

occ	Coef.	Std. Err.	z	P>|z|	[95% Conf. Interval]	
Menial						
ed	-.8307514	.1297238	-6.40	0.000	-1.085005	-.5764973
exper	-.0338038	.0192045	-1.76	0.078	-.071444	.0038364
_cons	10.34842	1.779603	5.82	0.000	6.860465	13.83638
BlueCol						
ed	-.9225522	.1085452	-8.50	0.000	-1.135297	-.7098075
exper	-.031449	.0150766	-2.09	0.037	-.0609987	-.0018994
_cons	12.27337	1.507683	8.14	0.000	9.318368	15.22838
Craft						
ed	-.6876114	.0952882	-7.22	0.000	-.8743729	-.50085
exper	-.0002589	.0131021	-0.02	0.984	-.0259385	.0254207
_cons	9.017976	1.36333	6.61	0.000	6.345897	11.69005
WhiteCol						
ed	-.4196403	.0956209	-4.39	0.000	-.6070539	-.2322268
exper	.0008478	.0147558	0.06	0.954	-.0280731	.0297687
_cons	4.972973	1.421146	3.50	0.000	2.187578	7.758368

(occ==Prof is the base outcome)

Next we compute coefficients for nonwhites:

```
. mlogit occ ed exper if white==0, base(5) nolog
Multinomial logistic regression              Number of obs   =        28
                                             LR chi2(8)      =     17.79
                                             Prob > chi2     =    0.0228
Log likelihood = -32.779416                  Pseudo R2       =    0.2135
```

occ	Coef.	Std. Err.	z	P>|z|	[95% Conf. Interval]	
Menial						
ed	-.7012628	.3331146	-2.11	0.035	-1.354155	-.0483701
exper	-.1108415	.0741488	-1.49	0.135	-.2561705	.0344876
_cons	12.32779	6.053743	2.04	0.042	.4626714	24.19291
BlueCol						
ed	-.560695	.3283292	-1.71	0.088	-1.204208	.0828185
exper	-.0261099	.0682348	-0.38	0.702	-.1598477	.1076279
_cons	8.063397	6.008358	1.34	0.180	-3.712768	19.83956
Craft						
ed	-.882502	.3359805	-2.63	0.009	-1.541012	-.2239924
exper	-.1597929	.0744172	-2.15	0.032	-.305648	-.0139378
_cons	16.21925	6.059753	2.68	0.007	4.342356	28.09615
WhiteCol						
ed	-.5311514	.369815	-1.44	0.151	-1.255976	.1936728
exper	-.0520881	.0838967	-0.62	0.535	-.2165227	.1123464
_cons	7.821371	6.805367	1.15	0.250	-5.516904	21.15965

```
(occ==Prof is the base outcome)
```

The two sets of coefficients for `ed` are placed in `mnlbeta`:

```
. matrix mnlbeta = (-.8307514, -.9225522, -.6876114, -.4196403  \
        -.7012628, -.560695 , -.882502 , -.5311514)
```

Rows of the matrix correspond to the variables (i.e., `ed` for whites and `ed` for nonwhites) since this was the easiest way to enter the coefficients. For `mlogplot`, the columns must correspond to variables, so we transpose the matrix:

```
. matrix mnlbeta = mnlbeta´
```

We assign names to the columns using `mnlname` and to the rows using `mnlcatnm` (where the last element is the name of the reference outcome):

```
. global mnlname = "White NonWhite"
. global mnlcatnm = "Menial BlueCol Craft WhiteCol Prof"
```

We named the coefficients for `ed` for whites, `White`, and the coefficients for `ed` for nonwhites, `NonWhite`, as this will make the plot clearer. Next we compute the standard deviation of `ed`:

```
. summarize ed
```

Variable	Obs	Mean	Std. Dev.	Min	Max
ed	337	13.09496	2.946427	3	20

and enter the information into `mnlsd`:

```
. matrix mnlsd = (2.946427,2.946427)
```

The same value is entered twice because we want to use the overall standard deviation in education for both groups. To create the plot, we use the command

```
. mlogplot, vars(White NonWhite) packed
      or matrix std(ss)
      note("Racial Differences in Effects of Education")
```

which leads to

Racial Differences in Effects of Education

Factor Change Scale Relative to Category Prof

	.07	.1	.16	.26	.4	.64	1
White–std	B M		C	W			P
NonWhite–std		C	M	BW			P
	–2.72	–2.27	–1.81	–1.36	–.91	–.45	0

Logit Coefficient Scale Relative to Category Prof

Given the limitations of our dataset (e.g., there were only 28 cases in the logit for nonwhites) and our simple model, these results do not represent serious research on racial differences in occupational outcomes, but they show the flexibility of the `mlogplot` command.

6.7   Multinomial probit model with IIA

The multinomial probit regression command `mprobit` is the normal error counterpart to the multinomial logit model fitted by `mlogit` in the same way that `probit` is the normal counterpart to `logit`. However, `mprobit` uses a normalization that can obscure this fact. To understand this point, we need to consider how logit and probit models can be motivated as discrete-choice models in which a person maximizes her utility.

Let u_{im} be the utility that person i receives from alternative m. The utility is assumed to be determined by a linear combination of observed characteristics $\mathbf{x}_i$ and random error ε_{im}:

$$u_{im} = \mathbf{x}_i \boldsymbol{\beta}_m + \varepsilon_{im}$$

Since the utility associated with each alternative m is partly determined by chance through ε, the model is also called a random utility model (RUM). A person chooses alternative j if the utility associated with that alternative is larger than that for any other alternative. Accordingly, the probability of alternative m being chosen is

$$\Pr(y_i = m) = \Pr(u_{im} > u_{ij} \text{ for all } j \neq m)$$

The choice that a person makes under these assumptions will not change if the utility associated with each alternative changes by some fixed amount, say, δ. That is, if $u_{im} > u_{ij}$, then $u_{im} + \delta > u_{ij} + \delta$. Thus the choice is based on the difference in the utilities between alternatives. We can incorporate this idea into the model by taking the difference in the utilities for two alternatives. To illustrate this, assume that there are three alternatives. We can consider the utility of each alternative relative to some base alternative. It does not matter which alternative is chosen as the base, so we assume that each utility is compared with alternative 1. Accordingly, we have

$$u_{i1} - u_{i1} = 0$$
$$u_{i2} - u_{i1} = \mathbf{x}_i \left(\boldsymbol{\beta}_2 - \boldsymbol{\beta}_1 \right) + \left(\varepsilon_{i2} - \varepsilon_{i1} \right)$$
$$u_{i3} - u_{i1} = \mathbf{x}_i \left(\boldsymbol{\beta}_3 - \boldsymbol{\beta}_1 \right) + \left(\varepsilon_{i3} - \varepsilon_{i1} \right)$$

If we define $u_{im}^* \equiv u_{im} - u_{i1}$, $\varepsilon_{im}^* \equiv \varepsilon_{im} - \varepsilon_{i1}$ and $\boldsymbol{\beta}_{m|1} \equiv \boldsymbol{\beta}_m - \boldsymbol{\beta}_1$, the model can be written as:

$$u_{i2}^* = \mathbf{x}_i \boldsymbol{\beta}_{2|1} + \varepsilon_{i2}^*$$
$$u_{i3}^* = \mathbf{x}_i \boldsymbol{\beta}_{3|1} + \varepsilon_{i3}^*$$

The specific form of the model depends on the distribution of the error terms. Assuming that the εs have an extreme value distribution with mean 0 and variance $\pi^2/6$ leads to the MNLM that we discussed with respect to `mlogit`. Assuming that the εs have a normal distribution leads to a probit-type model. To understand the model fitted by `mprobit` and how it relates to the usual binary probit model, we need to pay careful attention to the assumed variance of the errors. The binary probit model fitted by `probit` makes the usual assumption that $\text{Var}(\varepsilon_j) = 1/2$, so $\text{Var}\left(\varepsilon_j^*\right) = \text{Var}(\varepsilon_j) + \text{Var}(\varepsilon_1) = 1$. Since we assume that the errors are uncorrelated, $\text{Cov}(\varepsilon_j, \varepsilon_1) = 0$. Using our earlier example for labor force participation, we can fit the binary probit model:

(Continued on next page)

```
. use http://www.stata-press.com/data/lf2/binlfp2, clear
(Data from 1976 PSID-T Mroz)

. probit lfp k5 k618 age wc hc lwg inc, nolog
Probit regression                                Number of obs   =        753
                                                 LR chi2(7)      =     124.36
                                                 Prob > chi2     =     0.0000
Log likelihood = -452.69496                      Pseudo R2       =     0.1208
```

lfp	Coef.	Std. Err.	z	P>|z|	[95% Conf. Interval]	
k5	-.8747112	.1135583	-7.70	0.000	-1.097281	-.6521411
k618	-.0385945	.0404893	-0.95	0.340	-.117952	.0407631
age	-.0378235	.0076093	-4.97	0.000	-.0527375	-.0229095
wc	.4883144	.1354873	3.60	0.000	.2227642	.7538645
hc	.0571704	.1240052	0.46	0.645	-.1858754	.3002161
lwg	.3656287	.0877792	4.17	0.000	.1935847	.5376727
inc	-.020525	.0047769	-4.30	0.000	-.0298875	-.0111626
_cons	1.918422	.3806536	5.04	0.000	1.172355	2.66449

The coefficients are for the comparison of alternative 1 (being in the labor force) to alternative 0 (not being in the labor force), so we are estimating $\beta_{1|0}$. Using the same data with `mprobit`, we obtain

```
. mprobit lfp k5 k618 age wc hc lwg inc, nolog baseoutcome(0)
Multinomial probit regression                    Number of obs   =        753
                                                 Wald chi2(7)    =     107.38
Log likelihood = -452.69496                      Prob > chi2     =     0.0000
```

lfp	Coef.	Std. Err.	z	P>|z|	[95% Conf. Interval]	
inLF						
k5	-1.237028	.1605958	-7.70	0.000	-1.55179	-.9222664
k618	-.0545809	.0572605	-0.95	0.340	-.1668094	.0576477
age	-.0534905	.0107612	-4.97	0.000	-.0745821	-.0323989
wc	.6905808	.191608	3.60	0.000	.315036	1.066126
hc	.0808511	.1753699	0.46	0.645	-.2628677	.4245699
lwg	.5170771	.1241385	4.17	0.000	.2737701	.7603841
inc	-.0290268	.0067555	-4.30	0.000	-.0422673	-.0157862
_cons	2.713059	.5383259	5.04	0.000	1.65796	3.768158

```
(lfp=NotInLF is the base outcome)
```

The `baseoutcome(0)` option indicates that category 0 (not being in the labor force) is the base category, so that we are estimating the coefficients $\beta_{1|0}$. If we used the `baseoutcome(1)` option, we would be estimating $\beta_{0|1}$. Comparing the `mprobit` and `probit` output, we see that the z's are identical, but the coefficients for `mprobit` are larger than those for `probit`. The reason is that `mprobit` assumes that $\text{Var}(\varepsilon_j) = 1$, so that $\text{Var}(\varepsilon_j^*) = 2$. Or, in standard deviations, $\text{SD}(\varepsilon_j) = 1$ and $\text{SD}(\varepsilon_j^*) = \sqrt{2} \approx 1.414$. This leads to a change in scale (just like changing the units for income from dollars to pennies in a linear regression) so that the coefficients from `mprobit` will be larger by a factor of $\sqrt{2}$. For example, comparing the coefficients for k5 for the two models, we see that $-1.237028 = \sqrt{2} \times -.8747112$. Although this can be confusing, it does not really matter as long as you understand what `mprobit` is doing. If you compare the coefficients from `mprobit` with other analyses based on the usual probit model, you will be incorrect if you do not take into account the difference in scales. That is, you will incorrectly conclude that the coefficients are substantively larger in the data analyzed with `mprobit`. To avoid hand calculations to convert the coefficients from `mprobit` to the usual scale used with probit models, you can use the `probitparam` option to `mprobit`.[3] For example, if we entered the command:

```
. mprobit lfp k5 k618 age wc hc lwg inc, nolog baseoutcome(0)
```

the estimated coefficients for `mprobit` will match those from `probit`.

As shown by Long (1997), the scale of coefficients is based on an arbitrary identification assumption that does not affect the predicted probabilities. To illustrate this, we can compare predicted probabilities for `probit` and `mprobit`:

```
. use http://www.stata-press.com/data/lf2/binlfp2, clear
(Data from 1976 PSID-T Mroz)
. probit lfp k5 k618 age wc hc lwg inc, nolog
  (output omitted )
. predict p_probit1
(option p assumed; Pr(lfp))
```

With `probit`, only the probability for outcome 1 is computed. Next we use `mprobit`:

```
. mprobit lfp k5 k618 age wc hc lwg inc, nolog baseoutcome(0)
  (output omitted )
. predict p_mprobit0 p_mprobit1
(option pr assumed; predicted probabilities)
```

With `mprobit`, `predict` computes a predicted probability for each outcome category, so we need to provide names for two new variables. If we correlate the predictions, we find that `probit` and `mprobit` compute the same predicted probabilities:

```
. pwcorr p_probit1 p_mprobit1 p_mprobit0 p_mprobit1p p_mprobit0p
```

	p_prob~1	p_mpro~1	p_mpro~0	p_mpr~1p	p_mpr~0p
p_probit1	1.0000				
p_mprobit1	1.0000	1.0000			
p_mprobit0	-1.0000	-1.0000	1.0000		
p_mprobit1p	1.0000	1.0000	-1.0000	1.0000	
p_mprobit0p	-1.0000	-1.0000	1.0000	-1.0000	1.0000

The model fit by `mprobit` assumes that the errors are normal. With normal errors, it is possible for the errors to be correlated across alternatives, thus potentially removing the IIA assumption. Indeed, researchers usually discuss the multinomial probit model for the case when errors are correlated, since this is the only real advantage of the multinomial probit over multinomial logit. But `mprobit` assumes that the *errors are uncorrelated*. Accordingly, `mprobit` estimates an exact counterpart to the multinomial logit model fitted by `mlogit`—meaning that it also assumes IIA. If you use both `mprobit`

---

3. This option was added with version 1.0.7 of `mprobit`.

and `mlogit` with the same model and data, you will get nearly identical predictions. For example, we can compare predictions by estimating both `mlogit` and `mprobit` for our model and computing the predicted probabilities of observing each outcome category. Here are the commands we use, without showing the output.

```
. mprobit occ white ed exper, baseoutcome(1)
. predict mpp1 mpp2 mpp3 mpp4 mpp5
. mlogit occ white ed exper, baseoutcome(1)
. predict mlp1 mlp2 mlp3 mlp4 mlp5
```

We then correlate the predicted probabilities for the first outcome:

```
. correlate *p1
(obs=337)
```

	mpp1	mlp1
mpp1	1.0000	
mlp1	0.9979	1.0000

Clearly, there is not much difference. Train (2003, 39) points out that the thicker tails of the extreme value distribution used for the MNLM compared with the normal allow for "slightly more aberrant behavior", but he also notes that it is unlikely that this difference will be empirically distinguishable.

Finally, although the models fit by `mprobit` and `mlogit` produce nearly identical predictions, it is much harder to fit models with `mprobit` since it must compute integrals by Gaussian quadrature (this method approximates the integral by using a function computed at a limited number of evaluation or quadrature points). For our example of occupational attainment, estimation with `mlogit` took .05 seconds, compared with 12 seconds using `mprobit`. We have not seen enough empirical consequence to justify the extra computational time required by `mprobit`. Further, probit models cannot be interpreted using odds ratios. Nonetheless, the SPost commands `listcoef`, `fitstat`, and `prvalue` can be used with `mprobit`.

6.8   Stereotype logistic regression[4]

In the last chapter, we postponed discussion of the stereotype logistic model (SLM) because this model is easier to understand once you are familiar with multinomial logit. The SLM, proposed by Anderson (1984), is more flexible than the OLM since it does not require the proportional odds assumptions. Yet, it can be more parsimonious than the MNLM.[5] Although more parsimonious for the number of parameters, we will show that the full interpretation of the SLM is as complicated as that of the MNLM. If the full complexity of the model is not considered, valuable information can be lost and incorrect conclusions can be made.

There is also some confusion about whether the SLM requires the dependent variable to be ordered. In its simplest form, the SLM orders the dependent variable along one dimension, but the outcomes are not necessarily ordered the way you think they are! For example, you might think that your outcomes are ordered 1, 2, 3, and 4, but the SLM might determine that the ordering should be 2, 1, 3, and 4, a point we illustrate below. In higher-dimensional SLMs, categories are ordered on more than one dimension and the idea of ordinality is lost as the SLM becomes identical to the MNLM.

Multinomial logit and ordinal outcomes

We introduce the SLM by considering the results from a MNLM of attitudes toward working mothers that was our running example for ordinal outcomes.

(Continued on next page)

---

4. In keeping with standard terminology used to discuss this model, we use the term "outcome category" or "category" rather than "alternative".

5. Anderson developed the stereotype model in reaction to the limitations of the ordinal logit model, which he referred to as the "grouped continuous regression model". The term "grouped continuous" was used since the ordered logit model can be motivated in terms of a continuous, latent variable that is divided by thresholds that lead to the observed categories (i.e., you group a continuous variable). In contrast, he thinks of the outcome categories in the stereotype model as "assessed". Each respondent is considered to have stereotypes that characterize the outcome categories. The respondent, referred to as a judge by Anderson, assesses each category and then picks that category whose stereotype most closely matches the judge's views on the question being asked. Although this explains the name of the model, there is no reason to limit the application of the model to outcomes that are generated in an assessed fashion.

```
. mlogit warm yr89 male white age ed prst, nolog baseoutcome(4)
Multinomial logistic regression                 Number of obs   =      2293
                                                LR chi2(18)     =    349.54
                                                Prob > chi2     =    0.0000
Log likelihood = -2820.9982                     Pseudo R2       =    0.0583
```

warm	Coef.	Std. Err.	z	P>|z|	[95% Conf. Interval]	
SD						
yr89	-1.160197	.1810497	-6.41	0.000	-1.515048	-.8053457
male	1.226454	.167691	7.31	0.000	.8977855	1.555122
white	.834226	.2641771	3.16	0.002	.3164485	1.352004
age	.0316763	.0052183	6.07	0.000	.0214487	.041904
ed	-.1435798	.0337793	-4.25	0.000	-.209786	-.0773736
prst	-.0041656	.0070026	-0.59	0.552	-.0178904	.0095592
_cons	-.722168	.4928708	-1.47	0.143	-1.688177	.2438411
D						
yr89	-.4255712	.1318065	-3.23	0.001	-.6839071	-.1672352
male	1.326716	.137554	9.65	0.000	1.057115	1.596317
white	.4126344	.1872718	2.20	0.028	.0455885	.7796804
age	.0292275	.0042574	6.87	0.000	.0208832	.0375718
ed	-.0513285	.0283399	-1.81	0.070	-.1068737	.0042167
prst	-.0130318	.0055446	-2.35	0.019	-.023899	-.0021645
_cons	-.3088357	.3938354	-0.78	0.433	-1.080739	.4630676
A						
yr89	-.0625534	.1228908	-0.51	0.611	-.3034149	.1783082
male	.8666833	.1310965	6.61	0.000	.6097389	1.123628
white	.3002409	.1710551	1.76	0.079	-.0350211	.6355028
age	.0066719	.0041053	1.63	0.104	-.0013744	.0147181
ed	-.0330137	.0274376	-1.20	0.229	-.0867904	.020763
prst	-.0017323	.0052199	-0.33	0.740	-.0119631	.0084985
_cons	.3932277	.3740361	1.05	0.293	-.3398697	1.126325

(warm==SA is the base outcome)

If warm is truly an ordinal outcome with respect to the explanatory variables, what pattern would we expect for the βs across equations? (Remember that each panel in the output for mlogit corresponds to a different equation. The top panel, labeled SD, presents coefficients for the equation comparing outcome SD to the base category SA; the panel labeled D presents coefficients comparing D with SA, and so on.) If an increase in an explanatory variable is associated with an increase in the odds of answering A versus SA, we would also expect the variable to be associated with an increase in the odds of D versus SA, as well as an increase in the odds of SD versus SA. Consequently, we would expect $\beta_{k,A|SA}$, $\beta_{k,D|SA}$, and $\beta_{k,SD|SA}$ to be in the same direction. More than this, we would expect the coefficient to be largest when comparing categories SD and SA, which are furthest apart, and least for adjacent categories, such as A and SA. In the mlogit output shown above, this means that if warm is ordinal, we would expect the coefficients in the top panel labeled SD to be the largest (either positive or negative), followed by those in the D panel, with the smallest found in the A panel. Indeed, this pattern is found for all the variables except for prst, which is not significant, and male, which shows only a slight deviation from the ordinal pattern.

A further possibility is that the magnitudes of $\beta_{k,A|SA}$, $\beta_{k,D|SA}$, and $\beta_{k,SD|SA}$ are not just consistently ordered for each independent variable but that the *relative magnitudes* of these coefficients are the same for all independent variables. The implication is that the "distance" or "difficulty" in moving from SD to D compared with moving from SD to SA is the same for all the independent variables. For example, if $\beta_{yr89,SD|SA}$ is 2.7 times larger than $\beta_{yr89,D|SA}$, then $\beta_{male,SD|SA}$ would be 2.7 times larger than $\beta_{male,D|SA}$, and so on, for each variable and each ratio of coefficients. If this pattern adequately characterizes the data generating process, the more parsimonious SLM should fit the data nearly as well as the MNLM. This simpler model would provide a set of coefficients β_k and scaling parameters ϕ_j such that

$$\beta_{k,SD|SA} = \phi_{SD}\beta_k$$
$$\beta_{k,D|SA} = \phi_D\beta_k$$
$$\beta_{k,A|SA} = \phi_A\beta_k$$

If these constraints are applied to the MNLM, you have the one-dimensional stereotype logit model.

6.8.1   Formal statement of the one-dimensional SLM

With these ideas in mind, we can present the model more formally, beginning with a comparison to the MNLM. To simplify the presentation, we assume that there are three outcome categories and two independent variables. For the MNLM (see page 227), the model with base category three can be written as

$$\Pr(y = m \mid \mathbf{x}) = \frac{\exp\left(\beta_{0,m|3} + \beta_{1,m|3}x_1 + \beta_{2,m|3}x_2\right)}{\sum_{j=1}^{3} \exp\left(\beta_{0,j|3} + \beta_{1,j|3}x_1 + \beta_{2,j|3}x_2\right)} \quad \text{for } m = 1, 3 \qquad (6.2)$$

We can write the SLM using notation similar to the MNLM (later we will change the notation to match that used in the Stata manual):

$$\Pr(y = m \mid \mathbf{x}) = \frac{\exp\left(\phi_m\widetilde{\beta}_0 x_0 + \phi_m\widetilde{\beta}_1 x_1 + \phi_m\widetilde{\beta}_2 x_2\right)}{\sum_{j=1}^{3} \exp\left(\phi_j\widetilde{\beta}_0 x_0 + \phi_j\widetilde{\beta}_1 x_1 + \phi_j\widetilde{\beta}_2 x_2\right)} \quad \text{for } m = 1, 3 \qquad (6.3)$$

The only difference between (6.2) and (6.3) is that $\beta_{k,m|3}$ is replaced by $\phi_m\widetilde{\beta}_k$. This replacement forces the ratio of coefficients to be equal across variables. Specifically,

$$\frac{\phi_j\widetilde{\beta}_1}{\phi_m\widetilde{\beta}_1} = \frac{\phi_j\widetilde{\beta}_2}{\phi_m\widetilde{\beta}_2} = \frac{\phi_j}{\phi_m}$$

By comparison, in the MNLM, the ratio $\widehat{\beta}_{1,j|3}/\widehat{\beta}_{1,m|3}$ might be similar to $\widehat{\beta}_{2,j|3}/\widehat{\beta}_{2,m|3}$, but the model does not require this.

Some of the parameters in the SLM as specified in (6.3) are not identified. Accordingly, we must add identifying constraints before the parameters can be estimated. To understand the constraints that are used by Stata to identify the parameters, we find it helpful to compare them with the identifying constraints used for the MNLM. (Although these constraints are the defaults used by `slogit`, `slogit` also allows you to use other identifying constraints.) In the MNLM, we assume that $\beta_{k,J|J} = 0$, where J is the base category. This constraint simply says that a change in x_k does not change the odds of J compared with J. The corresponding constraint in the SLM is $\phi_3 \tilde{\beta}_k = 0$. We assume that $\phi_3 = 0$ since we do not want to require $\tilde{\beta}_k = 0$, which would eliminate the effect of x_k for all pairs of outcomes. This is our first identification constraint. To understand the next constraint, we compare $\beta_{k,1|3}$ and $\beta_{k,2|3}$ for x_k in the MNLM with the corresponding pairs of coefficients $\phi_1 \tilde{\beta}_k$ and $\phi_2 \tilde{\beta}_k$ in the SLM. There are two free parameters, $\beta_{k,1|3}$ and $\beta_{k,2|3}$, in the MNLM, but three parameters, ϕ_1, ϕ_2, and $\tilde{\beta}_k$, in the SLM. To eliminate the "extra" SLM parameter, we assume that $\phi_1 = 1$. With these constraints, the SLM is identified.

The notation we have used so far highlights the similarities between the MNLM and the SLM but differs from the notation used by Stata. To switch notations, we define $\theta_m \equiv \phi_m \tilde{\beta}_0$ where $\theta_3 = 0$, and $\phi_m \beta_1 \equiv -\phi_m \tilde{\beta}_1$ where $\phi_3 = 0$ and $\phi_1 = 1$. Since the sign has changed, a positive coefficient in the MNLM corresponds to a negative coefficient in the SLM. With this new notation, we can write the model as

$$\Pr(y = m \mid \mathbf{x}) = \frac{\exp(\theta_m - \phi_m \beta_1 x_1 - \phi_m \beta_2 x_2)}{\sum_{j=1}^{3} \exp(\theta_j - \phi_j \beta_1 x_1 - \phi_j \beta_2 x_2)} \tag{6.4}$$

We can generalize this equation to J outcomes and K independent variables

$$\Pr(y = m \mid \mathbf{x}) = \frac{\exp(\theta_m - \phi_m \mathbf{x} \boldsymbol{\beta})}{\sum_{j=1}^{J} \exp(\theta_j - \phi_j \mathbf{x} \boldsymbol{\beta})} \tag{6.5}$$

where $\theta_J = 0$, $\phi_J = 0$ and $\phi_1 = 1$ are used to identify the model.

6.8.2 Fitting the SLM with slogit

The stereotype logistic model is fitted with the following command and its basic options:

slogit *depvar* [*indepvars*] [*if*] [*in*] [*weight*] [, dimension(*#*)
    baseoutcome(*#*) constraints(*#*) nocorner robust cluster(*varname*)]

Options

dimension(*#*) specifies the dimension of the model. The default is dimension(1). The maximum is either one less than the number of categories in the dependent variable or the number of explanatory variables, whichever is fewest. The dimension of an SLM is discussed below.

baseoutcome(*#*) specifies the outcome category whose associated θ and ϕ estimates will be constrained to zero. By default, this is the highest numbered category.

constraints(#) and nocorner allow users to specify alternative identifying constraints or to add more constraints. nocorner tells Stata that the default constraints should not be used.

robust and cluster(*varname*) are used as in other models.

Example

For our example on attitudes toward working mothers, we fit the SLM as follows:

```
. slogit warm yr89 male white age ed prst, nolog
Stereotype logistic regression                    Number of obs   =        2293
                                                  Wald chi2(6)    =      185.45
Log likelihood = -2845.5947                       Prob > chi2     =      0.0000
 ( 1)  [phi1_1]_cons = 1
```

warm	Coef.	Std. Err.	z	P>|z|	[95% Conf. Interval]	
yr89	.9440522	.1527945	6.18	0.000	.6445805	1.243524
male	-1.256064	.151801	-8.27	0.000	-1.553588	-.9585394
white	-.6390139	.2149275	-2.97	0.003	-1.060264	-.2177637
age	-.0384116	.0044971	-8.54	0.000	-.0472258	-.0295974
ed	.1093335	.0292606	3.74	0.000	.0519838	.1666832
prst	.0114819	.0057911	1.98	0.047	.0001316	.0228322
/phi1_1	1	.	.	.	.	.
/phi1_2	.7488452	.0541062	13.84	0.000	.6427991	.8548913
/phi1_3	.3183653	.04977	6.40	0.000	.220818	.4159127
/phi1_4	0	(base outcome)				
/theta1	-1.060064	.4113478	-2.58	0.010	-1.86629	-.2538366
/theta2	.1323735	.3127481	0.42	0.672	-.4806014	.7453484
/theta3	.6272993	.1425944	4.40	0.000	.3478194	.9067792
/theta4	0	(base outcome)				

(warm=SA is the base outcome)

The top panel contains estimates of the βs the next panel contains estimates of the ϕs where the constraints are shown for $\phi_1 = 1$ and $\phi_4 = 0$. Notice that $\phi_1 > \widehat{\phi}_2 > \widehat{\phi}_3 > \phi_4$, a point we discuss below. The intercepts θ are shown in the last panel, including the constraint $\theta_4 = 0$.

6.8.3   Interpretation using predicted probabilities

From (6.5), predict computes predictions for each case in the sample. For example,

```
. slogit warm yr89 male white age ed prst, nolog
  (output omitted )
. predict slpr1 slpr2 slpr3 slpr4
(option pr assumed; predicted probabilities)
```

```
. list warm slpr1-slpr4 in 1/4
```

	warm	slpr1	slpr2	slpr3	slpr4
1.	SD	.1028566	.3021472	.4070647	.1879315
2.	SD	.2440264	.4385175	.2544496	.0630064
3.	SD	.1081623	.3103383	.401561	.1799384
4.	SD	.2134578	.4207702	.2861099	.0796621

Fitting the same model with `mlogit`, we can also compute predicted probabilities:

```
. mlogit warm yr89 male white age ed prst, nolog
(output omitted)
. predict mlpr1 mlpr2 mlpr3 mlpr4
(option pr assumed; predicted probabilities)
```

To compare the SLM and MNLM, we can correlate the predicted probabilities:

```
. pwcorr slpr1 mlpr1
```

	slpr1	mlpr1
slpr1	1.0000	
mlpr1	0.9070	1.0000

```
. pwcorr slpr2 mlpr2
```

	slpr2	mlpr2
slpr2	1.0000	
mlpr2	0.9293	1.0000

```
. pwcorr slpr3 mlpr3
```

	slpr3	mlpr3
slpr3	1.0000	
mlpr3	0.8748	1.0000

```
. pwcorr slpr4 mlpr4
```

	slpr4	mlpr4
slpr4	1.0000	
mlpr4	0.9459	1.0000

The probabilities are highly but not perfectly correlated across the two models, providing some support that the constraints imposed by the SLM hold.

Next we use SPost commands to make predictions at specific values of the independent variables. For example, to compare the predicted probabilities for men and women who are average on all other characteristics, we use `prvalue` with the `save` and `diff` options:

```
. quietly prvalue, x(male=0) rest(mean) save
. prvalue, x(male=1) rest(mean) diff
slogit: Change in Predictions for warm
                     Current      Saved     Change
   Pr(y=SD|x):        0.1619     0.0852     0.0767
   Pr(y=D|x):         0.3775     0.2723     0.1052
   Pr(y=A|x):         0.3426     0.4243    -0.0818
   Pr(y=SA|x):        0.1181     0.2181    -0.1001

                 yr89       male    white       age        ed       prst
   Current=   .39860445        1  .8765809  44.935456  12.218055  39.585259
     Saved=   .39860445        0  .8765809  44.935456  12.218055  39.585259
      Diff=       0            1     0          0          0          0
```

This shows that the predicted probabilities for each outcome category differ by roughly
.10 between men and women, with women having more positive attitudes toward working mothers.[6] Or, we can use **prchange** to compute discrete changes:

```
. prchange male, rest(mean)
slogit: Changes in Probabilities for warm
male
             Avg|Chg|            SD          D          A          SA
   0->1     .09093904      .07668686   .1051912  -.08178753  -.10009058

                  SD          D          A          SA
Pr(y|x)     .11714774  .32349858  .39201239  .16734134

              yr89      male     white       age        ed      prst
    x=      .398604   .464893   .876581   44.9355   12.2181   39.5853
sd(x)=      .489718   .498875   .328989   16.779    3.16083   14.4923
```

6.8.4 Interpretation using odds ratios

Since the SLM is a logit model (i.e., the outcome is the log of the odds), we can take the
log of (6.5) for two outcome categories to specify the SLM as log linear in the odds:

$$\ln \frac{\Pr(y=q|\mathbf{x})}{\Pr(y=r|\mathbf{x})} = (\theta_q - \theta_r) - (\phi_q - \phi_r)\mathbf{x}\boldsymbol{\beta}$$

Taking the exponential, we have a model that is multiplicative in the odds:

$$\Omega_{q|r}(\mathbf{x}) = \frac{\Pr(y=q|\mathbf{x})}{\Pr(y=r|\mathbf{x})} \tag{6.6}$$

$$= \frac{\exp(\theta_q - \phi_q \mathbf{x}\boldsymbol{\beta})}{\exp(\theta_r - \phi_r \mathbf{x}\boldsymbol{\beta})}$$

$$= \exp\left\{(\theta_q - \theta_r) - (\phi_q - \phi_r)\mathbf{x}\boldsymbol{\beta}\right\}$$

---

6. **prvalue** does not compute standard errors in predictions for the stereotype model.

Equation (6.6) can be used to estimate the factor change in the odds for a unit change in x_k, holding all other variables constant. To do this, we take the ratio of the odds after x_k increases by one to the odds before the change. Using basic algebra,

$$\frac{\Omega_{q|r}(\mathbf{x}, x_k + 1)}{\Omega_{q|r}(\mathbf{x}, x_k)} = \exp\left\{(\phi_r - \phi_q)\beta_k\right\} \tag{6.7}$$

$$= \left(\frac{e^{\phi_q}}{e^{\phi_r}}\right)^{-\beta_k}$$

This equation shows that the effect of x_k on the odds of outcome q versus outcome r differs across outcome comparisons according to the scaling coefficients ϕ_q and ϕ_r. Using the same approach as for the MNLM, we can interpret the effect of x_k on the odds as follows:

> For a unit increase in x_k, the odds of outcomes q versus r change by a factor of $\exp\left\{(\phi_r - \phi_q)\beta_k\right\}$, holding all other variables constant.

Shortly, we will provide examples of such interpretations using the model of attitudes toward working mothers.

Using the general formula in (6.7), we can compute the odds ratios for all pairs of outcomes. Although this formula uses the three coefficients ϕ_r, ϕ_q, and β_k to compute the odds ratio, the identification constraints simplify computation of the odds of the highest numbered category compared with the lowest numbered category (assuming that you are using the default identification assumptions and letting slogit determine the base category). Here the base category is 4 = SA so that $\phi_4 = \phi_{\mathrm{SA}} = 0$ and $\phi_1 = \phi_{\mathrm{SD}} = 1$. Then

$$\frac{\Omega_{\mathrm{SA}|\mathrm{SD}}(\mathbf{x}, x_k + 1)}{\Omega_{\mathrm{SA}|\mathrm{SD}}(\mathbf{x}, x_k)} = \exp\left\{(\phi_{\mathrm{SD}} - \phi_{\mathrm{SA}})\beta_k\right\}$$

$$= \exp\left\{(1 - 0)\beta_k\right\}$$

$$= e^{\beta_k}$$

This shows that the βs estimated by slogit can be interpreted directly in terms of the odds of the base category versus the first category (again, keep in mind this is assuming that you are using the default options). Although this makes it simple to examine the effects of x_k on the odds of the base category compared with the lowest numbered category, if you stop there you can easily overlook critical aspects of your data. In fact, you might not notice that the ordering of your outcome variable differs from what you expect, a point that is illustrated on page 288.

The easiest way to examine the effects of each variable on the odds of all pairs of outcomes is to use listcoef, expand, where the expand option indicates that you want to see the expanded set of comparisons for all pairs of outcomes. We will break the output from listcoef into parts to explain how it can be interpreted.

```
. listcoef, expand
slogit (N=2293): Factor Change in Odds
  Odds of: SA vs SD
```

warm	b	z	P>|z|	e^b	e^bStdX	SDofX
yr89	0.94405	6.179	0.000	2.5704	1.5878	0.4897
male	-1.25606	-8.274	0.000	0.2848	0.5344	0.4989
white	-0.63901	-2.973	0.003	0.5278	0.8104	0.3290
age	-0.03841	-8.541	0.000	0.9623	0.5249	16.7790
ed	0.10933	3.737	0.000	1.1155	1.4128	3.1608
prst	0.01148	1.983	0.047	1.0115	1.1810	14.4923
phi1_1	1.00000	.	.			
phi1_2	0.74885	13.840	0.000			
phi1_3	0.31837	6.397	0.000			
theta1	-1.06006	-2.577	0.010			
theta2	0.13237	0.423	0.672			
theta3	0.62730	4.399	0.000			

(output omitted)

This output is similar to that produced by slogit. The biggest difference is that the exponentials of the estimated β_ks and $\beta_k s_k$s are also printed. These can be interpreted as follows:

> From 1977 to 1989, the odds of strongly agreeing versus strongly disagreeing increased by a factor of 2.57, holding all other variables constant.

Or, using percent change:

> In 1989 compared with 1977, the odds of strongly agreeing versus strongly disagreeing increased by 157% (= 100(2.57 − 1)), holding all other variables constant.

Or, for age we can say:

> For a standard deviation increase in age, roughly 17 years, the odds of strongly agreeing versus strongly disagreeing decreased by a factor of .52, holding all other variables constant.

All comparisons are for the odds of the extreme categories of strongly agreeing and strongly disagreeing, which corresponds to the βs estimated above. The expand option computes odds ratios for all comparisons. Here are the results for yr89:

```
slogit (N=2293): Factor Change in the Odds of warm
Variable: yr89 (sd=.48971781)
       Odds comparing|
    Group 1 vs Group 2|      b        z      P>|z|     e^b     e^bStdX

    SD       -D       | -0.23710   -3.177   0.001    0.7889    0.8904
    SD       -A       | -0.64350   -5.377   0.000    0.5255    0.7297
    SD       -SA      | -0.94405   -6.179   0.000    0.3890    0.6298
    D        -SD      |  0.23710    3.177   0.001    1.2676    1.1231
    D        -A       | -0.40640   -5.640   0.000    0.6660    0.8195
    D        -SA      | -0.70695   -6.661   0.000    0.4931    0.7074
    A        -SD      |  0.64350    5.377   0.000    1.9031    1.3704
    A        -D       |  0.40640    5.640   0.000    1.5014    1.2202
    A        -SA      | -0.30055   -4.762   0.000    0.7404    0.8631
    SA       -SD      |  0.94405    6.179   0.000    2.5704    1.5878
    SA       -D       |  0.70695    6.661   0.000    2.0278    1.4137
    SA       -A       |  0.30055    4.762   0.000    1.3506    1.1586
```

(*output omitted*)

We can interpret these results as follows:

> From 1977 to 1989, the odds of more positive attitudes increased. Holding
> all other variables constant, the odds of SA compared with A increased by
> a factor of 1.4, the odds of A compared with D by a factor of 1.5, and the
> odds of D versus SD by a factor of 1.3. All coefficients are significant at the
> .01 level for a two-tailed test.

6.8.5   Distinguishability and the ϕ parameters

When interpreting the SLM, it is also useful to examine the estimates of the ϕs, which
indicate the distance between categories. More formally, we say that the ϕs measure
the distinguishability of the categories. If two ϕs are equal, then the model does not
distinguish between the corresponding categories and they can be collapsed. To see why
this is the case, look at the formula for the effect of a change in x_k on the odds of any
two outcomes:

$$\frac{\Omega_{m|n}(\mathbf{x}, x_k + 1)}{\Omega_{m|n}(\mathbf{x}, x_k)} = \exp\left\{ (\phi_n - \phi_m)\, \beta_k \right\}$$

If $\phi_n = \phi_m$, then

$$\frac{\Omega_{m|n}(\mathbf{x}, x_k + 1)}{\Omega_{m|n}(\mathbf{x}, x_k)} = \exp\left(0\beta_k\right) = 1$$

When the ϕs are equal, none of the independent variables affects the odds of category
m versus n. That is, the two categories are indistinguishable.

A LR test can be used to test if categories can be combined by comparing the
unconstrained model to a model that adds the constraint $\phi_n = \phi_m$. Two examples
show how this is done. First, we fit the SLM without the equality constraint. When

you examine this output, notice the identification constraint listed at the top ((1) [phi1_1]_cons = 1) and that the estimates for /phi1_1 and /phi1_2 differ:

```
. slogit warm yr89 male white age ed prst, nolog
Stereotype logistic regression              Number of obs   =        2293
                                            Wald chi2(6)    =      185.45
Log likelihood = -2845.5947                 Prob > chi2     =      0.0000
 ( 1)  [phi1_1]_cons = 1
```

warm	Coef.	Std. Err.	z	P>|z|	[95% Conf. Interval]	
yr89	.9440522	.1527945	6.18	0.000	.6445805	1.243524
male	-1.256064	.151801	-8.27	0.000	-1.553588	-.9585394
white	-.6390139	.2149275	-2.97	0.003	-1.060264	-.2177637
age	-.0384116	.0044971	-8.54	0.000	-.0472258	-.0295974
ed	.1093335	.0292606	3.74	0.000	.0519838	.1666832
prst	.0114819	.0057911	1.98	0.047	.0001316	.0228322
/phi1_1	1	.	.	.	.	.
/phi1_2	.7488452	.0541062	13.84	0.000	.6427991	.8548913
/phi1_3	.3183653	.04977	6.40	0.000	.220818	.4159127
/phi1_4	0	(base outcome)				
/theta1	-1.060064	.4113478	-2.58	0.010	-1.86629	-.2538366
/theta2	.1323735	.3127481	0.42	0.672	-.4806014	.7453484
/theta3	.6272993	.1425944	4.40	0.000	.3478194	.9067792
/theta4	0	(base outcome)				

```
(warm=SA is the base outcome)
```

Next we store the results, which allows us to later compute a LR test using this model.

```
. est store slbase
```

We can add the constraint $\phi_1 = \phi_2$, which implies that SD and D can be combined, with the command:

```
. constraint define 1 [phi1_1]_cons=[phi1_2]_cons
```

[phi1_1]_cons is how Stata labels ϕ_1, where / is *not* included in the name of the parameter. Ditto for ϕ_2. Next we fit the model with this constraint imposed using the constraint() option:

(*Continued on next page*)

```
. slogit warm yr89 male white age ed prst, nolog constraint(1)
Stereotype logistic regression                    Number of obs    =       2293
                                                  Wald chi2(6)     =     203.13
Log likelihood = -2853.5867                       Prob > chi2      =     0.0000
 ( 1)  [phi1_1]_cons = 1
 ( 2)  [phi1_1]_cons - [phi1_2]_cons = 0
```

warm	Coef.	Std. Err.	z	P>|z|	[95% Conf. Interval]	
yr89	.6887246	.1182899	5.82	0.000	.4568806	.9205686
male	-1.116511	.1306907	-8.54	0.000	-1.37266	-.8603624
white	-.4802162	.176398	-2.72	0.006	-.8259499	-.1344826
age	-.0330558	.0038039	-8.69	0.000	-.0405113	-.0256002
ed	.0765727	.0245146	3.12	0.002	.0285251	.1246204
prst	.0116892	.0048988	2.39	0.017	.0020877	.0212906
/phi1_1	1	.	.	.	.	.
/phi1_2	1	.	.	.	.	.
/phi1_3	.399509	.0540204	7.40	0.000	.293631	.505387
/phi1_4	0	(base outcome)				
/theta1	-1.006048	.3507383	-2.87	0.004	-1.693483	-.3186139
/theta2	-.1163713	.3478987	-0.33	0.738	-.7982402	.5654976
/theta3	.5558963	.1509611	3.68	0.000	.2600179	.8517747
/theta4	0	(base outcome)				

```
(warm=SA is the base outcome)
```

Both constraints are listed at the top of the output. To test if the new constraint is reasonable, we use `lrtest`, where we use the shortcut `.` to refer to the last model fitted:

```
. lrtest slbase .
Likelihood-ratio test                             LR chi2(1)  =      15.98
(Assumption: . nested in slbase)                  Prob > chi2 =     0.0001
```

We conclude that categories 1 and 2 are distinguishable. Testing the distinguishability of categories 2 and 3 can be done as follows:

```
. constraint define 2 [phi1_2]_cons=[phi1_3]_cons
. slogit warm yr89 male white age ed prst, nolog constraint(2)
  (output omitted)
. lrtest slbase .
Likelihood-ratio test                             LR chi2(1)  =      89.20
(Assumption: . nested in slbase)                  Prob > chi2 =     0.0000
```

The LR test again rejects that the categories can be combined.

6.8.6   Ordinality in the one-dimensional SLM

The SLM assumes that the dependent categories can be ordered, but the ordering is not necessarily the same as how the outcome categories are numbered (or how you assume that they are ordered). Looking at the formula for the odds ratios

$$\frac{\Omega_{m|n}(\mathbf{x}, x_k + 1)}{\Omega_{m|n}(\mathbf{x}, x_k)} = \exp\left\{ (\phi_n - \phi_m)\,\beta_k \right\}$$

we see that the magnitude of the odds ratios will increase as category values m and n are further apart only if $\phi_1 > \phi_2 > \ldots \phi_{J-1} > \phi_J$. But `slogit` does not impose this constraint when fitting the model. If you look at our estimates above from the model of attitudes toward working mothers, you will see that the estimates of the ϕs are ordered such that $\phi_1 > \widehat{\phi}_2 > \widehat{\phi}_3 > \phi_4$. But what would have happened if our dependent variable was not ordered from 1 to 4? Imagine that we create a second variable, named `warm2`, that is equal to `warm` except that we switch the middle two categories so that agree $= 3$ instead of 2 and disagree $= 2$ instead of 3. Although a real survey would not order the categories like this, poorly worded alternatives (among other things) can sometimes lead to unexpected inversions of the middle categories from what might semantically seem like their "correct" ordering. In any case, we can fit a model with `warm2` as the outcome:

```
. slogit warm2 yr89 male white age ed prst, nolog
  (output omitted )
. est store warm2
```

We want to compare our results with those for `warm`, so we fit that model, save the results, and use `estimates table` to compare the results (output has been edited to make the table simpler).

```
. slogit warm yr89 male white age ed prst, nolog
  (output omitted )
. est store warm
```

(Continued on next page)

```
. estimates table warm warm2, b(%9.3f) t(%6.2f)
```

Variable	warm	warm2
dim1		
yr89	0.944	0.944
	6.18	6.18
male	-1.256	-1.256
	-8.27	-8.27
white	-0.639	-0.639
	-2.97	-2.97
age	-0.038	-0.038
	-8.54	-8.54
ed	0.109	0.109
	3.74	3.74
prst	0.011	0.011
	1.98	1.98
phi1_1		
_cons	1.000	1.000
	.	.
phi1_2		
_cons	0.749	0.318
	13.84	6.40
phi1_3		
_cons	0.318	0.749
	6.40	13.84
theta1		
_cons	-1.060	-1.060
	-2.58	-2.58
theta2		
_cons	0.132	0.627
	0.42	4.40
theta3		
_cons	0.627	0.132
	4.40	0.42

```
                            legend: b/t
```

The estimates for the $\exp(\beta_k)$s are *exactly* the same for both models. $\exp(\beta_k)$ indicates the factor change in the odds of outcomes for the highest numbered and lowest numbered categories, and these are the same for both `warm` and `warm2`. To consider the consequences of changes for the middle categories, however, we need to examine the ϕs. Here we find that in the model for `warm` the ϕs are ordered $1 = \hat{\phi}_1^{[warm]} > \hat{\phi}_2^{[warm]} > \hat{\phi}_3^{[warm]} > \hat{\phi}_4^{[warm]} = 0$. But, in the model for `warm2`, they are ordered $1 = \hat{\phi}_1^{[warm2]} > \hat{\phi}_3^{[warm2]} > \hat{\phi}_2^{[warm2]} > \hat{\phi}_4^{[warm2]} = 0$. This means that the SLM orders `warm2` as 1, 3, 2, and 4, which is what we would expect given that we reversed the middle categories when creating the new variable. This simply shows that the one-dimensional SLM orders the categories of the outcome, but the order does not need to correspond to how the outcome has been numbered. More formally, `slogit` does not impose the constraint $\phi_1 > \phi_2 > \ldots \phi_{J-1} > \phi_J$. By contrast, the ordered logit and probit models assume that the outcome categories are numbered in their proper order, and if you switch the middle categories around, you can get quite different results. The critical point regarding interpretation is that you need to examine the effects of the independent variables on all pairs of outcomes, not just those between the largest and smallest categories. This is where `listcoef, expand` comes in handy. And this is why the SLM is really not any simpler to interpret than the MNLM. You still need to examine the comparisons among all pairs of outcomes.

Higher-dimension SLM

The MNLM requires no ordering for the outcome variable, whereas the one-dimensional SLM orders the outcomes along one dimension (even if it is not the dimension that you expected). Between ordinal and fully nominal variables are variables that can be ordered in more than one dimension (but fewer dimensions than the number of categories minus one). For example, the *direction* of opinion about working mothers varies from less positive to more positive. This is one dimension by which the variable can be ordered. Attitudes can also be ordered according to *extremity* of an attitude, varying between strongly held attitudes (either strongly agree or strongly disagree) and less strongly held attitudes (either agree or disagree). Ordering on more than one dimension is possible with higher-dimensional SLMs. Although these can be estimated by `slogit`, we consider them only briefly here and they are not included in the SPost commands.

The loglinear model for a one-dimensional SLM is

$$\ln \frac{\Pr(y = q | \mathbf{x})}{\Pr(y = r | \mathbf{x})} = (\theta_q - \theta_r) - (\phi_q - \phi_r) \mathbf{x} \boldsymbol{\beta}$$

For a two-dimensional model, we add another set of ϕs and another set of βs:

$$\ln \frac{\Pr(y = q | \mathbf{x})}{\Pr(y = r | \mathbf{x})} = (\theta_q - \theta_r) - (\phi_q^{[1]} - \phi_r^{[1]}) \mathbf{x} \boldsymbol{\beta}^{[1]} - (\phi_q^{[2]} - \phi_r^{[2]}) \mathbf{x} \boldsymbol{\beta}^{[2]}$$

Suppose that we thought that only sex and age were important. Our model might be

$$\ln \frac{\Pr(y = q | \mathbf{x})}{\Pr(y = r | \mathbf{x})} = (\theta_q - \theta_r) - (\phi_q^{[1]} - \phi_r^{[1]}) \left(\beta_{age}^{[1]} age + \beta_{male}^{[1]} male \right)$$
$$- (\phi_q^{[2]} - \phi_r^{[2]}) \left(\beta_{age}^{[2]} age + \beta_{male}^{[2]} male \right)$$

With two dimensions, you can find that variables are significant in some, but not all, of the equations. And the pattern of ϕs from the two dimensions can differ so that the ordering for the first dimension (the [1] parameters) is different than for the second dimension (the [2] parameters). An SLM with $J - 1$ dimensions is simply a different way to parameterize the MNLM.

7   Models for nominal outcomes with alternative-specific data

So far we have focused on models where case-specific variables are associated with the likelihood of observing a given outcome. For example, we considered how individual characteristics such as education and experience affect a person's occupation. But many datasets contain alternative-specific variables that measure characteristics of the alternatives that vary across cases. Consider the following examples:

- The outcome is the mode of transportation that a person uses to get to work: car, bus, or train (see, e.g., Hensher 1986). We want to estimate the effect of travel time on the respondent's choice, but the amount of time for a given mode of travel can differ for each traveler.

- The outcome is the type of car an individual purchases: European, American, or Japanese (see [R] **clogit**). We want to estimate the effect of the number of dealerships, but the number of dealerships of each type varies among buyers since they live in different cities.

- The outcome is which candidate a respondent votes for in a multiparty election (e.g., Alvarez and Nagler 1998). We want to estimate how the distance between a respondent and a candidate on particular issues (e.g., taxation, defense, gun control) affects voter choice, but the distance from each candidate to each respondent varies across respondents.

In these examples, the alternative-specific variables are the time it would take a respondent to get to work for each mode of transportation, the number of car dealerships for each city's carmakers, and the distance between a respondent and each candidate on a political issue. While you may be used to thinking of cases having one value per variable, what is needed here is a way to have as many values for a variable as there are alternatives.

We begin by showing how to arrange the data to fit models with alternative-specific explanatory variables. Then we introduce the conditional logit model (CLM), which is equivalent to the multinomial logit model (MNLM) when you have only case-specific explanatory variables. The CLM also provides a flexible means of combining case-specific and alternative-specific variables. But, like the MNLM, the CLM imposes the assumption of the independence of irrelevant alternatives (IIA) discussed in the last chapter. The alternative-specific multinomial probit model (ASMNPM) can relax the IIA assumption by using information from alternative-specific variables. Last, we consider the rank-ordered logit model (ROLM) for situations in which you have a ranking of all alternatives, as opposed to knowing only which single alternative is most preferred.

7.1   Alternative-specific data organization

Models for alternative-specific variables in Stata require that the data be arranged differently from the other models we consider in this book. We illustrate this with an example from Greene and Hensher (1995). We have data on $N = 152$ groups of people traveling for their vacation, choosing between $J = 3$ alternative modes of travel: train, bus, or car. Within each group of travelers, there are three rows of data corresponding to the three alternatives faced by each group. Accordingly, we have $N \times J = 152 \times 3 = 456$ observations (i.e., rows or records in the dataset). For each group of travelers, the first row is for the alternative of taking a train; the second for taking a bus; and the third for taking a car. Two dummy variables are used to indicate the mode of travel corresponding to a given row. Variable `train` is 1 if the row contains information about taking the train, else `train` is 0. `bus` is 1 if the row contains information about taking a bus, else `bus` is 0. And if both `train` and `bus` are 0, the row has information about driving a car. The actual choice made by a group is indicated with the dummy variable `choice` that equals 1 in the row corresponding to the alternative selected by the group. For example, let's look at the data for the first two groups, corresponding to the first six rows of the dataset:

```
. use http://www.stata-press.com/data/lf2/travel2.dta, clear
(Greene & Hensher 1997 data on travel mode choice)
. list id mode choice train bus time invc in 1/6, nolabel sepby(id)
```

	id	mode	choice	train	bus	time	invc
1.	1	1	0	1	0	406	31
2.	1	2	0	0	1	452	25
3.	1	3	1	0	0	180	10
4.	2	1	0	1	0	398	31
5.	2	2	0	0	1	452	25
6.	2	3	1	0	0	255	11

Groups are indicated by `id`. Each group of travelers, represented by three rows in the dataset, is a case. Each row corresponds to a particular alternative as shown by `mode`, which equals 1 for train, 2 for bus, and 3 for car. Both groups listed above chose to travel by car, which is shown by `choice` equal to 1 in the third row for each group. Variables `train` and `bus` reproduce the information in `mode`. Variable `time` indicates how long a group thinks it will take them to travel using a given mode of transportation. `time` is an alternative-specific variable since it varies by the alternative and the group. The first group anticipates that their trip will take 406 minutes by train, 452 minutes by bus, and 180 minutes by car. In contrast, the second group believes that their trip will take 398 minutes by train, 452 minutes by bus, and 255 minutes by car. `invc`, which contains the in-vehicle cost of the trip, is also an alternative-specific variable. Later, we show how case-specific variables, such as the size of the travel party or the household income, can be used.

Some datasets with alternative-specific variables are already arranged with one row per alternative per case, as shown above. If this is your situation, you can immediately use `clogit`, `asmprobit`, and `rologit` without rearranging your data. And you will not need the information in the rest of this section. Other datasets, however, may have one row per case. Data in `travel2` could have been arranged with all data for a group contained in one row. For example, data on groups 1 and 2 (listed above) would look like this:

```
. use http://www.stata-press.com/data/lf2/travel2case.dta, clear
(Greene & Hensher 1997 data in one-row-per-case format)
. list id time1 time2 time3 invc1 invc2 invc3 choice in 1/2, nolabel
```

	id	time1	time2	time3	invc1	invc2	invc3	choice
1.	1	406	452	180	31	25	10	3
2.	2	398	452	255	31	25	11	3

In `travel2case.dta`, the variable `choice` indicates which mode of travel is chosen. The three rows of information in `travel2` for the three alternatives are replaced with three variables in one row: `time1` indicates time expected by train (which is mode 1), `time2` indicates time expected by bus (which is mode 2), and `time3` indicates time expected by car (which is mode 3). The variables `invc1`, `invc2`, and `invc3` likewise indicate the in-vehicle cost for train, bus, and car, respectively. You should convince yourself that `travel2case.dta` contains the same information as `travel2.dta`, just arranged differently.

Although it is straightforward to use data-management commands in Stata to convert data from a case-based arrangement to an alternative-based one (e.g., to move from `travel2case.dta` to `travel2.dta`), we have written the command `case2alt` to make the process easier and less error prone.

To convert the variables on time and in-vehicle costs from the case-specific format of `travel2case.dta` into the alternative-specific format of `travel2.dta`, we use the command:

```
. case2alt, alt(time invc) case(id) choice(choice) gen(choice2) altnum(mode)
choice indicated by: choice2
case identifier: id
case-specific interactions: train* bus* car*
alternative-specific variables: time invc
```

The option `alt` contains the names of variables with alternative-specific information. For example, `alt(time invc)` tells `case2alt` to convert the variables `time1`, `time2`, and `time3` into one, alternative-specific variable named `time` and to convert `invc1`, `invc2` and `invc3` into one variable named `invc`. `case(id)` means that the variable `id` indicates the case number. `choice(choice)` specifies that variable `choice` indicates the choice made by the group, and `gen(choice2)` specifies the name to be used for the choice variable in the new data. Our converted data look like this:

```
. list id choice2 train bus time invc in 1/6, nolabel sepby(id)
```

	id	choice2	train	bus	time	invc
1.	1	0	1	0	406	31
2.	1	0	0	1	452	25
3.	1	1	0	0	180	10
4.	2	0	1	0	398	31
5.	2	0	0	1	452	25
6.	2	1	0	0	255	11

The only difference from our listing of `travel2.dta` is that we have named the choice variable `choice2` (to avoid confusion with the case-specific variable `choice`).

7.1.1   Syntax for case2alt

The syntax for `case2alt` is

case2alt, {<u>c</u>hoice(*varname*) | <u>r</u>ank(*stubname*)} [<u>a</u>lt(*stubnames*)

   <u>casev</u>ars(*varlist*) case(*varname*) <u>g</u>enerate(*newvar*) <u>rep</u>lace

   altnum(*varname*) <u>non</u>ames]

Options

choice(*varname*) or rank(*stubname*) is required. For data without rankings, `choice()` specifies the variable that indicates the value of the chosen alternative. For data with ranked outcomes, `rank()` specifies the prefix of the variable that contains the information about rank (e.g., if *stubname* is `outcome`, then `outcome1` would contain the rank information for alternative 1, `outcome2` the rank information for alternative 2, and so on.) For both rank and nonrank data, `case2alt` assumes that the values of the outcome variable are positive integers. We recommend that the outcome values be consecutive integers beginning with 1.

alt(*stubnames*) contains the *stubnames* for alternative-specific variables. This requires that the variables *stubname#* exist for the value corresponding to each alternative. For example, if *stubnames* contains `time invc` and the three alternatives are identified by the values 1, 2, and 3, then variables `time1`, `time2`, `time3`, `invc1`, `invc2`, and `invc3` must contain information for the corresponding alternatives in the case-specific dataset.

`casevars`(*varlist*) contains the case-specific variables. These variables are characteristics of the case rather than characteristics of the alternatives. This list should not include the outcome variable or the ID variable.

`case`(*varname*) specifies the variable that identifies individual cases (e.g., the travel groups in the `travel2.dta` data). If *varname* exists, it will be used as the identifying variable. If *varname* does not exist, it will be created. If *varname* is unspecified, a new variable named `_id` will be created.

`generate`(*newvar*) and `replace` are used to name the variable that will indicate which alternative is chosen. The variable contains 1 for the chosen alternative and 0 for alternatives that are not chosen. The new variable will be named in one of three ways: (1) if *newvar* is specified, the new variable will be named *newvar*; (2) if `replace` is specified, it will be given the name specified in `choice()`; and (3) if neither `replace` nor `generate()` is specified, the new variable will be named `choice`.

`altnum`(*varname*) specifies the name of a new variable that will identify the alternatives. If not specified, the name `_altnum` will be used.

`nonames` indicates the case-specific interactions should be named `y#` instead of using the value labels of the outcome variable.

In the rest of this chapter, we consider models that required alternative-specific data. These include the conditional logit model fitted by `clogit`, the alternative-specific multinomial probit model fitted by `asmprobit`, and the rank-ordered logit model fitted by `rologit`.

7.2   The conditional logit model

The conditional logit model allows us to fit how the choice among nominal alternatives is affected by characteristics of the alternatives that vary across cases. In the CLM, the predicted probability of observing outcome m is

$$\Pr\left(y_i = m \mid \mathbf{z}_i\right) = \frac{\exp\left(\mathbf{z}_{im}\boldsymbol{\gamma}\right)}{\sum_{j=1}^{J} \exp\left(\mathbf{z}_{ij}\boldsymbol{\gamma}\right)} \quad \text{for } m = 1 \text{ to } J$$

where $\mathbf{z}_{im}$ contains values of the independent variables for alternative m for case i. In the example of the CLM that we use, there are three alternatives for transportation: train, bus, and car. Suppose that we have a single independent variable, z_{im}, that is the amount of time it would take respondent i to travel using mode of transportation m. Then γ is a parameter indicating the effect of time on the probability of choosing one mode over another. In general, for each variable z_k, there are J values of the variable for each case, but only the single parameter γ_k.

7.2.1   Fitting the conditional logit model

The basic syntax for clogit is

clogit *depvar* [*indepvars*] [*if*] [*in*] [*weight*], group(*varname*)

  [constraints(*constraints*) robust cluster(*varname*) level(#) or]

Options

group(*varname*) is required and specifies the variable that identifies the cases.

constraints(*clist*) specifies the linear constraints to be applied during estimation. The default is to perform unconstrained estimation. Constraints are defined with the constraint command.

robust indicates that robust variance estimates are to be used. When cluster() is specified, robust standard errors are automatically used. See chapter 3 for more details.

cluster(*varname*) specifies that the observations be independent across the groups specified by unique values of *varname* but not necessarily independent within the groups. See chapter 3 for more details.

level(#) specifies the level, as a percentage, for confidence intervals. The default is level(95).

or requests that odds ratios $\exp(\widehat{\gamma}_k)$ be reported instead of $\widehat{\gamma}_k$.

Example of the clogit model

For our travel example, the dependent variable is choice, a binary variable indicating which mode of transportation was chosen. The independent variables include the dummy variables train and bus that identify each alternative mode of transportation and the alternative-specific variables time and invc. To fit the model, we use the option group(id) to specify that the id variable identifies the cases:

```
. use http://www.stata-press.com/data/lf2/travel2.dta, clear
(Greene & Hensher 1997 data on travel mode choice)

. clogit choice train bus time invc, group(id) nolog
Conditional (fixed-effects) logistic regression   Number of obs   =        456
                                                  LR chi2(4)      =     172.06
                                                  Prob > chi2     =     0.0000
Log likelihood = -80.961135                       Pseudo R2       =     0.5152
```

choice	Coef.	Std. Err.	z	P>|z|	[95% Conf. Interval]	
train	2.671238	.4531611	5.89	0.000	1.783058	3.559417
bus	1.472335	.4007152	3.67	0.000	.6869474	2.257722
time	-.0191453	.0024509	-7.81	0.000	-.0239489	-.0143417
invc	-.0481658	.0119516	-4.03	0.000	-.0715905	-.0247411

7.2.2   Interpreting odds ratios from clogit

In the results that we just obtained, the coefficients for `time` and `invc` are negative. This indicates that the longer it takes to travel by a given mode, the less likely that alternative is to be chosen. Similarly, the more it costs, the less likely an alternative is to be chosen. More specific interpretations are possible by using `listcoef` to transform the estimates into odds ratios:

```
. listcoef, help
clogit (N=456): Factor Change in Odds

  Odds of: 1 vs 0
```

choice	b	z	P>|z|	e^b
train	2.67124	5.895	0.000	14.4579
bus	1.47233	3.674	0.000	4.3594
time	-0.01915	-7.812	0.000	0.9810
invc	-0.04817	-4.030	0.000	0.9530

```
    b = raw coefficient
    z = z-score for test of b=0
 P>|z| = p-value for z-test
   e^b = exp(b) = factor change in odds for unit increase in X
 SDofX = standard deviation of X
```

For the alternative-specific variables `time` and `invc`, each odds ratio is the multiplicative effect of a unit change in that independent variable on the odds of any given alternative mode of travel. For example,

> Increasing the time of travel by 1 minute for a given mode of transportation decreases the odds of using that mode by a factor of .98 (2%), holding the values for the other alternatives constant.

That is, if the time it takes to travel by car increases by 1 minute while the time it takes to travel by train and bus remain constant, the odds of traveling by car decrease by 2%. The odds ratios for the alternative-specific constants `bus` and `train` indicate the relative likelihood of selecting these alternatives versus traveling by car (the omitted category), assuming that cost and time are the same for all modes. For example,

> If cost and time were equal, groups of travelers would be 4.36 times more likely to travel by bus than by car and they would be 14.46 times more likely to travel by train than by car.

7.2.3   Interpreting probabilities from clogit

Using predict

You can use Stata's `predict` command to compute predicted probabilities for each alternative for each group, where the predicted probabilities sum to 1 for each group.

For example,

```
. predict prob
(option pc1 assumed; conditional probability for single outcome within group)
```

The message in parentheses indicates that, by default, conditional probabilities are being computed. To see what was done, we can list the variables in the model along with the predicted probabilities for the first two cases:

```
. list id prob choice train bus time invc in 1/6, nolabel sepby(id)
```

	id	prob	choice	train	bus	time	invc
1.	1	.0642477	0	1	0	406	31
2.	1	.0107205	0	0	1	452	25
3.	1	.9250318	1	0	0	180	10
4.	2	.2535607	0	1	0	398	31
5.	2	.0363011	0	0	1	452	25
6.	2	.7101382	1	0	0	255	11

For the first case (i.e., the first travel group), the predicted probability of traveling by car (the option chosen) is .925, although the predicted probability of traveling by train is only .064. The choice actually made corresponds to the cheapest and quickest mode of transportation. For the second group, the predicted probability of traveling by car (the option chosen) is .710, although the predicted probability of traveling by train is .254. Again the choice corresponds to choosing the cheapest and quickest mode of transportation. Let's consider another group who chose to travel by train:

```
. list id prob choice train bus time invc in 16/18, nolabel sepby(id)
```

	id	prob	choice	train	bus	time	invc
16.	6	.5493771	1	1	0	385	20
17.	6	.0643481	0	0	1	452	13
18.	6	.3862748	0	0	0	284	12

The probability of choosing train was estimated to be .549, although the probability of driving was .386. This group chose to travel by train, even though it was neither cheapest nor fastest.

Using asprvalue

For the CLM, the predictions depend on the values of all independent variables *for each alternative*. Therefore, the SPost commands prvalue, prtab, prcounts, and prgen do not work with clogit (similarly, they do not work with asmprobit or rologit). We have written the command asprvalue to compute predicted probabilities for models that allow alternative-specific variables. The syntax for asprvalue is asprvalue [*varlist*] [, x(*variables_and_values*) <u>rest</u>(*stat*) <u>base</u>(*name*)

   <u>cat</u>(*varnames*) <u>save</u> diff <u>br</u>ief]

Options

x(*variables_and_values*) assigns values to the independent variables for computing the
  predictions. For case-specific variables, only one value can be provided. For example, x(psize=2). For alternative-specific variables, one value can be provided, which
  will be assigned to all the alternatives, or J values can be provided to assign sequentially to the J alternatives. For example, with three alternatives we might specify
  x(time=200 300 400 invc=10). Variable time is given three values corresponding
  to different times for the three alternative modes of travel; invc was given one value,
  so all alternatives are assumed to have the same value. For clogit and rologit
  (discussed in section 7.5), values for alternative-specific variables should be ordered
  the same way as they appear in the model (or are specified using cat()), with the
  value to be assigned to the reference category given last. For asmprobit (discussed
  in section 7.3), the different alternatives are specified using one variable rather than
  a series of dummy variables, and values for alternative-specific variables should be
  ordered to correspond with the ascending values of the variable. The values assigned
  to each alternative are included in the output, and you should always confirm the
  assignment of the proper value to the proper alternative.

rest(*stat*) assigns values to the independent variables not specified in x(). The default is mean, which assigns to each variable its overall mean. asmean (standing
  for <u>a</u>lternative-<u>s</u>pecific <u>mean</u>) assigns the overall mean to case-specific variables but
  assigns the alternative-specific means to alternative-specific variables. For example,
  for time, mean would compute the mean over all cases and alternatives and assign
  this same value to each alternative, whereas asmean would compute the mean for
  time within each alternative (i.e., separate means for bus, train, and car) and assign
  each alternative the mean for that alternative.

base(*name*) provides the name of the reference category to be used to label the output.
  If base() is not specified, the name base is assumed. This option is not required
  for asmprobit.

cat(*varnames*) provides the names of the dummy variables for the alternatives in
  the model. The base category should not be included. If you fit a clogit or
  rologit model with no case-specific variables, cat() must be specified. Otherwise,
  asprvalue will automatically determine the names of the alternatives.

save and diff save predicted probabilities and compares current probabilities with
  saved probabilities, respectively.

brief provides minimal output.

When case-specific variables are included along with alternative-specific variables using `clogit` or `rologit`, the names of the case-specific variables must be named following the conventions used by `case2alt` (you can create the variables yourself so long as you name them properly). We will discuss this more when we use `asprvalue` with case-specific variables below.

To illustrate how to use `asprvalue`, suppose that we were interested in the predicted probabilities of traveling by train, bus, and car when `invc` and `time` have the same values for each alternative. We can compute these probabilities by typing

```
. asprvalue, cat(train bus) base(car)
clogit: Predictions for choice
              prob
train   .72955889
  bus   .21998006
  car   .05046107
alternative-specific variables
             train        bus        car
time   643.44079   643.44079   643.44079
invc   48.618421   48.618421   48.618421
```

The option `rest(mean)` is assumed by default. The results indicate that if the trip took the same amount of time and had the same in-vehicle costs for all modes of travel, groups would have a probability of .73 of choosing to travel by train and only a .05 probability of choosing to travel by car. With the CLM, this interpretation is correct regardless of the specific values of `time` and `invc` we choose, so long as the values are the same for each mode of travel. We can confirm this by using `x()`, where we specify that the average time is 300 minutes and the average cost is $20 for all alternatives:

```
. asprvalue, x(time=300 invc=20) cat(train bus) base(car)
clogit: Predictions for choice
              prob
train   .72955889
  bus   .21998006
  car   .05046107
alternative-specific variables
          train    bus    car
time        300    300    300
invc         20     20     20
```

Next we can compute the predicted probabilities of taking each mode of transportation for a trip that has the average time and average cost for each mode of travel—in other words, a trip that by car takes the average length of a trip by car, by bus takes the average length of a bus trip, and by train the average length of a train trip. We can obtain these averages by specifying the `rest(asmean)` option:

```
. asprvalue, rest(asmean) cat(train bus) base(car)
clogit: Predictions for choice
              prob
train    .43451011
  bus    .15197465
  car    .41351524

alternative-specific variables
             train        bus        car
time    643.44079   674.61842   578.26974
invc    48.618421   33.144737   20.092105
```

The probability of traveling by train is much lower than before and the probability of traveling by car much higher, because traveling by car requires less time and less in-vehicle costs on average.

We can use the `save` and `diff` options to compute how the predicted probability of selecting a given alternative changes when the value of an alternative-specific variable changes. Imagine that we are interested in how much the predicted probability of traveling by train decreases if we add 10 minutes to an average trip by train while holding the other times constant. First, we compute the predictions with all variables at their means, using `quietly` to suppress the output and specifying the average travel times for train, car, and bus with the `x()` option:

```
. quietly asprvalue, x(time=643.4 674.6 578.3) rest(asmean) ///
> cat(train bus) base(car) save
```

Next we increase the time for train from 643.4 to 653.4:

```
. asprvalue, x(time=653.4 674.6 578.3) rest(asmean) ///
> cat(train bus) base(car) diff
clogit: Predictions for choice
            Current       Saved        Diff
train     .38845369   .43478274   -.04632905
  bus     .16446434   .15200497    .01245937
  car     .44708198   .41321227    .03386971

alternative-specific variables
                    train        bus        car
Current:time        653.4      674.6      578.3
Current:invc    48.618421   33.144737   20.092105
  Saved:time        643.4      674.6      578.3
  Saved:invc    48.618421   33.144737   20.092105
   Dif:time           10          0          0
   Dif:invc            0          0          0
```

The results indicate that if the average time of a train trip increased by 10 minutes, the probability of choosing to travel by train would decrease by .046, holding the times and costs of other alternatives constant at their means. The probability of traveling by bus would increase by .012, and the probability of traveling by car would increase by .034. To ensure that you understand the meaning of the alternative-specific parameter estimated by `clogit`, note that the change in predicted probabilities for each these alternative corresponds to a factor change in the odds of .83. That is, $\{.3885/(1-.3885)\}/\{.4348/(1-.4348)\} = .826$ and $\exp(10 \times \hat{\gamma}_{time}) = \exp(-.191) = .826$. Next we make the same changes in the anticipated travel times for bus:

```
. asprvalue, x(time=643.4 684.6 578.3) rest(asmean) cat(train bus) base(car) diff
clogit: Predictions for choice
            Current       Saved       Diff
train     .44661152    .43478274    .01182878
  bus     .12893429    .15200497   -.02307068
  car     .42445418    .41321227    .01124191

alternative-specific variables

                   train        bus         car
Current:time       643.4      684.6       578.3
Current:invc   48.618421  33.144737   20.092105
Saved:time         643.4      674.6       578.3
Saved:invc     48.618421  33.144737   20.092105
Dif:time               0         10           0
Dif:invc               0          0           0
```

We could do the same thing for car with the command:

```
. asprvalue, x(time=643.4 674.6 588.3) rest(asmean) cat(train bus) base(car) diff
```

7.2.4   Fitting the multinomial logit model using clogit

Any multinomial logit model can be fitted using clogit by expanding the dataset (explained below) and respecifying the independent variables as a set of interactions. For the details on the mathematics behind this process, see Long (1997, 181). This is of more than academic interest for two reasons. First, it opens up the possibility of mixed models that include both case-specific and alternative-specific variables (see section 7.2.5). Second, it is possible to impose constraints on parameters in clogit that are not possible with mlogit (see mclgen and mclest by Hendrickx (2000) for further details). The first step is to convert the case-specific data to alternative-specific data (required by clogit). We do this with case2alt.

Setting up the data with case2alt

Even though the nomocc2.dta dataset has no alternative-specific variables, we still need to rearrange the data to use clogit. First, we load the data and look at the outcome variable occ:

sequences, Train (2003, 233) warns that his experiments using Halton sequences have occasionally produced anomalous results and that caution should be used.

7.3.3 Alternative-based data with uncorrelated errors

If we assume that the errors are uncorrelated, the ASMNPM model is the normal counterpart to the CLM fitted by clogit. By comparing the commands and results for clogit and asmprobit, we can start to understand how asmprobit specifies options. We present an abbreviated description of the asmprobit syntax and options below. Readers may wish to skip the description of options for now given that we discuss them in detail below.

asmprobit *depvar* [*indepvars*] [*if*] [*in*] [*weight*], case(*varname*)

    alternatives(*varname*) [casevars(*varlist*) constraints(*constraints*)

    correlation(*correlation*) stddev(*stddev*) structural noconstant

    basealternative(#|*lbl*|*str*) scalealternative(#|*lbl*|*str*)] intpoints(#)

Options

case(*varname*) specifies the variable that identifies each case. case() is required.

alternatives(*varname*) specifies the variable that identifies the alternatives for each case. The number of alternatives can vary with each case. alternatives() is required.

casevars(*varlist*) specify the case-specific variables that are constant for each case(). If there are a maximum of J alternatives, there will be $J - 1$ sets of coefficients associated with casevars().

constraints(*constraints*) specifies linear constraints on the coefficients.

correlation(*correlation*) specifies the correlation structure of the latent-variable errors. correlation() implies the structural covariance parameterization discussed in section 7.4. The types of covariance structures include correlation(unstructured), the most general, which results in $J(J - 3)/2 + 1$ correlation estimates; correlation(exchangeable), which provides for one correlation coefficient common to all latent variables, with the exception of the latent variable associated with the basealternative(); and correlation(independent), which assumes that all correlations are zero.

stddev(*stddev*) specifies the variance structure of the latent-variable errors. stddev() implies the structural covariance parameterization. stddev(heteroskedastic), the default, has $J - 2$ estimable parameters. The standard deviations of the latent-variable errors for the alternatives specified in basealternative() and scalealternative() are fixed to one. stddev(homoskedastic) constrains all the standard deviations to equal one.

```
. use http://www.stata-press.com/data/lf2/nomocc2, clear
(1982 General Social Survey)
. tab occ
```

Occupation	Freq.	Percent	Cum.
Menial	31	9.20	9.20
BlueCol	69	20.47	29.67
Craft	84	24.93	54.60
WhiteCol	41	12.17	66.77
Prof	112	33.23	100.00
Total	337	100.00	

When converting the data, we will use the labels of the outcome categories for naming new variables. If we use the current labels, the new variable names will get long. Accordingly, we start by changing the value labels. We use describe to find out what the current value label is, use label define to change that label, and then run case2alt:

```
. describe occ
```

variable name	storage type	display format	value label	variable label
occ	byte	%10.0g	occlbl	Occupation

```
. label define occlbl 1 "m" 2 "b" 3 "c" 4 "w" 5 "p", modify
. tab occ
```

Occupation	Freq.	Percent	Cum.
m	31	9.20	9.20
b	69	20.47	29.67
c	84	24.93	54.60
w	41	12.17	66.77
p	112	33.23	100.00
Total	337	100.00	

```
. case2alt, casevars(ed exper white) choice(occ) gen(choice)
(note: variable _id used since case() not specified)
(note: variable _altnum used since altnum() not specified)
choice indicated by: choice
case identifier: _id
case-specific interactions: m* b* c* w* p*
```

In the rearranged data, the alternatives are indicated by dummy variables whose names are generated from the new value labels for occ. For the first two cases (10 observations) let's look at the dummy variables created to indicate the occupation corresponding to a given row and the variable occ that shows the occupation a person had. (Of course, we need only four of the five dummy variables that uniquely identify the alternatives, but case2alt automatically creates all five.) The variable _id is created by case2alt to indicate which rows belong to the same case:

```
. list _id choice m b c w p in 1/10, sepby(_id)
```

	_id	choice	m	b	c	w	p
1.	1	1	1	0	0	0	0
2.	1	0	0	1	0	0	0
3.	1	0	0	0	1	0	0
4.	1	0	0	0	0	1	0
5.	1	0	0	0	0	0	1
6.	2	1	1	0	0	0	0
7.	2	0	0	1	0	0	0
8.	2	0	0	0	1	0	0
9.	2	0	0	0	0	1	0
10.	2	0	0	0	0	0	1

`case2alt` also creates interaction terms between each of the dummy variables and each of the independent variables. Consider the interactions with `ed` for the first two cases:

```
. list _id m b c w mXed bXed cXed wXed in 1/10, sepby(_id)
```

	_id	m	b	c	w	mXed	bXed	cXed	wXed
1.	1	1	0	0	0	11	0	0	0
2.	1	0	1	0	0	0	11	0	0
3.	1	0	0	1	0	0	0	11	0
4.	1	0	0	0	1	0	0	0	11
5.	1	0	0	0	0	0	0	0	0
6.	2	1	0	0	0	12	0	0	0
7.	2	0	1	0	0	0	12	0	0
8.	2	0	0	1	0	0	0	12	0
9.	2	0	0	0	1	0	0	0	12
10.	2	0	0	0	0	0	0	0	0

The interaction of `ed` with the indicator variable for a given alternative is equal to `ed` only for the row corresponding to that alternative; otherwise, the value is 0. The interactions with `exper` and `white` have the same structure, so we do not list them.

Fitting multinomial logit with clogit

The interactions and indicator variables for all but one of the alternatives are included as independent variables for `clogit`. The omitted alternative becomes the base category:

```
. clogit choice mXwhite mXed mXexper m bXwhite bXed bXexper b    ///
> cXwhite cXed cXexper c wXwhite wXed wXexper w,                 ///
> group(_id) nolog
Conditional (fixed-effects) logistic regression    Number of obs   =       1685
                                                   LR chi2(16)     =     231.16
                                                   Prob > chi2     =     0.0000
Log likelihood = -426.80048                        Pseudo R2       =     0.2131
```

choice	Coef.	Std. Err.	z	P>|z|	[95% Conf. Interval]	
mXwhite	-1.774306	.7550543	-2.35	0.019	-3.254186	-.2944273
mXed	-.7788519	.1146293	-6.79	0.000	-1.003521	-.5541826
mXexper	-.0356509	.018037	-1.98	0.048	-.0710028	-.000299
m	11.51833	1.849356	6.23	0.000	7.893659	15.143
bXwhite	-.5378027	.7996033	-0.67	0.501	-2.104996	1.029391
bXed	-.8782767	.1005446	-8.74	0.000	-1.07534	-.6812128
bXexper	-.0309296	.0144086	-2.15	0.032	-.05917	-.0026893
b	12.25956	1.668144	7.35	0.000	8.990061	15.52907
cXwhite	-1.301963	.647416	-2.01	0.044	-2.570875	-.0330509
cXed	-.6850365	.0892996	-7.67	0.000	-.8600605	-.5100126
cXexper	-.0079671	.0127055	-0.63	0.531	-.0328693	.0169351
c	10.42698	1.517943	6.87	0.000	7.451864	13.40209
wXwhite	-.2029212	.8693072	-0.23	0.815	-1.906732	1.50089
wXed	-.4256943	.0922192	-4.62	0.000	-.6064407	-.2449479
wXexper	-.001055	.0143582	-0.07	0.941	-.0291967	.0270866
w	5.279722	1.684006	3.14	0.002	1.979132	8.580313

The estimated parameters are identical to those produced by `mlogit` on page 231, so their interpretation is also the same.

7.2.5   Using clogit with case- and alternative-specific variables

The MNLM has case-specific variables, such as an individual's income. For case-specific variables, the value of a variable does not differ across outcomes, but we estimate $J - 1$ parameters for each case-specific variable. The CLM has alternative-specific variables, such as the time it takes to get to work with a given mode of transportation. For alternative-specific variables, values vary across alternatives, but we estimate one parameter for the effect of the variable. An interesting possibility is combining the two in one model, referred to as a *mixed model*. For example, in explaining the choice people make on mode of transportation, we might want to know if wealthier people are more likely to drive than take the bus. To create a mixed model, we combine the formulas for the MNLM and the CLM (see Long 1997, 178–182; Cameron and Trivedi 2005, 500–503):

$$\Pr(y_i = m \mid \mathbf{x}_i, \mathbf{z}_i) = \frac{\exp(\mathbf{z}_{im}\boldsymbol{\gamma} + \mathbf{x}_i\boldsymbol{\beta}_m)}{\sum_{j=1}^{J} \exp(\mathbf{z}_{ij}\boldsymbol{\gamma} + \mathbf{x}_i\boldsymbol{\beta}_j)} \qquad \text{where } \boldsymbol{\beta}_1 = 0 \qquad (7.1)$$

As in the CLM, $\mathbf{z}_{im}$ contains values of the alternative-specific variables for alternative m and case i, and $\boldsymbol{\gamma}$ contains the effects of the alternative-specific variables. As in the multinomial logit model, $\mathbf{x}_i$ contains case-specific independent variables for case i, and $\boldsymbol{\beta}_m$ contains coefficients for the effects on alternative m relative to the base alternative.

Example of a mixed model

This mixed model can be fitted using `clogit`. For the alternative-specific variables, the data are set up in the same way as for the conditional logit model above. For case-specific variables, interaction terms are created as illustrated in the last section. To apply this to the travel example, we add two case-specific variables to our model: `hinc` is household income, and `psize` is the number of people who will be traveling together. First, we create the interactions between the indicators of mode of travel and the case-specific variables.

```
. use http://www.stata-press.com/data/lf2/travel2, clear
(Greene & Hensher 1997 data on travel mode choice)
. gen busXhinc = bus*hinc
. gen trainXhinc = train*hinc
. gen busXpsize = bus*psize
. gen trainXpsize = train*psize
```

Then we fit the model with `clogit`:

```
. clogit choice busXhinc busXpsize bus trainXhinc trainXpsize train ///
> time invc, group(id) nolog
```

Conditional (fixed-effects) logistic regression			Number of obs	=	456
			LR chi2(8)	=	178.97
			Prob > chi2	=	0.0000
Log likelihood = -77.504846			Pseudo R2	=	0.5359

choice	Coef.	Std. Err.	z	P>|z|	[95% Conf. Interval]	
busXhinc	-.0080174	.0200322	-0.40	0.689	-.0472798	.031245
busXpsize	-.5141037	.4007015	-1.28	0.199	-1.299464	.2712569
bus	2.486465	.8803649	2.82	0.005	.7609815	4.211949
trainXhinc	-.0342841	.0158471	-2.16	0.031	-.0653438	-.0032243
trainXpsize	-.0038421	.3098075	-0.01	0.990	-.6110537	.6033695
train	3.499641	.7579665	4.62	0.000	2.014054	4.985228
time	-.0185035	.0025035	-7.39	0.000	-.0234103	-.0135966
invc	-.0402791	.0134851	-2.99	0.003	-.0667095	-.0138488

Interpretation of odds ratios using listcoef

The model we just fitted includes both alternative- and case-specific variables. Regardless of whether alternative-specific variables are included in the model, the odds ratios for case-specific variables can be interpreted the same as in the MNLM. Odds ratios for contrasts with the reference category can be obtained using `listcoef`, restricting our list of variables to only those that are case specific:

```
. listcoef busXhinc trainXhinc busXpsize trainXpsize, percent help
clogit (N=456): Percentage Change in Odds
  Odds of: 1 vs 0
```

choice	b	z	P>|z|	%
busXhinc	-0.00802	-0.400	0.689	-0.8
trainXhinc	-0.03428	-2.163	0.031	-3.4
busXpsize	-0.51410	-1.283	0.199	-40.2
trainXpsize	-0.00384	-0.012	0.990	-0.4

```
        b = raw coefficient
        z = z-score for test of b=0
    P>|z| = p-value for z-test
        % = percent change in odds for unit increase in X
    SDofX = standard deviation of X
```

Keeping in mind that car is the reference category for these coefficients, we can make the following interpretations:

> A unit increase in income decreases the odds of traveling by train versus traveling by car by 3.4%, holding all else constant.

Similarly:

> Each added member of the traveling party decreases the odds of traveling by bus versus traveling by car by 40.2%, holding all else constant.

Unfortunately, `listcoef` after `clogit` does not compute results for all contrasts like it does after `mlogit`. To get a comparison between, say, taking the bus and taking a train, you must refit the model with either `bus` or `train` as the reference category and include the variables `car`, `carXhinc`, and `carXpsize`.

The coefficients for alternative-specific variables can also be listed with `listcoef` (we obtain factor change coefficients rather than the percent change coefficients since we did not use the `percent` option):

```
. listcoef time invc bus train, help
clogit (N=456): Factor Change in Odds
  Odds of: 1 vs 0
```

choice	b	z	P>|z|	e^b
time	-0.01850	-7.391	0.000	0.9817
invc	-0.04028	-2.987	0.003	0.9605
bus	2.48647	2.824	0.005	12.0187
train	3.49964	4.617	0.000	33.1036

```
        b = raw coefficient
        z = z-score for test of b=0
    P>|z| = p-value for z-test
      e^b = exp(b) = factor change in odds for unit increase in X
    SDofX = standard deviation of X
```

We can interpret the coefficients for the alternative-specific variables `time` and `invc` just as we did in the model above where we did not include case-specific variables. As before, increasing the time of travel by 1 minute for a given mode of transportation decreases the odds of using that mode of travel by a factor of .98 (2%), holding the values for the other alternatives constant.

Interpretation of predicted probabilities using asprvalue

As we mentioned earlier, if you use `asprvalue` with a `clogit` (or `rologit`) model that has case-specific variables, these interactions must be named following the conventions used by `case2alt`. The interactions must be named *altname*X*csvname*, where *altname* is the name of an alternative (and the name of the corresponding dummy variable used to indicate the row for that alternative) and *csvname* is the name of the case-specific variable. For example, all the interactions of case-specific variables with the alternative `bus` must begin `busX` and all the interactions with case-specific variable `hinc` must end with `Xhinc`. As we discuss in the next section, any alternative by alternative-specific variable interactions must be similarly named (e.g., `busXtime`). A capital X cannot be used in any names of variables except to denote these interactions. A benefit of this requirement is that if there are case-specific variables in the model, you do not need to use the `cat()` option to specify the dummy variables indicating the alternatives in the model.

If we wanted to compute the predicted probabilities holding all case-specific and alternative-specific variables to their means, we type

```
. asprvalue, base(car)
clogit: Predictions for choice
              prob
   bus    .23370957
 train    .70268089
   car    .06360951
case-specific variables
           hinc       psize
x=   31.809211   1.8092105
alternative-specific variables
            bus       train         car
time   643.44079   643.44079   643.44079
invc   48.618421   48.618421   48.618421
```

Recall that the average cost and time for traveling by car is less than that of travel by train. Accordingly, you might be more interested in the predicted probabilities holding the alternatives equal to their respective means.

```
. asprvalue, rest(asmean) base(car)
clogit: Predictions for choice
              prob
   bus    .1513233
 train   .43435812
   car   .41431856

case-specific variables
            hinc       psize
x=    31.809211   1.8092105

alternative-specific variables
             bus        train          car
time   674.61842   643.44079    578.26974
invc   33.144737   48.618421    20.092105
```

The mean values of `time` and `invc` for each alternative are presented in the last two rows of the output above. For the predicted probabilities, we see that for a person of average income and in an average-sized party, and a trip that is average on all characteristics for all three alternatives, the probability of traveling by train is only slightly higher than the probability of traveling by car, whereas the probability of traveling by bus is less.

Often, of course, what we are most interested in is the change in the predicted probabilities corresponding to a discrete change in the independent variables. We can compute these using the `save` and `diff` options. For example, if we were interested in how the probabilities change as party size increases from one person to two:

```
. quietly asprvalue, x(psize=1) rest(asmean) base(car) save
. asprvalue, x(psize=2) rest(asmean) base(car) diff
clogit: Predictions for choice
              Current        Saved         Diff
   bus     .13919763    .21251462   -.07331699
 train     .44040644    .40365174     .0367547
   car     .42039591    .38383365    .03656226

case-specific variables
                 hinc       psize
Current    31.809211           2
  Saved    31.809211           1
   Diff            0           1

alternative-specific variables
                        bus        train          car
Current:time     674.61842   643.44079    578.26974
Current:invc     33.144737   48.618421    20.092105
  Saved:time     674.61842   643.44079    578.26974
  Saved:invc     33.144737   48.618421    20.092105
    Dif:time             0           0            0
    Dif:invc             0           0            0
```

For a person of average income and a trip with the average cost and time for each mode of transportation, increasing party size from one to two increases the probability of traveling by train and car each by .037, whereas the predicted probability of traveling by bus falls by a corresponding .073.

As before, we can look at how the predicted probabilities change as the alternative-specific variables change. For example, we can imagine an initiative to increase vastly the efficiency of bus travel so that the length of time for the average bus trip in the sample decreased by 20% (from 675 minutes to 540 minutes). We calculate the estimated effect that would have on the probability of traveling by bus as follows:

```
. quietly asprvalue, x(time=675 643 578) rest(asmean) base(car) save
. asprvalue, x(time=540 643 578) rest(asmean) base(car) diff
clogit: Predictions for choice
             Current       Saved         Diff
     bus     .6813612    .14957573    .53178547
   train   .16333339    .43592516   -.27259177
     car   .15530542     .4144991   -.25919369
case-specific variables
                  hinc       psize
Current   31.809211   1.8092105
  Saved   31.809211   1.8092105
   Diff           0           0
alternative-specific variables
                       bus       train         car
Current:time           540         643         578
Current:invc   33.144737   48.618421   20.092105
  Saved:time           675         643         578
  Saved:invc   33.144737   48.618421   20.092105
    Dif:time          -135           0           0
    Dif:invc             0           0           0
```

We can see the predicted probability of traveling by bus would increase dramatically, from .15 to .68. This is equivalent to a little more than a 12-fold increase in odds (from 1/6 to a little over 2), which is what we can calculate from the coefficient for `time`: $\exp(.01850 \times 135) = 12.15$.

Allowing the effects of alternative-specific variables to vary over the alternatives

When working with alternative-specific variables, it is easy to imagine scenarios in which the effects of those variables will vary over the alternatives. Here we might imagine that adding 1 minute of travel time by bus to a prospective trip does not have the same effect on the odds of selection as does adding a minute of travel time by car. In the CLM, we can model this simply by adding terms for the interaction of `time` with the dummy variables for the `bus` and `train` alternatives. In our example,

```
. gen busXtime = bus*time

. gen trainXtime = train*time

. clogit choice busXhinc busXpsize busXtime bus trainXhinc trainXpsize ///
> trainXtime train time invc, group(id) nolog
Conditional (fixed-effects) logistic regression    Number of obs   =        456
                                                   LR chi2(10)     =     181.05
                                                   Prob > chi2     =     0.0000
Log likelihood = -76.463933                        Pseudo R2       =     0.5421
```

choice	Coef.	Std. Err.	z	P>|z|	[95% Conf. Interval]	
busXhinc	-.0055958	.0197897	-0.28	0.777	-.0443829	.0331914
busXpsize	-.4860495	.4153594	-1.17	0.242	-1.300139	.32804
busXtime	-.0015935	.0014169	-1.12	0.261	-.0043706	.0011835
bus	3.28312	1.179105	2.78	0.005	.9721155	5.594124
trainXhinc	-.03692	.0163369	-2.26	0.024	-.0689397	-.0049004
trainXpsize	.0364256	.3203728	0.11	0.909	-.5914936	.6643447
trainXtime	-.001628	.0012597	-1.29	0.196	-.004097	.0008409
train	4.304662	1.011084	4.26	0.000	2.322974	6.286349
time	-.0185236	.0025318	-7.32	0.000	-.0234858	-.0135613
invc	-.0355277	.0145275	-2.45	0.014	-.0640011	-.0070543

The nonsignificant results for the busXtime and trainXtime interactions indicate that we have no evidence that the effect of time varies either between bus and car or between train and car.

Note We have considered the conditional logit model only in the context of choices among an unordered set of alternatives. The possible uses of clogit are much broader. The *Stata Base Reference Manual* entry for clogit contains more examples and references.

7.3 Alternative-specific multinomial probit[1]

In chapter 6, we motivated the multinomial probit model for case-specific data in terms of a person choosing among alternatives to maximize her utility. Her characteristics, such as age or education, affect the utility provided by each alternative. Here we extend the model to incorporate alternative-specific data. The inclusion of such data allows us to relax the assumption that the errors are uncorrelated, which eliminates the IIA restriction of clogit for alternative-specific data. Until recently, difficulties computing the multidimensional normal integrals needed to fit the multinomial probit model with correlated errors made the model impractical. But work begun by McFadden (1989) has

---

1. David Drukker at StataCorp was extremely helpful as we wrote this section and explored issues of identification. The examples in this section assume that you are using Stata 9.1 (a free update to Stata 9) and asmprobit version 2.0 or later. To determine the version of asmprobit that you are using, enter the command which asmprobit.

largely solved the computational problems by fitting the model using what is known as maximum simulated likelihood (see Train 2003, part II; Cameron and Trivedi 2005, 393–398). In Stata, this model is referred to as the *alternative-specific multinomial probit model* (ASMNPM) and can be fitted using `asmprobit`. The term "alternative-specific" in the name alludes to the fact that alternative-specific variables are necessary to identify the error correlations. If only case-based variables are available, the correlations are not identified. Accordingly, `asmprobit` solves the IIA problem for conditional logit models with alternative-specific data, but not for multinomial logit models with only case-specific data.

7.3.1 The model

Assume that $\mathbf{x}_{im}$ contains alternative-specific information about alternative m for case i and that ε_{im} is a random, normally distributed error. Let u_{im} be the utility that case i receives from alternative m where

$$u_{im} = \mathbf{x}_{im}\boldsymbol{\beta} + \varepsilon_{im} \qquad \text{for } m = 1, J$$

A person chooses alternative j when $u_{ij} > u_{im}$ for all $m \neq j$. Accordingly, with J choices, the probability of choice m is

$$\Pr(y_i = m) = \Pr(u_{im} > u_{ij} \text{ for all } j \neq m) \qquad (7.2)$$

Since the errors are normally distributed, we can allow them to be correlated across the equations for different alternatives. Suppose that there are four alternatives. The covariance matrix for the εs would be

$$\boldsymbol{\Sigma}_\varepsilon = \begin{bmatrix} \sigma_{11} & & & \\ \sigma_{21} & \sigma_{22} & & \\ \sigma_{31} & \sigma_{32} & \sigma_{33} & \\ \sigma_{41} & \sigma_{42} & \sigma_{43} & \sigma_{44} \end{bmatrix}$$

If this matrix is constrained so that $\boldsymbol{\Sigma}_\varepsilon = \mathbf{I}$ (that is, the errors have a unit variance and are uncorrelated), we have the normal error counterpart to the conditional logit model where the errors are assumed to have an extreme value distribution. And, just like the CLM, the model has the IIA property.

While allowing the errors to be correlated relaxes the IIA condition, Train (2003, 104–110) shows that the parameters in $\boldsymbol{\Sigma}_\varepsilon$ are not identified unless constraints are imposed. These constraints reflect that neither adding a constant to the utility for each alternative nor dividing each utility by a constant will affect the choice that is made according to (7.2). Since $u_{im} > u_{ij}$ implies that $u_{im} + \delta > u_{ij} + \delta$, the choice of alternative m over alternative j is not affected by the base level of utility. Similarly, since $u_{im} > u_{ij}$ implies that $u_{im}\tau > u_{ij}\tau$ (for all $\tau > 0$), the choice is also unaffected by the scale used to measure utility. Accordingly, we must normalize the model to eliminate the effects of the base level and scale of utility.

To remove the effect of level, we use the difference between each alternative's utility and the utility of the base alternative. Suppose that we select the first alternative as the base. The new equations specify how much utility an alternative provides beyond that provided by the first alternative:

$$u_{i1} - u_{i1} = 0$$
$$u_{i2} - u_{i1} = (\mathbf{x}_{i2} - \mathbf{x}_{i1})\,\boldsymbol{\beta} + (\varepsilon_{i2} - \varepsilon_{i1})$$
$$u_{i3} - u_{i1} = (\mathbf{x}_{i3} - \mathbf{x}_{i1})\,\boldsymbol{\beta} + (\varepsilon_{i3} - \varepsilon_{i1})$$
$$u_{i4} - u_{i1} = (\mathbf{x}_{i4} - \mathbf{x}_{i1})\,\boldsymbol{\beta} + (\varepsilon_{i4} - \varepsilon_{i1})$$

Defining $\varepsilon_{im}^* \equiv \varepsilon_{im} - \varepsilon_{i1}$, $u_{im}^* \equiv u_{im} - u_{i1}$, and $\mathbf{x}_{im}^* \equiv \mathbf{x}_{im} - \mathbf{x}_{i1}$, leads to

$$u_{i1}^* = 0$$
$$u_{i2}^* = \mathbf{x}_{i2}^*\boldsymbol{\beta} + \varepsilon_{i2}^*$$
$$u_{i3}^* = \mathbf{x}_{i3}^*\boldsymbol{\beta} + \varepsilon_{i3}^*$$
$$u_{i4}^* = \mathbf{x}_{i4}^*\boldsymbol{\beta} + \varepsilon_{i4}^*$$

By subtracting u_{i1} from each equation, we have reduced the number of errors by one since $\varepsilon_{i1}^* = \varepsilon_{i1} - \varepsilon_{i1} = 0$. The covariance matrix for the differenced errors is

$$\boldsymbol{\Sigma}_\varepsilon^* = \begin{bmatrix} \sigma_{22}^* & & \\ \sigma_{32}^* & \sigma_{33}^* & \\ \sigma_{42}^* & \sigma_{43}^* & \sigma_{44}^* \end{bmatrix}$$

To set the scale, we fix the value of one of the variances σ_{mm}^*. Which variance we fix does not matter, so we arbitrarily pick σ_{22}^*. Whereas some treatments of the ASMNPM fix the variance to 1, `asmprobit` fixes the value to 2 (see our discussion on page 274 regarding `mprobit`), which leads to

$$\boldsymbol{\Sigma}_\varepsilon^* = \begin{bmatrix} 2 & & \\ \sigma_{32}^* & \sigma_{33}^* & \\ \sigma_{42}^* & \sigma_{43}^* & \sigma_{44}^* \end{bmatrix}$$

Fixing a base alternative and the variance of one of the differenced errors exactly identifies the regression coefficients and the covariance matrix for the differenced errors. By default, `asmprobit` imposes these restrictions and estimates the parameters of $\boldsymbol{\Sigma}_\varepsilon^*$. We discuss this further below when we provide an example.

7.3.2 Informal explanation of estimation by simulation

The probability of choosing alternative m for a given set of values for the independent variables requires us to compute

$$\Pr(y_i = m) = \Pr(u_{im} > u_{ij} \text{ for all } j \neq m)$$
$$= \Pr(\mathbf{x}_{im}\boldsymbol{\beta} + \varepsilon_{im} > \mathbf{x}_{ij}\boldsymbol{\beta} + \varepsilon_{ij} \text{ for all } j \neq m)$$

which in turn requires integration of the multivariate normal distribution for the εs. Or, after adding the identification constraints, we need to compute

$$\Pr(y_i = m) = \Pr\left(u_{im}^* > u_{ij}^* \text{ for all } j \neq m\right)$$
$$= \Pr\left(\mathbf{x}_{im}^* \boldsymbol{\beta} + \varepsilon_{im}^* > \mathbf{x}_{ij}^* \boldsymbol{\beta} + \varepsilon_{ij}^* \text{ for all } j \neq m\right)$$

The integral must be approximated, because it has no closed form. Following Train (2003), `asmprobit` uses simulation to approximate the integral.[2] Since this has practical implications, it is worth taking a minute to explain how this works, albeit in a greatly simplified context.

Suppose that z is normally distributed with a mean of 0 and a variance of 1 and that we want to know $\Pr(z > 1.96)$. That is, what percent of the standardized, normal density is above 1.96? For multinomial probit, we want to compute the more complex quantity $\Pr(u_{im} > u_{ij} \text{ for all } j \neq m)$, but the principles are the same. Most readers will recognize 1.96 as the critical value for a two-tailed test at the .05 level, so the answer is .025 (to be more precise, .02499790). If we did not know the answer, we could use Stata's pseudorandom number generator `uniform()` to estimate the probability. Let's start by taking 100 random draws from a normal distribution. First, we set the seed so that we can get the same results if we run the program again. Then we set the number of observations to 100, and finally we draw the random numbers:

```
. clear
. set seed 11020
. set obs 100
obs was 0, now 100
. gen r_norm = invnormal(uniform())
```

To determine the percent of the values in `r_norm` that are larger than 1.96, we create a dummy equal to 1 if `r_norm` is greater than 1.96, else 0. Then the mean computed by `sum` tells us the proportion of random draws that are above 1.96:

```
. gen rgt196 = r_norm>1.96
. summarize rgt196
```

Variable	Obs	Mean	Std. Dev.	Min	Max
rgt196	100	.01	.1	0	1

With 100 random numbers, the estimate is .01, which is not very accurate. So this time let's draw 1,000 random numbers:

---

2. Estimation that maximizes the likelihood when using simulation to compute the integrals is known as maximum simulated likelihood (MSL). See Train (2003) or Cameron and Trivedi (2005, 393–398) for more information.

```
. clear
. set obs 1000
obs was 0, now 1000
. gen r_norm = invnormal(uniform())
. gen rgt196 = r_norm>1.96
. summarize rgt196
```

Variable	Obs	Mean	Std. Dev.	Min	Max
rgt196	1000	.028	.1650553	0	1

The estimate .028 is much closer. And with 1,000,000 numbers, the estimate is very accurate:

```
. clear
. set obs 1000000
obs was 0, now 1000000
. gen r_norm = invnormal(uniform())
. gen rgt196 = r_norm>1.96
. summarize rgt196
```

Variable	Obs	Mean	Std. Dev.	Min	Max
rgt196	1000000	.025162	.156617	0	1

This is essentially how simulation methods compute integrals, although the specific details get complicated.

One of the more important complications is that `asmprobit` uses Hammersley or Halton sequences instead of uniform variates. Above we used `uniform()` to generate pseudorandom numbers that were uniformly distributed. These were transformed with `invnormal()` to create normal variates. Such pseudorandom numbers are not uniform across their range of values. For example, if we create two variables with uniform random numbers and plot them, we get something like this:

(Continued on next page)

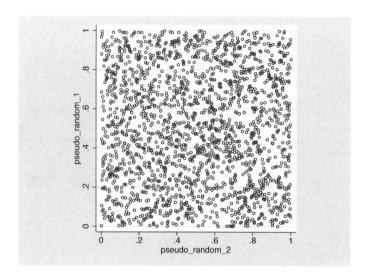

Some areas of the graph have many points that overlap, whereas other areas have empty spaces. Halton and Hammersley sequences use number theory to create a sequence of numbers that are more uniform. They are not random numbers, although they are sometimes referred to as quasirandom numbers. If we plot two Halton sequences, we see a much more uniform distribution of points:

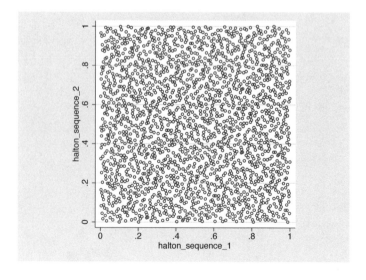

Using Halton or Hammersley sequences when computing integrals has been shown to require an order of magnitude fewer simulation points to obtain the same accuracy as uniform-based methods (Sandor and Andra 2004; Train 2003, 224–238), thus speeding convergence by a factor of roughly 10. Despite the speed advantages from using such sequences, Train (2003, 233) warns that his experiments using Halton sequences have occasionally produced anomalous results and that caution should be used.

7.3.3 Alternative-based data with uncorrelated errors

If we assume that the errors are uncorrelated, the ASMNPM model is the normal counterpart to the CLM fitted by `clogit`. By comparing the commands and results for `clogit` and `asmprobit`, we can start to understand how `asmprobit` specifies options. We present an abbreviated description of the `asmprobit` syntax and options below. Readers may wish to skip the description of options for now given that we discuss them in detail below.

`asmprobit` *depvar* [*indepvars*] [*if*] [*in*] [*weight*], `case`(*varname*)
    `alternatives`(*varname*) [`casevars`(*varlist*) `constraints`(*constraints*)
    `correlation`(*correlation*) `stddev`(*stddev*) `structural` `noconstant`
    `basealternative`(*# | lbl | str*) `scalealternative`(*# | lbl | str*)] `intpoints`(*#*)

Options

case(*varname*) specifies the variable that identifies each case. `case()` is required.

alternatives(*varname*) specifies the variable that identifies the alternatives for each case. The number of alternatives can vary with each case. `alternatives()` is required.

casevars(*varlist*) specify the case-specific variables that are constant for each `case()`. If there are a maximum of J alternatives, there will be $J-1$ sets of coefficients associated with `casevars()`.

constraints(*constraints*) specifies linear constraints on the coefficients.

correlation(*correlation*) specifies the correlation structure of the latent-variable errors. `correlation()` implies the structural covariance parameterization discussed in section 7.4. The types of covariance structures include `correlation(unstructured)`, the most general, which results in $J(J-3)/2+1$ correlation estimates; `correlation(exchangeable)`, which provides for one correlation coefficient common to all latent variables, with the exception of the latent variable associated with the `basealternative()`; and `correlation(independent)`, which assumes that all correlations are zero.

stddev(*stddev*) specifies the variance structure of the latent-variable errors. `stddev()` implies the structural covariance parameterization. `stddev(heteroskedastic)`, the default, has $J-2$ estimable parameters. The standard deviations of the latent-variable errors for the alternatives specified in `basealternative()` and `scalealternative()` are fixed to one. `stddev(homoskedastic)` constrains all the standard deviations to equal one.

structural requests the $J \times J$ structural covariance parameterization instead of the default $J - 1 \times J - 1$ differenced covariance parameterization (the covariance of the latent errors differenced with that of the base alternative). Section 7.4 discusses this option. The differenced covariance parameterization will achieve the same maximum simulated likelihood regardless of the choice of base() and scale(). On the other hand, text below describes how the structural covariance parameterization imposes additional normalizations that may bound the model away from its maximum likelihood and thus prevent convergence with some datasets or choices of base() and scale().

noconstant suppresses the $J - 1$ alternative-specific constant terms.

basealternative(#| lbl| str) specifies the alternative used to normalize the latent-variable location. Note that the base alternative may be specified as a number, label, or string. The standard deviation for the latent-variable error associated with the base alternative is fixed to one and its correlations with all other latent-variables' errors are set to zero. The default is the first alternative. basealternative() cannot be equal to scalealternative().

scalealternative(#| lbl| str) specifies the alternative used to normalize the latent-variable scale. Note that the scale alternative may be specified as a number, label, or string. The default is to use the second alternative. scalealternative() cannot be equal to basealternative().

intpoints(#) specifies the number of points to use in the pseudo- or quasi-Monte Carlo integration. Larger values of # provide better approximations to the log likelihood, but at the cost of added computational time.

Examples

Using the travel example, we fit the CLM and compute predicted probabilities:

```
. use http://www.stata-press.com/data/lf2/travel2, clear
(Greene & Hensher 1997 data on travel mode choice)
. clogit choice time invc train bus, group(id) nolog
Conditional (fixed-effects) logistic regression      Number of obs   =        456
                                                     LR chi2(4)      =     172.06
                                                     Prob > chi2     =     0.0000
Log likelihood = -80.961135                          Pseudo R2       =     0.5152
```

choice	Coef.	Std. Err.	z	P>|z|	[95% Conf. Interval]	
time	-.0191453	.0024509	-7.81	0.000	-.0239489	-.0143417
invc	-.0481658	.0119516	-4.03	0.000	-.0715905	-.0247411
train	2.671238	.4531611	5.89	0.000	1.783058	3.559417
bus	1.472335	.4007152	3.67	0.000	.6869474	2.257722

```
. predict clmpr
(option pc1 assumed; conditional probability for single outcome within group)
```

To fit the corresponding ASMNPM, we use the command:

```
. asmprobit choice time invc, case(id) alternatives(mode) basealternative(Car) ///
> correlation(independent) stddev(homoskedastic) nolog
```

The outcome variable and the two alternative-specific variables are specified as they were with `clogit`. Other options are specified differently.

1. The variable indicating which records belong to a given case are now indicated with the `case()` option, rather than with the `group()` option used for `clogit`.

2. The variable indicating the alternative corresponding to a given observation within a case is indicated with `alternatives()`.

3. For `clogit`, the base alternative was implicitly specified by not listing an indicator variable for that alternative. So, with `clogit` we include `train` and `bus`, implying that `car` is the base alternative. With `asmprobit`, the base alternative is specified with the option `basealternative()`.

4. The second line of the command specifies the correlation structure. The option `correlation()` indicates that the errors are uncorrelated, and `stddev()` restricts all the variances to be equal.

Here are the results from `asmprobit`:

```
Alternative-specific multinomial probit      Number of obs     =        456
Case variable: id                            Number of cases   =        152

Alternative variable: mode                   Alts per case: min =         3
                                                            avg =       3.0
                                                            max =         3
Integration sequence:       Hammersley
Integration points:              150         Wald chi2(2)      =      86.90
Log simulated-likelihood = -92.058576        Prob > chi2       =     0.0000
```

choice	Coef.	Std. Err.	z	P>|z|	[95% Conf.	Interval]
mode						
time	-.0095924	.0011094	-8.65	0.000	-.0117668	-.0074179
invc	-.0451214	.0084322	-5.35	0.000	-.0616482	-.0285947
Train						
_cons	1.818842	.3074054	5.92	0.000	1.216339	2.421346
Bus						
_cons	.711724	.2509273	2.84	0.005	.2199155	1.203532
Car	(base alternative)					

```
(mode=Car is the alternative normalizing location)
(mode=Train is the alternative normalizing scale)
```

When we list the covariance matrix for the errors, we see that the errors are uncorrelated with unit variance:

```
. estat covariance
```

	Train	Bus	Car
Train	1		
Bus	0	1	
Car	0	0	1

We can also compute predicted probabilities for each observation:

```
. predict mnppr
(option pr assumed; Pr(mode))
```

The results from `asmprobit` and `clogit` are similar in the relative magnitudes of their estimates and the levels of statistical significance. If we compare the predicted probabilities from the two models, we see that the predictions are highly correlated:

```
. pwcorr clmpr mnppr
```

	clmpr	mnppr
clmpr	1.0000	
mnppr	0.9689	1.0000

Although the correlation is large, it is smaller than the correlations we found for binary logit and probit in chapter 4. There, the predicted probabilities had a correlation of .9998. When comparing `mprobit` and `mlogit`, the correlations were above .99. Further, the values of the log likelihoods from `asmprobit` and `clogit` show a surprisingly large difference: -80.961 compared with -92.059.

We explored this difference extensively. We could not find a problem in how we set up the commands or in the data itself (which has become the classic example for this type of model). We also experimented with options that control the estimation process for `asmprobit`. We then created simulated data so that we could fit models with `clogit` and `asmprobit`, knowing that we had a correctly specified model. In such cases, the log likelihoods were extremely close. We have concluded that when one or both of the models are misspecified, they can converge to noticeably different log likelihoods. When this occurs, we recommend that great care should be taken in applying either model to the data (we are aware of how unfortunately vague this advice is). The current example is the only time that we have seen a large difference between results from a logit-type and the corresponding probit-type model. The travel data might be unusual or ASMNPM could be sensitive to errors in specification. Although we were able to artificially construct datasets that produced similar differences in log likelihoods with `logit` and `probit`, we have never seen this problem with binary models when using real data.

7.3.4 Alternative-based data with correlated errors

The main reason to use `asmprobit` rather than `clogit` is to allow correlated errors, which we illustrate using the travel example. First, we load the data and list the first two cases:

```
. use http://www.stata-press.com/data/lf2/travel2.dta, clear
(Greene & Hensher 1997 data on travel mode choice)
. list id mode choice train bus time invc in 1/6, nolabel sepby(id)
```

	id	mode	choice	train	bus	time	invc
1.	1	1	0	1	0	406	31
2.	1	2	0	0	1	452	25
3.	1	3	1	0	0	180	10
4.	2	1	0	1	0	398	31
5.	2	2	0	0	1	452	25
6.	2	3	1	0	0	255	11

As before, each case is indicated by id and has three observations corresponding to the three alternatives. mode indicates the alternative for a given observation, choice indicates which alternative was chosen, and time and invc are the alternative-specific variables. We use these variables in specifying the ASMNPM with the default correlation structure:

```
. asmprobit choice time invc, case(id) alternatives(mode)
Iteration 0:   log simulated-likelihood = -106.44614
  (output omitted)
Iteration 16:  log simulated-likelihood = -81.363746
```

Alternative-specific multinomial probit	Number of obs	=	456
Case variable: id	Number of cases	=	152
Alternative variable: mode	Alts per case: min	=	3
	avg	=	3.0
	max	=	3
Integration sequence: Hammersley			
Integration points: 150	Wald chi2(2)	=	19.71
Log simulated-likelihood = -81.363746	Prob > chi2	=	0.0001

choice	Coef.	Std. Err.	z	P>|z|	[95% Conf. Interval]	
mode						
time	-.0268044	.0060833	-4.41	0.000	-.0387275	-.0148813
invc	-.0817238	.0225907	-3.62	0.000	-.1260007	-.0374468
Train	(base alternative)					
Bus						
_cons	-1.776282	.5360452	-3.31	0.001	-2.826911	-.7256528
Car						
_cons	-4.503204	1.191181	-3.78	0.000	-6.837877	-2.168532
/lnl2_2	1.388244	.2833897	4.90	0.000	.8328104	1.943678
/l2_1	-.9430005	1.339645	-0.70	0.481	-3.568657	1.682656

```
(mode=Train is the alternative normalizing location)
(mode=Bus is the alternative normalizing scale)
```

Estimation required more iterations than `clogit`. Indeed, with large samples or many variables, `asmprobit` could take hours to converge. The `Integration sequence: Hammersley` line indicates how the sequence was generated for the Monte Carlo estimation. You need to consider this further only if you have trouble getting your model to converge. `Integration points: 150` indicates how many draws were used to compute the integrals used for estimation. If you have trouble getting your model to converge, you can try increasing the number of points by adding the option `intpoints(#)` with a value larger than the default of 150.

Before discussing the parameter estimates for `time` and `invc`, let's look at the estimates of Σ_ε^*:

```
. estat covariance
```

	Bus	Car
Bus	2	
Car	-1.333604	16.95176

Note: covariances are for alternatives differenced with Train

The note indicates that this is the covariance matrix for the differences in errors relative to alternative `Train`: $(\varepsilon_{i\text{Bus}} - \varepsilon_{i\text{Train}})$ and $(\varepsilon_{i\text{Car}} - \varepsilon_{i\text{Train}})$. The variance for $(\varepsilon_{i\text{Bus}} - \varepsilon_{i\text{Train}})$ is fixed at 2 to identify the model, whereas the two remaining elements were estimated. We can also obtain the estimated correlation matrix:

```
. estat correlation
```

	Bus	Car
Bus	1.000	
Car	-0.229	1.000

Note: correlations are for alternatives differenced with Train

It is tempting to use this information to describe the correlations among errors for the utilities. For example, you might *incorrectly* assume that positive errors in the utility for `Car` are associated with negative errors in the utility for `Bus`. This is incorrect since we are estimating these values after the constraints are imposed. That is, -0.229 is the estimated correlation between $(\varepsilon_{i\text{Bus}} - \varepsilon_{i\text{Train}})$ and $(\varepsilon_{i\text{Car}} - \varepsilon_{i\text{Train}})$, which is unlikely to be of much substantive interest.

As with `clogit`, the coefficients for `time` and `invc` show the negative effect of travel time and cost on choosing a mode of transportation. When comparing the magnitude of these values to those from other models (e.g., from `probit`) or from other programs, remember that by default `asmprobit` is identified with the assumption that the fixed variance in Σ_ε^* is 2, whereas other software packages might assume fixed variances of 1. This will make the coefficients from `asmprobit` larger by a factor of $\sqrt{2}$, but the z-scores will remain the same.

7.4   The structural covariance matrix[3]

By default, asmprobit estimates the covariance matrix $\boldsymbol{\Sigma}_\varepsilon^*$ for differences in errors relative to the base alternative. But, with the structural option, asmprobit can estimate the structural covariance matrix $\boldsymbol{\Sigma}_\varepsilon$ for the undifferenced errors. When using the structural option, it is *essential* to understand that not all the parameters in $\boldsymbol{\Sigma}_\varepsilon$ are identified and that more identification restrictions are required to estimate the parameters in the structural covariance matrix. In a model with J alternatives, there are $1/2\,J\,(J+1)$ distinct parameters in $\boldsymbol{\Sigma}_\varepsilon$, but the data identify only $\{1/2\,(J-1)\,J\}-1$ of these. Accordingly, to estimate the structural covariance matrix you must impose at least $J+1$ restrictions. By default, asmprobit ..., structural sets two error variances to 1, which provides two restrictions. $J-1$ additional restrictions are imposed by setting the error covariances between the base alternative and the other alternatives to 0. For example, with $J=3$:

$$\boldsymbol{\Sigma}_\varepsilon = \begin{bmatrix} 1 & & \\ 0 & 1 & \\ 0 & \sigma_{32} & \sigma_{33} \end{bmatrix} \tag{7.3}$$

To illustrate the structural option, we use the travel example and specify the options base(1) to specify that $\sigma_{11}=1$ and $\sigma_{j1}=0$ and scale(2) to specify that $\sigma_{22}=1$. The output is

(*Continued on next page*)

---

3. The material in this section is more advanced than in the rest of the chapter. However, if you want to use the structural option for asmprobit, it is important to study this section carefully.

```
. asmprobit choice time invc, nolog case(id) alternatives(mode) ///
> base(1) scale(2) structural
```

| Alternative-specific multinomial probit | | | Number of obs | = | 456 |
| Case variable: id | | | Number of cases | = | 152 |

Alternative variable: mode			Alts per case: min =	3
			avg =	3.0
			max =	3

```
Integration sequence:      Hammersley
Integration points:              150            Wald chi2(2)    =    19.71
Log simulated-likelihood = -81.363746           Prob > chi2     =   0.0001
```

choice	Coef.	Std. Err.	z	P>|z|	[95% Conf. Interval]	
mode						
time	-.0268047	.0060834	-4.41	0.000	-.0387279	-.0148814
invc	-.0817245	.022591	-3.62	0.000	-.126002	-.0374469
Train	(base alternative)					
Bus						
_cons	-1.776277	.5360479	-3.31	0.001	-2.826912	-.7256426
Car						
_cons	-4.503222	1.19119	-3.78	0.000	-6.837912	-2.168533
/lnsigma3	1.384793	.2869218	4.83	0.000	.8224362	1.947149
/atanhr3_2	-.6688769	.7249952	-0.92	0.356	-2.089841	.7520876
sigma1	1	(base alternative)				
sigma2	1	(scale alternative)				
sigma3	3.993998	1.145965			2.276038	7.008678
rho3_2	-.5842407	.4775274			-.9698546	.6363927

```
(mode=Train is the alternative normalizing location)
(mode=Bus is the alternative normalizing scale)
```

The estimated structural covariance matrix, corresponding to (7.3), is

```
. estat covariance
```

	Train	Bus	Car
Train	1		
Bus	0	1	
Car	0	-2.333456	15.95202

And the correlations among errors are

```
. estat correlation
```

	Train	Bus	Car
Train	1.000		
Bus	0.000	1.000	
Car	0.000	-0.584	1.000

Bunch (1991) showed that the $J-1$ zero-covariance restrictions such as those imposed by `structural` place restrictions on the variances of the differenced covariance matrix. If the variances of the differenced covariance matrix are not large enough to accommodate the zero covariance restrictions, more constraints are placed on the model, which causes what should be an exactly identified model to be overidentified. This result is practically important since with some datasets (e.g., `travel.dta`) the default restrictions imposed by `asmprobit ...,` `structural` result in an overidentified model, which may differ from the model you think you are estimating. Here is an example. First, we estimate our travel example using the `structural` option:

```
. asmprobit choice time invc, nolog case(id) alternatives(mode) structural
Alternative-specific multinomial probit      Number of obs      =      456
Case variable: id                            Number of cases    =      152

Alternative variable: mode                   Alts per case: min =        3
                                                            avg =      3.0
                                                            max =        3
Integration sequence:       Hammersley
Integration points:                150       Wald chi2(2)       =    19.71
Log simulated-likelihood = -81.363746        Prob > chi2        =   0.0001
```

choice	Coef.	Std. Err.	z	P>|z|	[95% Conf.	Interval]
mode						
time	-.0268047	.0060834	-4.41	0.000	-.0387279	-.0148814
invc	-.0817245	.022591	-3.62	0.000	-.126002	-.0374469
Train	(base alternative)					
Bus						
_cons	-1.776277	.5360479	-3.31	0.001	-2.826912	-.7256426
Car						
_cons	-4.503222	1.19119	-3.78	0.000	-6.837912	-2.168533
/lnsigma3	1.384793	.2869218	4.83	0.000	.8224362	1.947149
/atanhr3_2	-.6688769	.7249952	-0.92	0.356	-2.089841	.7520876
sigma1	1	(base alternative)				
sigma2	1	(scale alternative)				
sigma3	3.993998	1.145965			2.276038	7.008678
rho3_2	-.5842407	.4775274			-.9698546	.6363927

```
(mode=Train is the alternative normalizing location)
(mode=Bus is the alternative normalizing scale)
```

The log simulated-likelihood and the estimated coefficients for `time` and `invc` are identical to those from the model fitted without `structural`.

What if we had used different alternatives to set the base and scale? By default, the `structural` option implies the options `base(1)` and `scale(2)`. We can change these by explicitly adding the following options:

```
. asmprobit choice time invc, nolog case(id) alternatives(mode) ///
> base(1) scale(3) structural
```

Alternative-specific multinomial probit			Number of obs	=	456
Case variable: id			Number of cases	=	152

Alternative variable: mode			Alts per case: min =	3
			avg =	3.0
			max =	3

Integration sequence:	Hammersley				
Integration points:	150		Wald chi2(2)	=	64.14
Log simulated-likelihood = -87.572171			Prob > chi2	=	0.0000

choice	Coef.	Std. Err.	z	P>|z|	[95% Conf. Interval]
mode						
time	-.0104124	.0013766	-7.56	0.000	-.0131105	-.0077143
invc	-.0413258	.0082817	-4.99	0.000	-.0575576	-.025094
Train	(base alternative)					
Bus						
_cons	-1.014891	.2696893	-3.76	0.000	-1.543473	-.4863102
Car						
_cons	-1.843542	.3190822	-5.78	0.000	-2.468931	-1.218152
/lnsigma2	-.5633886	.3749217	-1.50	0.133	-1.298222	.1714443
/atanhr3_2	-7.736511	547.6298	-0.01	0.989	-1081.071	1065.598
sigma1	1	(base alternative)				
sigma2	.5692767	.2134342			.2730169	1.187018
sigma3	1	(scale alternative)				
rho3_2	-.9999996	.0004175			-1	1

```
(mode=Train is the alternative normalizing location)
(mode=Car is the alternative normalizing scale)
```

The log simulated-likelihood is -87.572, which is more negative than the value for the model without structural or for the model with the `base(1)` `scale(2)` `structural` options. The estimated coefficients for `time` and `invc` also differ. This is an example of the problem noted by Bunch (1991) where the constraints to set the base and scale alternatives result in an overidentified model. This can be seen when we look at the estimated correlation matrix:

For the zero-truncated models `ztp` and `ztnb` (see section 8.4), predictions that are conditional on $y > 0$ are computed:

name`Crate` The conditional predicted rate $E(y \mid y > 0)$.

name`Cpr`k The predicted probability $\Pr(y = k \mid y > 0)$ for $k = 1$ to *max*.

name`Cprgt` The predicted probability $\Pr(y > k \mid y > 0)$.

name`Ccu`k The predicted cumulative probability $\Pr(y \le k \mid y > 0)$ for $k = 1$ to *max*.

When the `plot` option is specified, $max + 1$ observations (for counts 0 through *max*) are generated for the following variables:

name`val` The value k of the count y ranging from 0 to *max*.

name`obeq` The *observed* probability $\Pr(y = k)$. These values are the same as the ones you could obtain by running `tabulate` on the count variable (e.g., `tabulate art`).

name`oble` The *observed* cumulative probability $\Pr(y \le k)$.

name`preq` The average *predicted* probability $\Pr(y = k)$.

name`prle` The average *predicted* cumulative probability $\Pr(y \le k)$.

For the `ztp` and `ztnb` models, conditional predictions are also computed:

name`Cpreq` The average predicted probability $\Pr(y = k \mid y > 0)$.

name`Cprle` The average predicted cumulative probability $\Pr(y \le k \mid y > 0)$.

Which observations are used to compute the averages? By default, `prcounts` computes averages for all observations in memory, which could include observations that were not used in the estimation. For example, if your model was `poisson art if fem==1`, the averages computed by `prcounts` would be based on all observations, including those where `fem` is not 1. To restrict the averages to the sample used in estimation, you need to add the condition `if e(sample)==1`. For example, `prcounts isfem if e(sample)==1, plot`.

8.1.3   Comparing observed and predicted counts with prcounts

If the `plot` option was used with `prcounts`, it is simple to construct a graph that compares the observed probabilities for each value of the count variable with the predicted probabilities from fitting the Poisson distribution. For example,

```
. prcounts psn, plot max(9)
. label var psnobeq "Observed Proportion"
. label var psnpreq "Poisson Prediction"
. label var psnval "# of Articles"
. list psnval psnobeq psnpreq in 1/10
```

	psnval	psnobeq	psnpreq
1.	0	.3005464	.1839859
2.	1	.2688525	.311469
3.	2	.1945355	.2636423
4.	3	.0918033	.148773
5.	4	.073224	.0629643
6.	5	.0295082	.0213184
7.	6	.0185792	.006015
8.	7	.0131148	.0014547
9.	8	.0010929	.0003078
10.	9	.0021858	.0000579

The listed values are the observed and predicted probabilities for observing scientists with 0–9 publications. These can be plotted with `graph`:

```
. graph twoway connected psnobeq psnpreq psnval, ///
> ytitle("Probability") ylabel(0(.1).4)          ///
> xlabel(0(1)9)
```

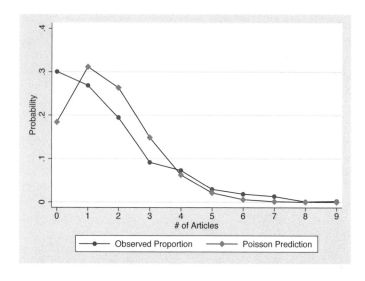

The graph clearly shows that the fitted Poisson distribution (represented by ◆) un-derpredicts 0s and overpredicts counts 1, 2, and 3. This pattern of overprediction and underprediction is characteristic of fitting a count model that does not take into account *heterogeneity* among sample members in their rate, μ. As fitting the univariate Pois-son distribution assumes that all scientists have the same rate of productivity, which is clearly unrealistic, our next step is to allow heterogeneity in μ based on observed characteristics of the scientists.

Advanced: plotting Poisson distributions Earlier we plotted the Poisson distribution for four values of μ. The trick to doing this is to construct artificial data with
   a given mean rate of productivity. Here are the commands we used to generate
   the graph on page 350:

```
. clear
. set obs 25
obs was 0, now 25
. gen ya = .8
. poisson ya, nolog
note: you are responsible for interpretation of noncount dep. variable
```

```
Poisson regression                          Number of obs   =         25
                                            LR chi2(0)      =      -0.00
                                            Prob > chi2     =          .
Log likelihood = -22.685774                 Pseudo R2       =    -0.0000
```

ya	Coef.	Std. Err.	z	P>|z|	[95% Conf. Interval]
_cons	-.2231435	.2236068	-1.00	0.318	-.6614048 .2151177

```
. prcounts pya, plot max(20)
. gen yb = 1.5
. poisson yb, nolog
note: you are responsible for interpretation of noncount dep. variable
```

```
Poisson regression                          Number of obs   =         25
                                            LR chi2(0)      =       0.00
                                            Prob > chi2     =          .
Log likelihood =  -29.41213                 Pseudo R2       =     0.0000
```

yb	Coef.	Std. Err.	z	P>|z|	[95% Conf. Interval]
_cons	.4054651	.1632993	2.48	0.013	.0854043 .7255259

```
. prcounts pyb, plot max(20)
. gen yc = 2.9
```

```
. poisson yc, nolog
note: you are responsible for interpretation of noncount dep. variable
Poisson regression                                Number of obs   =         25
                                                  LR chi2(0)      =       0.00
                                                  Prob > chi2     =          .
Log likelihood = -36.997981                       Pseudo R2       =     0.0000
```

yc	Coef.	Std. Err.	z	P>|z|	[95% Conf. Interval]
_cons	1.064711	.117444	9.07	0.000	.8345247 1.294897

```
. prcounts pyc, plot max(20)
. gen yd = 10.5
. poisson yd, nolog
note: you are responsible for interpretation of noncount dep. variable
Poisson regression                                Number of obs   =         25
                                                  LR chi2(0)      =       0.00
                                                  Prob > chi2     =          .
Log likelihood = -52.564007                       Pseudo R2       =     0.0000
```

yd	Coef.	Std. Err.	z	P>|z|	[95% Conf. Interval]
_cons	2.351375	.0617213	38.10	0.000	2.230404 2.472347

```
. prcounts pyd, plot max(20)
. label var pyapreq "mu=0.8"
. label var pybpreq "mu=1.5"
. label var pycpreq "mu=2.9"
. label var pydpreq "mu=10.5"
. label var pyaval "y=# of Events"
. graph twoway connected pyapreq pybpreq pycpreq pydpreq pyaval, ///
> ytitle("Probability") ylabel(0(.1).5) xlabel(0(2)20)
```

8.2   The Poisson regression model

The Poisson regression model (PRM) extends the Poisson distribution by allowing each observation to have a different value of μ. More formally, the PRM assumes that the observed count for observation i is drawn from a Poisson distribution with mean μ_i, where μ_i is estimated from observed characteristics. This is sometimes referred to as incorporating *observed heterogeneity* and leads to the structural equation

$$\mu_i = E\left(y_i \mid \mathbf{x}_i\right) = \exp\left(\mathbf{x}_i\boldsymbol{\beta}\right)$$

Taking the exponential of $\mathbf{x}\boldsymbol{\beta}$ forces μ to be positive, which is necessary because counts can be only 0 or positive. To see how this works, consider the PRM with one independent variable, $\mu = \exp\left(\alpha + \beta x\right)$, which can be plotted as

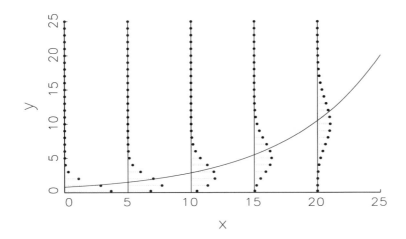

In this graph the mean μ, shown by the curved line, increases as x increases. For each value of μ, the distribution around the mean is shown by the dots, which should be thought of as coming out of the page and which represent the probability of each count. Interpretation of the model involves assessing how changes in the independent variables affect the conditional mean and the probabilities of various counts. Details on interpretation are given after we consider estimation.

8.2.1   Fitting the PRM with poisson

The PRM is fitted with the command

poisson *depvar* [*indepvars*] [*if*] [*in*] [*weight*] [, no̲co̲n̲stant

exposure(*varname*) c̲onstraints(*constraints*) r̲obust c̲luster(*varname*)

l̲evel(#) i̲rr no̲log]

In our experience, poisson converges quickly, and difficulties are rare.

Variable lists

depvar is the dependent variable. poisson does not require this to be an integer. But if you have noninteger values, you obtain the warning

    note: you are responsible for interpretation of noncount dep. variable.

indepvars is a list of independent variables. If *indepvars* is not included, a model with only an intercept is fitted, which corresponds to fitting a univariate Poisson distribution, as shown in the last section.

Specifying the estimation sample

if and in qualifiers can be used to restrict the estimation sample. For example, if you want to fit a model for only women, you could specify `poisson art mar kid5 phd ment if fem==1`.

Listwise deletion Stata excludes observations in which there are missing values for any of the variables in the model. Accordingly, if two models are fitted using the same data but have different independent variables, it is possible to have different samples. We recommend that you use `mark` and `markout` (discussed in chapter 3) to explicitly remove observations with missing data.

Weights

`poisson` can be used with `fweights`, `pweights`, and `iweights`. See chapter 3 for details.

Options

`noconstant` suppresses the constant term (interecept) in the model.

`exposure(varname)` specifies a variable indicating the amount of time during which an observation was "at risk" of the event occurring. Details are given in an example below.

`constraints(constraints)` specifies linear constraints on the coefficients.

`robust` requests that robust variance estimates be used. See chapter 3 for details.

`cluster(varname)` specifies that the observations are independent across the groups specified by unique values of *varname* but not necessarily within the groups. When `cluster()` is specified, robust standard errors are automatically used. See chapter 3 for details.

`level(#)` specifies the level of the confidence interval for estimated parameters. By default, a 95% interval is used. You can change the default level to, say, 90%, with the command `set level 90`.

`irr` reports estimated coefficients transformed to incidence-rate ratios defined as $\exp(\beta)$. These are discussed in section 8.2.3.

`nolog` suppresses the iteration history.

8.2.2   Example of fitting the PRM

If scientists who differ in their rates of productivity are combined, the univariate distribution of articles will be overdispersed (i.e., the variance is greater than the mean). Differences among scientists in their rates of productivity could be due to factors such as sex, marital status, number of young children, prestige of the graduate program, and the number of articles written by a scientist's mentor. To account for these differences, we add these variables as independent variables:

```
. use http://www.stata-press.com/data/lf2/couart2, clear
(Academic Biochemists / S Long)
. poisson art fem mar kid5 phd ment, nolog
```

Poisson regression				Number of obs	=	915
				LR chi2(5)	=	183.03
				Prob > chi2	=	0.0000
Log likelihood = -1651.0563				Pseudo R2	=	0.0525

art	Coef.	Std. Err.	z	P>|z|	[95% Conf. Interval]	
fem	-.2245942	.0546138	-4.11	0.000	-.3316352	-.1175532
mar	.1552434	.0613747	2.53	0.011	.0349512	.2755356
kid5	-.1848827	.0401272	-4.61	0.000	-.2635305	-.1062349
phd	.0128226	.0263972	0.49	0.627	-.038915	.0645601
ment	.0255427	.0020061	12.73	0.000	.0216109	.0294746
_cons	.3046168	.1029822	2.96	0.003	.1027755	.5064581

The way you interpret a count model depends on whether you are interested in the expected value of the count variable or in the distribution of counts. If your interest is in the expected count, several methods can be used to compute the change in the expectation for a change in an independent variable. If your interest is in the distribution of counts or perhaps just the probability of a specific count, the probability of a count for a given level of the independent variables can be computed.

8.2.3   Interpretation using the rate, μ

In the PRM,
$$\mu = E\left(y \mid \mathbf{x}\right) = \exp\left(\mathbf{x}\boldsymbol{\beta}\right)$$

Changes in μ for changes in the independent variable can be interpreted in several ways.

Factor change in E(y | x)

Perhaps the most common method of interpretation is the factor change in the rate. If we define $E\left(y \mid \mathbf{x}, x_k\right)$ as the expected count for a given $\mathbf{x}$, where we explicitly note the value of x_k, and define $E\left(y \mid \mathbf{x}, x_k + \delta\right)$ as the expected count after increasing x_k by δ units, then

$$\frac{E\left(y \mid \mathbf{x}, x_k + \delta\right)}{E\left(y \mid \mathbf{x}, x_k\right)} = e^{\beta_k \delta} \tag{8.1}$$

Therefore, the parameters can be interpreted that for a change of δ in x_k, the expected count increases by a factor of $\exp(\beta_k \times \delta)$, holding all other variables constant.

For example,

Factor change: for a unit change in x_k, the expected count changes by a factor of $\exp(\beta_k)$, holding all other variables constant.

Standardized factor change: for a standard deviation change in x_k, the expected count changes by a factor of $\exp(\beta_k \times s_k)$, holding all other variables constant.

Incidence-rate ratio

In some discussions of count models, μ is referred to as the *incidence rate*, and (8.1) for $\delta = 1$ is called the *incidence-rate ratio*. These coefficients can be computed by adding the option `irr` to the estimation command. They can also be computed with our `listcoef`, which is illustrated below.

Percent change in E(y | x)

The percentage change in the expected count for a δ unit change in x_k, holding other variables constant, can also be computed as

$$100 \times \frac{E\left(y \mid \mathbf{x}, x_k + \delta\right) - E\left(y \mid \mathbf{x}, x_k\right)}{E\left(y \mid \mathbf{x}, x_k\right)} = 100 \times \left\{\exp\left(\beta_k \times \delta\right) - 1\right\}$$

Example of factor and percent change

Factor change coefficients can be computed using `listcoef`:

```
. listcoef fem ment, help
poisson (N=915): Factor Change in Expected Count
  Observed SD: 1.926069
```

art	b	z	P>|z|	e^b	e^bStdX	SDofX
fem	-0.22459	-4.112	0.000	0.7988	0.8940	0.4987
ment	0.02554	12.733	0.000	1.0259	1.2741	9.4839

```
       b = raw coefficient
       z = z-score for test of b=0
   P>|z| = p-value for z-test
     e^b = exp(b) = factor change in expected count for unit increase in X
 e^bStdX = exp(b*SD of X) = change in expected count for SD increase in X
   SDofX = standard deviation of X
```

For example, the coefficients for `fem` and `ment` can be interpreted as follows:

Being a female scientist decreases the expected number of articles by a factor of .80, holding all other variables constant.

For a standard deviation increase in the mentor's productivity, roughly 9.5 articles, a scientist's mean productivity increases by a factor of 1.27, holding other variables constant.

To compute *percent change*, we add the option `percent`:

```
. listcoef fem ment, percent help
poisson (N=915): Percentage Change in Expected Count
  Observed SD: 1.926069
```

art	b	z	P>|z|	%	%StdX	SDofX
fem	-0.22459	-4.112	0.000	-20.1	-10.6	0.4987
ment	0.02554	12.733	0.000	2.6	27.4	9.4839

```
        b = raw coefficient
        z = z-score for test of b=0
    P>|z| = p-value for z-test
        % = percent change in expected count for unit increase in X
    %StdX = percent change in expected count for SD increase in X
    SDofX = standard deviation of X
```

For example, the percent change coefficients for `fem` and `ment` can be interpreted as follows:

Being a female scientist decreases the expected number of articles by 20%, holding all other variables constant.

For every additional article by the mentor, a scientist's predicted mean productivity increases by 2.6%, holding other variables constant.

The standardized percent change coefficient can be interpreted as follows:

For a standard deviation increase in the mentor's productivity, a scientist's mean productivity increases by 27%, holding all other variables constant.

Marginal change in E(y | x)

Another method of interpretation is the marginal change in $E(y \mid \mathbf{x})$:

$$\frac{\partial E(y \mid \mathbf{x})}{\partial x_k} = E(y \mid \mathbf{x}) \beta_k$$

For $\beta_k > 0$, the larger the current value of $E(y \mid \mathbf{x})$, the larger the rate of change; for $\beta_k < 0$, the smaller the rate of change. The marginal with respect to x_k depends on both β_k and $E(y \mid \mathbf{x})$. Thus the value of the marginal depends on the levels of all variables in the model. In practice, this measure is often computed with all variables held at their means.

Example of marginal change using prchange

As the marginal is not appropriate for binary independent variables, we request only the change for the continuous variables phd and ment. The marginal effects are in the column that is labeled MargEfct:

```
. prchange phd ment, rest(mean)

poisson: Changes in Rate for art
          min->max      0->1      -+1/2     -+sd/2  MargEfct
   phd    0.0794      0.0200     0.0206     0.0203    0.0206
   ment   7.9124      0.0333     0.0411     0.3910    0.0411

exp(xb):    1.6101

              fem        mar       kid5        phd       ment
     x=   .460109    .662295    .495082    3.10311    8.76721
  sd(x)=  .498679    .473186     .76488    .984249    9.48392
```

Example of marginal change using mfx

By default, mfx computes the marginal change with variables held at their means:

```
. mfx

Marginal effects after poisson
      y  = predicted number of events (predict)
         =  1.6100936
```

variable	dy/dx	Std. Err.	z	P>|z|	[	95% C.I.	]	X
fem*	-.3591461	.08648	-4.15	0.000	-.528643	-.189649		.460109
mar*	.2439822	.09404	2.59	0.009	.059671	.428293		.662295
kid5	-.2976785	.06414	-4.64	0.000	-.423393	-.171964		.495082
phd	.0206456	.04249	0.49	0.627	-.062635	.103926		3.10311
ment	.0411262	.00317	12.97	0.000	.034912	.04734		8.76721

```
(*) dy/dx is for discrete change of dummy variable from 0 to 1
```

The estimated marginal effects for phd and ment match those given above. For dummy variables, mfx, by default, computes the discrete change, as the variable changes from 0 to 1, a topic we will now consider.

Discrete change in E(y | x)

You can also compute the discrete change in the expected count for a change in x_k from x_S to x_E,

$$\frac{\Delta E\left(y \mid \mathbf{x}\right)}{\Delta x_k} = E\left(y \mid \mathbf{x}, x_k = x_E\right) - E\left(y \mid \mathbf{x}, x_k = x_S\right)$$

which can be interpreted as follows:

For a change in x_k from x_S to x_E, the expected count changes by $\Delta E\left(y \mid \mathbf{x}\right)/\Delta x_k$, holding all other variables at the specified values.

As with earlier chapters, the discrete change can be computed several ways depending on your purpose:

1. The total possible effect of x_k is found by letting x_k change from its minimum to its maximum.

2. The effect of a binary variable x_k is computed by letting x_k change from 0 to 1. This is the quantity computed by `mfx` for binary variables.

3. The *uncentered* effect of a unit change in x_k at the mean is computed by changing from $\overline{x}_k$ to $\overline{x}_k + 1$. The *centered* discrete change is computed by changing from $\left(\overline{x}_k - 1/2\right)$ to $\left(\overline{x}_k + 1/2\right)$.

4. The *uncentered* effect of a standard deviation change in x_k at the mean is computed by changing from $\overline{x}_k$ to $\overline{x}_k + s_k$. The *centered* change is computed by changing from $\left(\overline{x}_k - s_k/2\right)$ to $\left(\overline{x}_k + s_k/2\right)$.

5. The *uncentered* effect of a change of δ units in x_k from $\overline{x}_k$ to $\overline{x}_k + \delta$. The *centered* change is computed by changing from $\left(\overline{x}_k - \delta/2\right)$ to $\left(\overline{x}_k + \delta/2\right)$.

Discrete changes are computed with `prchange`. By default, changes are computed centered on the values specified with `x()` and `rest()`. To compute changes that begin at the specified values, such as a change from $\overline{x}_k$ to $\overline{x}_k + 1$, you must specify the `uncentered` option. By default, `prchange` computes results for changes in the independent variables of 1 unit and a standard deviation. With the `delta(#)` option, you can request changes of # units. When using discrete change, remember that the magnitude of the change in the expected count depends on the levels of all variables in the model.

Example of discrete change using prchange

Here we set all variables to their mean:

```
. prchange fem ment, rest(mean)
poisson: Changes in Rate for art
           min->max      0->1      -+1/2     -+sd/2   MargEfct
   fem     -0.3591    -0.3591    -0.3624    -0.1804    -0.3616
   ment     7.9124     0.0333     0.0411     0.3910     0.0411

exp(xb):     1.6101

               fem        mar       kid5        phd       ment
     x=    .460109    .662295    .495082    3.10311    8.76721
  sd(x)=   .498679    .473186     .76488    .984249    9.48392
```

Examples of interpretation are the following:

Being a female scientist decreases the expected productivity by .36 articles, holding all other variables at their means.

A standard deviation increase in the mentor's articles increases the scientist's rate of productivity by .39, holding all other variables at their mean.

To illustrate the use of the `uncentered` option, suppose that we want to know the effect of a change from 1 to 2 young children:

```
. prchange kid5, uncentered x(kid5=1)
poisson: Changes in Rate for art
           min->max        0->1       +1        +sd   MargEfct
kid5    -0.7512      -0.2978    -0.2476    -0.1934    -0.2711

exp(xb):    1.4666

              fem       mar      kid5      phd      ment
     x=   .460109   .662295         1   3.10311   8.76721
  sd(x)=  .498679   .473186    .76488   .984249   9.48392
```

The rate of productivity decreases by .25 as the number of young children increases from 1 to 2.

Example of discrete change with confidence intervals

Although `prchange` is an easy way to compute discrete changes, it does not include confidence intervals. These can be computed using the `save` and `diff` options with `prvalue`. To see how this is done, let's start by using `prchange` to compute the change in the expected number of publications when the number of young children increases from 1 to 3:

```
. prchange kid5, uncentered x(kid5=1) delta(2)
poisson: Changes in Rate for art
(Note: delta = 2)
           min->max        0->1     +delta      +sd   MargEfct
kid5    -0.7512      -0.2978    -0.4533    -0.1934    -0.2711

exp(xb):    1.4666

              fem       mar      kid5      phd      ment
     x=   .460109   .662295         1   3.10311   8.76721
  sd(x)=  .498679   .473186    .76488   .984249   9.48392
```

We find that as the number of young children increases from 1 to 3, the expected number of articles decreases by .45.

To estimate the bounds for the confidence interval around this prediction, we begin by `quietly` computing predictions for an average scientist with one young child:

```
. quietly prvalue, x(kid5=1) save
```

The `quietly` prefix suppresses output, whereas `save` keeps the results in memory. The next `prvalue` computes predictions for those with three children, using the `diff` option to compute the change from the saved results:

```
. prvalue, x(kid5=3) diff
poisson: Change in Predictions for art
Confidence intervals by delta method
                       Current    Saved    Change   95% CI for Change
     Rate:              1.0133    1.4666   -.45333  [-0.5938,  -0.3128]
         (output omitted)
                   fem        mar        kid5       phd       ment
   Current=  .46010929  .66229508           3  3.1031093  8.7672131
     Saved=  .46010929  .66229508           1  3.1031093  8.7672131
      Diff=          0          0           2          0          0
```

The results show a decrease of .45 in the expected number of articles as the number of young children in the family increases from 1 to 3, with estimated bounds for the 95% confidence interval at .31 and .59.

8.2.4   Interpretation using predicted probabilities

The estimated parameters can also be used to compute predicted probabilities using the following formula:

$$\widehat{\Pr}(y = m \mid \mathbf{x}) = \frac{e^{-\mathbf{x}\widehat{\boldsymbol{\beta}}} \left(\mathbf{x}\widehat{\boldsymbol{\beta}}\right)^{m}}{m!}$$

Predicted probabilities at specified values can be computed using `prvalue`. Predictions at the observed values for all observations can be made using `prcounts`, or `prgen` can be used to compute predictions that can be plotted.

Example of predicted probabilities using prvalue

`prvalue` computes predicted probabilities for values of the independent variables specified with `x()` and `rest()`. For example, to compare the predicted probabilities for married and unmarried women without young children, we first compute the predicted counts for single women without children by specifying `x(mar=0 fem=1 kid5=0)` and `rest(mean)`. We suppress the output with `quietly` but `save` the results for later use:

```
. * single women without children
. quietly prvalue, x(mar=0 fem=1 kid5=0) rest(mean) save
```

Next we compute the predictions for married women without children and use the `diff` option to compare these results with those we just saved:

```
. * compared to married women without children
. prvalue, x(mar=1 fem=1 kid5=0) rest(mean) diff

poisson: Change in Predictions for art
Confidence intervals by delta method
                     Current     Saved      Change   95% CI for Change
       Rate:          1.6471    1.4102      .23684   [ 0.0519,    0.4217]
       Pr(y=0|x):     0.1926    0.2441     -0.0515   [-0.0913,   -0.0116]
       Pr(y=1|x):     0.3172    0.3442     -0.0270   [-0.0484,   -0.0055]
       Pr(y=2|x):     0.2613    0.2427      0.0186   [ 0.0038,    0.0333]
       Pr(y=3|x):     0.1434    0.1141      0.0293   [ 0.0067,    0.0520]
       Pr(y=4|x):     0.0591    0.0402      0.0188   [ 0.0039,    0.0337]
       Pr(y=5|x):     0.0195    0.0113      0.0081   [ 0.0014,    0.0148]
       Pr(y=6|x):     0.0053    0.0027      0.0027   [ 0.0004,    0.0050]
       Pr(y=7|x):     0.0013    0.0005      0.0007   [ 0.0001,    0.0014]
       Pr(y=8|x):     0.0003    0.0001      0.0002   [ 0.0000,    0.0003]
       Pr(y=9|x):     0.0000    0.0000      0.0000   [-0.0000,    0.0001]

                    fem         mar        kid5         phd        ment
       Current=       1           1           0   3.1031093   8.7672131
         Saved=       1           0           0   3.1031093   8.7672131
          Diff=       0           1           0           0           0
```

The results show that married women are less likely than unmarried women to have one or no publications and are more likely to have two or more publications. Overall, their rate of productivity is .24 publications higher.

To examine the effects of the number of young children, we can use a series of calls to `prvalue`, where the `brief` option limits the amount of output:

```
. prvalue, x(mar=1 fem=1 kid5=0) rest(mean) brief maxcnt(4)

poisson: Predictions for art
                              95% Conf. Interval
       Rate:          1.6471   [ 1.4815,    1.8127]
       Pr(y=0|x):     0.1926   [ 0.1607,    0.2245]
       Pr(y=1|x):     0.3172   [ 0.2966,    0.3379]
       Pr(y=2|x):     0.2613   [ 0.2520,    0.2705]
       Pr(y=3|x):     0.1434   [ 0.1239,    0.1630]
       Pr(y=4|x):     0.0591   [ 0.0451,    0.0730]

. prvalue, x(mar=1 fem=1 kid5=1) rest(mean) brief maxcnt(4)

poisson: Predictions for art
                              95% Conf. Interval
       Rate:           1.369   [ 1.2304,    1.5077]
       Pr(y=0|x):     0.2544   [ 0.2191,    0.2896]
       Pr(y=1|x):     0.3482   [ 0.3352,    0.3612]
       Pr(y=2|x):     0.2384   [ 0.2231,    0.2536]
       Pr(y=3|x):     0.1088   [ 0.0908,    0.1267]
       Pr(y=4|x):     0.0372   [ 0.0273,    0.0472]
```

```
. prvalue, x(mar=1 fem=1 kid5=2) rest(mean) brief maxcnt(4)

poisson: Predictions for art

                              95% Conf. Interval
       Rate:            1.138  [ .9662,    1.3097]
       Pr(y=0|x):      0.3205  [ 0.2654,   0.3755]
       Pr(y=1|x):      0.3647  [ 0.3571,   0.3723]
       Pr(y=2|x):      0.2075  [ 0.1805,   0.2345]
       Pr(y=3|x):      0.0787  [ 0.0566,   0.1008]
       Pr(y=4|x):      0.0224  [ 0.0127,   0.0321]

. prvalue, x(mar=1 fem=1 kid5=3) rest(mean) brief maxcnt(4)

poisson: Predictions for art

                              95% Conf. Interval
       Rate:           .94587  [ .73935,   1.1524]
       Pr(y=0|x):      0.3883  [ 0.3081,   0.4685]
       Pr(y=1|x):      0.3673  [ 0.3630,   0.3717]
       Pr(y=2|x):      0.1737  [ 0.1337,   0.2137]
       Pr(y=3|x):      0.0548  [ 0.0302,   0.0793]
       Pr(y=4|x):      0.0130  [ 0.0043,   0.0216]
```

These values could be presented in a table or plotted. Overall, the probabilities of having no publications increases as the number of young children increases, which is expected since the regression coefficient for kid5 was negative. We also see that having more young children increases the chances of having only one publication and decreases the probabilities of more publications.

Example of predicted probabilities using prgen

prgen computes a series of predictions by holding all variables but one constant. The resulting predictions can then be plotted. Here we plot the predicted probability of not publishing for married men and married women with different numbers of children. First, we compute the predictions for women using the prefix fprm to indicate predictions for women from the PRM:

```
. prgen kid5, x(fem=1 mar=1) rest(mean) from(0) to(3) gen(fprm) n(4)

poisson: Predicted values as kid5 varies from 0 to 3.

           fem       mar      kid5       phd      ment
    x=       1         1  .49508197  3.1031093  8.7672131
```

Next we compute predictions for men, using the prefix mprm:

```
. prgen kid5, x(fem=0 mar=1) rest(mean) from(0) to(3) gen(mprm) n(4)

poisson: Predicted values as kid5 varies from 0 to 3.

           fem       mar      kid5       phd      ment
    x=       0         1  .49508197  3.1031093  8.7672131
```

In both calls of prgen, we requested four values with the n(4) option. This creates predictions for 0, 1, 2, and 3 children. To plot these predictions, we begin by adding value labels to the newly generated variables. Then we use the graph command:

```
. label var fprmp0 "Married Women"
. label var mprmp0 "Married Men"
. label var mprmx  "Number of Children"
. graph twoway connected fprmp0 mprmp0 mprmx,   ///
> ylabel(0(.1).4) yline(.1 .2 .3) xlabel(0(1)3) ///
> ytitle("Probability of No Articles")
```

This leads to the following graph, where the points marked with circles and diamonds
are placed at the tick marks for the number of children:

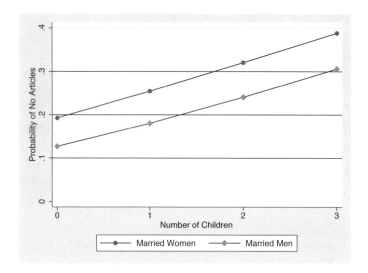

If you compare the values plotted for women with those computed with `prvalue` in the
prior section, you will see that they are the same—just computed differently.

Example of predicted probabilities using prcounts

`prcounts` computes predictions for all observations in the dataset. Also the predictions
are averaged across observations:

$$\overline{\Pr}(y = m) = \frac{1}{N} \sum_{i=1}^{N} \widehat{\Pr}(y_i = m \mid \mathbf{x}_i)$$

To show how this command can be used to compare predictions from different models,
we begin by fitting a univariate Poisson distribution and computing predictions with
`prcounts`:

```
. poisson art, nolog
Poisson regression                                    Number of obs   =        915
                                                      LR chi2(0)      =       0.00
                                                      Prob > chi2     =          .
Log likelihood = -1742.5735                           Pseudo R2       =     0.0000
```

art	Coef.	Std. Err.	z	P>|z|	[95% Conf. Interval]	
_cons	.5264408	.0254082	20.72	0.000	.4766416	.57624

```
. prcounts psn, plot max(9)
. label var psnpreq "Univariate Poisson Dist."
```

Because we specified the `plot` option and the prefix `psn`, the command `prcounts` created a new variable called `psnpreq` that contains the average predicted probabilities of counts 0–9 from a univariate Poisson distribution. We then fit the PRM with independent variables and again compute predictions with `prcounts`:

```
. poisson art fem mar kid5 phd ment, nolog
Poisson regression                                    Number of obs   =        915
                                                      LR chi2(5)      =     183.03
                                                      Prob > chi2     =     0.0000
Log likelihood = -1651.0563                           Pseudo R2       =     0.0525
```

art	Coef.	Std. Err.	z	P>|z|	[95% Conf. Interval]	
fem	-.2245942	.0546138	-4.11	0.000	-.3316352	-.1175532
mar	.1552434	.0613747	2.53	0.011	.0349512	.2755356
kid5	-.1848827	.0401272	-4.61	0.000	-.2635305	-.1062349
phd	.0128226	.0263972	0.49	0.627	-.038915	.0645601
ment	.0255427	.0020061	12.73	0.000	.0216109	.0294746
_cons	.3046168	.1029822	2.96	0.003	.1027755	.5064581

```
. prcounts prm, plot max(9)
. label var prmpreq "PRM"
. label var prmobeq "Observed"
```

In addition to the new variable `prmpreq`, `prcounts` also generates `prmobeq`, which contains the observed probability of counts 0–9. Another new variable, `prmval`, contains the value of the count. We now plot the values of `prmobeq`, `psnpreq`, and `prmpreq` with `prmval` on the x-axis:

```
. graph twoway connected prmobeq psnpreq prmpreq prmval, ///
> ytitle("Probability of Count") ylabel(0(.1).4)         ///
> xlabel(0(1)9)
```

(Continued on next page)

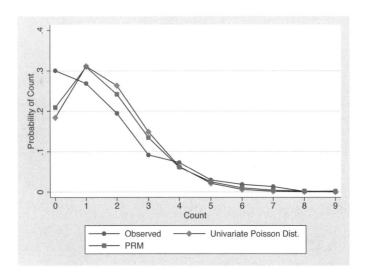

This graph shows that, even though many of the independent variables have significant effects on the number of articles published, there is only a modest improvement in the predictions made by the PRM over the univariate Poisson distribution, with somewhat more 0s predicted and slightly fewer 2s and 3s. Although this suggests the need for an alternative model, we will first discuss how different periods of exposure can be incorporated into count models.

8.2.5 Exposure time*

So far we have implicitly assumed that each observation was at risk of an event occurring for the same amount of time. For our example, this means that, for each person in the sample, we counted their articles over the same period. Often when collecting data, however, different observations have different *exposure* times. For example, the sample of scientists might have received their degrees in different years, and our outcome might have been total publications from Ph.D. to the date of the survey. Amount of time in the career clearly affects the number of publications.

Different exposure times can be easily incorporated into count models. Let t_i be the amount of time that observation i is at risk. If the rate (i.e., the expected number of observations for one unit of time) for that case is μ_i, then we would expect $t_i\mu_i$ to be the expected count over a period of length t_i. Then assuming only two independent variables for simplicity, our count equation becomes

$$\mu_i t_i = \{\exp\left(\beta_0 + \beta_1 x_1 + \beta_2 x_2\right)\} \times t_i$$

Because $t = \exp\left(\ln t\right)$, the equation can be rewritten as

$$\mu_i t_i = \exp\left(\beta_0 + \beta_1 x_1 + \beta_2 x_2 + \ln t_i\right)$$

This shows that the effect of different exposure times can be included as the log of the exposure time with a regression coefficient constrained to equal 1. Although we do not have data with different exposure times, we have artificially constructed three variables to illustrate this issue. profage is a scientist's professional age, which corresponds to the time a scientist has been "exposed" to the possibility of publishing; lnage is the natural log of profage; and totalarts is the total number of articles during the career (to see how these were created, you can examine the sample file st9ch8count.do). To fit the model including exposure time, we use the exposure() option:[1]

```
. poisson totalarts fem mar kid5 phd ment, nolog exposure(profage)
Poisson regression                            Number of obs   =         915
                                              LR chi2(5)      =      819.86
                                              Prob > chi2     =      0.0000
Log likelihood = -6222.1369                   Pseudo R2       =      0.0618
```

totalarts	Coef.	Std. Err.	z	P>|z|	[95% Conf. Interval]	
fem	-.1830187	.0221493	-8.26	0.000	-.2264306	-.1396067
mar	.0931759	.024757	3.76	0.000	.044653	.1416988
kid5	-.1312361	.0161656	-8.12	0.000	-.1629201	-.0995521
phd	.0437065	.0108052	4.04	0.000	.0225287	.0648844
ment	.0220197	.0008279	26.60	0.000	.020397	.0236424
_cons	.2576664	.0420177	6.13	0.000	.1753133	.3400195
profage	(exposure)					

The results can be interpreted using the same methods discussed above.

To show you what the exposure() option is doing, we can obtain the same results by adding lnage as an independent variable and constraining the coefficient for lnage to 1:

```
. constraint define 1 lnage=1
. poisson totalarts fem mar kid5 phd ment lnage, nolog constraint(1)
Poisson regression                            Number of obs   =         915
                                              Wald chi2(5)    =      959.53
Log likelihood = -6222.1369                   Prob > chi2     =      0.0000
 ( 1)  [totalarts]lnage = 1
```

totalarts	Coef.	Std. Err.	z	P>|z|	[95% Conf. Interval]	
fem	-.1830187	.0221493	-8.26	0.000	-.2264306	-.1396067
mar	.0931759	.024757	3.76	0.000	.044653	.1416988
kid5	-.1312361	.0161656	-8.12	0.000	-.1629201	-.0995521
phd	.0437065	.0108052	4.04	0.000	.0225287	.0648844
ment	.0220197	.0008279	26.60	0.000	.020397	.0236424
lnage	1	.	.	.	.	.
_cons	.2576664	.0420177	6.13	0.000	.1753133	.3400195

---

1. The output that follows differs from that presented in earlier editions because of a change in the random seed used in creating the variable for professional age.

You can also obtain the same result with `offset()` instead of `exposure()`, except that with `offset()` you specify a variable that is equal to the log of the exposure time. For example,

```
. poisson totalarts fem mar kid5 phd ment, nolog offset(lnage)
```

Although the `exposure()` and `offset()` are not considered further in this chapter, they can be used with the other models we discuss.

8.3 The negative binomial regression model

The PRM accounts for observed heterogeneity (i.e., observed differences among sample members) by specifying the rate, μ_i, as a function of observed x_ks. In practice, the PRM rarely fits, due to *overdispersion*. That is, the model underfits the amount of dispersion in the outcome. The negative binomial regression model (NBRM) addresses the failure of the PRM by adding a parameter, α, that reflects *unobserved* heterogeneity among observations.[2] For example, with three independent variables, the PRM is

$$\mu_i = \exp\left(\beta_0 + \beta_1 x_{i1} + \beta_2 x_{i2} + \beta_3 x_{i3}\right)$$

The NBRM adds an error, ε, that is assumed to be uncorrelated with the x's,

$$
\begin{aligned}
\widetilde{\mu}_i &= \exp\left(\beta_0 + \beta_1 x_{i1} + \beta_2 x_{i2} + \beta_3 x_{i3} + \varepsilon_i\right) \\
&= \exp\left(\beta_0 + \beta_1 x_{i1} + \beta_2 x_{i2} + \beta_3 x_{i3}\right) \exp\left(\varepsilon_i\right) \\
&= \exp\left(\beta_0 + \beta_1 x_{i1} + \beta_2 x_{i2} + \beta_3 x_{i3}\right) \delta_i
\end{aligned}
$$

where the second step follows by basic algebra, and the last step simply defines $\delta \equiv \exp(\varepsilon)$. To identify the model, we assume that

$$E(\delta) = 1$$

which corresponds to the assumption $E(\varepsilon) = 0$ in the LRM. With this assumption, it is easy to show that

$$E(\widetilde{\mu}) = \mu E(\delta) = \mu$$

Thus *the PRM and the NBRM have the same mean structure*. That is, if the assumptions of the NBRM are correct, the expected rate for a given level of the independent variables will be the same in both models. However, the standard errors in the PRM will be biased downward, resulting in spuriously large z-values and spuriously small p-values (Cameron and Trivedi 1986, 31).

The distribution of observations given both the values of the xs and δ is still Poisson in the NBRM. That is,

$$\Pr(y_i \mid \mathbf{x}_i, \delta_i) = \frac{e^{-\widetilde{\mu}_i}\widetilde{\mu}_i^{y_i}}{y_i!}$$

---

2. The NBRM can also be derived through a process of contagion where the occurrence of an event changes the probability of further events—an approach not considered further here.

Because δ is unknown, we cannot compute $\Pr(y \mid \mathbf{x})$. This limitation is resolved by assuming that δ is drawn from a gamma distribution (see Long 1997, 231–232 or Cameron and Trivedi 1998, 70–79 for details). Then we can compute $\Pr(y \mid \mathbf{x})$ as a weighted combination of $\Pr(y \mid \mathbf{x}, \delta)$ for all values of δ, where the weights are determined by $\Pr(\delta)$. The mathematics for this mixing of values of $\Pr(y \mid \mathbf{x}, \delta)$ are complex (and not particularly helpful for understanding the interpretation of the model) but lead to the negative binomial distribution

$$\Pr(y \mid \mathbf{x}) = \frac{\Gamma(y + \alpha^{-1})}{y!\Gamma(\alpha^{-1})} \left(\frac{\alpha^{-1}}{\alpha^{-1} + \mu} \right)^{\alpha^{-1}} \left(\frac{\mu}{\alpha^{-1} + \mu} \right)^{y}$$

where $\Gamma()$ is the gamma function.

In the negative binomial distribution, the parameter α determines the degree of dispersion in the predictions, as shown by the following figure:

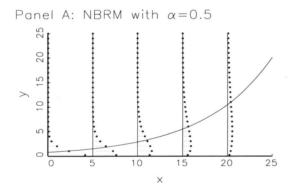

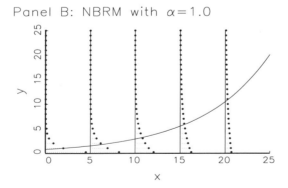

In both panels, the dispersion of predicted counts for a given value of x is larger than in the PRM. In particular, note the greater probability of a 0 count. Further, the larger value of α in panel B results in greater spread in the data. Indeed, if $\alpha = 0$, the NBRM reduces to the PRM, which turns out to be the key to testing for overdispersion. This is discussed in section 8.3.3.

8.3.1 Fitting the NBRM with nbreg

The NBRM is fitted with the following command and its basic options:

nbreg *depvar* [*indepvars*] [*if*] [*in*] [*weight*] [, no̲cons̲tant

    dispersion([me̲an | c̲onstant]) exposure(*varname*) constr̲aints(*constraints*)

    r̲obust c̲luster(*varname*) le̲vel(#) irr no̲log]

where the options are the same as those for poisson, with the exception of the option dispersion() that is discussed in the next section. Because of differences in how poisson and nbreg are implemented in Stata, models fitted with nbreg take substantially longer to converge.

NB1 and NB2 variance functions

The dispersion() option controls how the dispersion function is specified. This option determines whether Stata estimates the NB1 model or the NB2 model. To understand what this means, it is useful to begin with the variance or dispersion function from the PRM. In that model,

$$\text{Var}\left(y_i \mid \mathbf{x}_i\right) = E\left(y_i \mid \mathbf{x}_i\right) = \mu_i$$

which is referred to as equidispersion because the conditional mean equals the conditional variance. In real-world data, counts are usually overdispersed, meaning that the conditional variance is larger than the conditional mean. The NBRM addresses the problem by allowing overdispersion through the α parameter. Cameron and Trivedi (1986) show that a variety of variance functions is possible with the NBRM. The most commonly used function is

$$\text{Var}(y_i \mid \mathbf{x}) = \mu_i + \alpha\mu_i^2 \qquad (8.2)$$

which Cameron and Trivedi refer to as the NB2 model in reference to the squared term μ^2. This is the model fitted with the default option dispersion(mean). This is the only version of the NBRM available in Stata prior to version 9. The dispersion(constant) option specifies what Cameron and Trivedi refer to as the NB1 model. Here the variance function equals

$$\text{Var}(y_i \mid \mathbf{x}) = \mu_i + \alpha\mu_i \qquad (8.3)$$

The term NB1 refers to the power of μ in the $\alpha\mu$ term. Since the NB2 model is used most often in applied research, we use this form of the model throughout the book.

8.3.2   Example of fitting the NBRM

Here we use the same example as for the PRM above:

```
. use http://www.stata-press.com/data/lf2/couart2, clear
(Academic Biochemists / S Long)
. nbreg art fem mar kid5 phd ment, nolog
```

Negative binomial regression				Number of obs	=	915
				LR chi2(5)	=	97.96
Dispersion = mean				Prob > chi2	=	0.0000
Log likelihood = -1560.9583				Pseudo R2	=	0.0304

art	Coef.	Std. Err.	z	P>|z|	[95% Conf. Interval]	
fem	-.2164184	.0726724	-2.98	0.003	-.3588537	-.0739832
mar	.1504895	.0821063	1.83	0.067	-.0104359	.3114148
kid5	-.1764152	.0530598	-3.32	0.001	-.2804105	-.07242
phd	.0152712	.0360396	0.42	0.672	-.0553652	.0859075
ment	.0290823	.0034701	8.38	0.000	.0222811	.0358836
_cons	.256144	.1385604	1.85	0.065	-.0154294	.5277174
/lnalpha	-.8173044	.1199372			-1.052377	-.5822318
alpha	.4416205	.0529667			.3491069	.5586502

```
Likelihood-ratio test of alpha=0:   chibar2(01) =  180.20 Prob>=chibar2 = 0.000
```

The output is similar to that of poisson, except for the results at the bottom of the output, which initially can be confusing. Although the model was defined in terms of the parameter α, nbreg estimates $\ln(\alpha)$ with the estimate given in the line /lnalpha. This is done because estimating $\ln(\alpha)$ forces the estimated α to be positive. The value of $\widehat{\alpha}$ is given on the next line. Test statistics are not given because they require special treatment, as discussed in section 8.3.3.

Comparing the PRM and NBRM using estimates table

We can use estimates table to combine the results from poisson and nbreg:

```
. poisson art fem mar kid5 phd ment, nolog
  (output omitted )
. estimates store PRM
. nbreg art fem mar kid5 phd ment, nolog
  (output omitted )
. estimates store NBRM
```

```
. estimates table PRM NBRM, b(%9.3f) t label varwidth(32) drop(lnalpha:_cons)
> stats(alpha N)
```

Variable	PRM	NBRM
Gender: 1=female 0=male	-0.225	-0.216
	-4.11	-2.98
Married: 1=yes 0=no	0.155	0.150
	2.53	1.83
Number of children < 6	-0.185	-0.176
	-4.61	-3.32
PhD prestige	0.013	0.015
	0.49	0.42
Article by mentor in last 3 yrs	0.026	0.029
	12.73	8.38
Constant	0.305	0.256
	2.96	1.85
alpha		0.442
N	915.000	915.000

```
                                              legend: b/t
```

The estimates of the corresponding parameters from the PRM and the NBRM are close, but the z-values for the NBRM are consistently smaller than those for the PRM. This is the expected consequence of overdispersion.

8.3.3   Testing for overdispersion

If there is overdispersion, estimates from the PRM are inefficient with standard errors that are biased downward, even if the model includes the correct variables. Accordingly, it is important to test for overdispersion. Because the NBRM reduces to the PRM when $\alpha = 0$, we can test for overdispersion by testing H_0: $\alpha = 0$. There are two points to remember. First, nbreg estimates $\ln(\alpha)$ rather than α. A test of H_0: $\ln(\alpha) = 0$ corresponds to testing H_0: $\alpha = 1$, which is *not* the test we want. Second, as α must be greater than or equal to 0, the asymptotic distribution of $\widehat{\alpha}$ when $\alpha = 0$ is only half of a normal distribution. That is, all values less than 0 have a probability of 0. This requires an adjustment to the usual significance level of the test.

To test the hypothesis H_0: $\alpha = 0$, Stata provides an LR test that is listed after the estimates of the parameters:

```
     Likelihood-ratio test of alpha=0:   chibar2(01) =   180.20 Prob>=chibar2 = 0.000
```

This output differs from that of lrtest. The test statistic chibar2(01) is computed by the same formula shown in chapter 3:

$$G^2 = 2\left(\ln L_{\text{NBRM}} - \ln L_{\text{PRM}}\right)$$
$$= 2\left(-1560.96 - -1651.06\right) = 180.2$$

The significance level of the test is adjusted to account for the truncated sampling distribution of $\hat{\alpha}$. For details, you can click on `chibar2(01)`, which will be listed in blue in the Results window (a clickable link). Here the results are very significant and provide strong evidence of overdispersion. You can summarize this by saying that because there is significant evidence of overdispersion ($G^2 = 180.2$, $p < .01$), the negative binomial regression model is preferred to the Poisson regression model.

8.3.4 Interpretation using the rate μ

As the mean structure for the NBRM is identical to that for the PRM, the same methods of interpretation based on $E(y \mid \mathbf{x})$ can be used based on the equation

$$\frac{E(y \mid \mathbf{x}, x_k + \delta)}{E(y \mid \mathbf{x}, x_k)} = e^{\beta_k \delta}$$

This leads to the interpretation that for a change of δ in x_k, the expected count increases by a factor of $\exp(\beta_k \times \delta)$, holding all other variables constant.

Factor and percent change coefficients can be obtained using `listcoef`. For example,

```
. listcoef fem ment, help
nbreg (N=915): Factor Change in Expected Count
 Observed SD: 1.926069
```

art	b	z	P>|z|	e^b	e^bStdX	SDofX
fem	-0.21642	-2.978	0.003	0.8054	0.8977	0.4987
ment	0.02908	8.381	0.000	1.0295	1.3176	9.4839
ln alpha	-0.81730					
alpha	0.44162	SE(alpha) = 0.05297				

```
LR test of alpha=0: 180.20    Prob>=LRX2 = 0.000
```

```
        b = raw coefficient
        z = z-score for test of b=0
    P>|z| = p-value for z-test
      e^b = exp(b) = factor change in expected count for unit increase in X
  e^bStdX = exp(b*SD of X) = change in expected count for SD increase in X
    SDofX = standard deviation of X
```

(*Continued on next page*)

```
. listcoef fem ment, help percent
nbreg (N=915): Percentage Change in Expected Count
 Observed SD: 1.926069
```

art	b	z	P>|z|	%	%StdX	SDofX
fem	-0.21642	-2.978	0.003	-19.5	-10.2	0.4987
ment	0.02908	8.381	0.000	3.0	31.8	9.4839
ln alpha	-0.81730					
alpha	0.44162	SE(alpha) = 0.05297				

```
 LR test of alpha=0: 180.20    Prob>=LRX2 = 0.000
```

```
        b = raw coefficient
        z = z-score for test of b=0
    P>|z| = p-value for z-test
        % = percent change in expected count for unit increase in X
    %StdX = percent change in expected count for SD increase in X
    SDofX = standard deviation of X
```

These coefficients can be interpreted as follows:

Being a female scientist decreases the expected number of articles by a factor of .805, holding all other variables constant. Equivalently, being a female scientist decreases the expected number of articles by 19.5%, holding all other variables constant.

For every additional article by the mentor, a scientist's expected mean productivity increases by 3.0%, holding other variables constant.

For a standard deviation increase in the mentor's productivity, a scientist's expected mean productivity increases by 32%, holding all other variables constant.

Interpretations for marginal and discrete change can be computed and interpreted using the methods discussed for the PRM.

8.3.5   Interpretation using predicted probabilities

The methods from the PRM can also be used for interpreting predicted probabilities. The *only* difference is that the predicted probabilities are computed with the formula

$$\widehat{\Pr}\left(y \mid \mathbf{x}\right) = \frac{\Gamma(y + \alpha^{-1})}{y!\Gamma(\alpha^{-1})} \left(\frac{\widehat{\alpha}^{-1}}{\widehat{\alpha}^{-1} + \widehat{\mu}}\right)^{\widehat{\alpha}^{-1}} \left(\frac{\widehat{\mu}}{\widehat{\alpha}^{-1} + \widehat{\mu}}\right)^{y}$$

where $\widehat{\mu} = \exp\left(\mathbf{x}\widehat{\boldsymbol{\beta}}\right)$. As before, predicted probabilities can be computed using prgen, prchange, prcounts, and prvalue. As there is nothing new in how to use these commands, we provide only two examples that are designed to illustrate key differences and similarities between the PRM and the NBRM. First, we use prvalue to compute

predicted values for an "average" respondent. For the PRM, the output looks like the
following:

```
. quietly poisson art fem mar kid5 phd ment
. prvalue
poisson: Predictions for art
Confidence intervals by delta method
                                   95% Conf. Interval
     Rate:                1.6101   [ 1.5265,   1.6937]
     Pr(y=0|x):           0.1999   [ 0.1832,   0.2166]
     Pr(y=1|x):           0.3218   [ 0.3116,   0.3320]
     Pr(y=2|x):           0.2591   [ 0.2538,   0.2643]
     Pr(y=3|x):           0.1390   [ 0.1290,   0.1491]
     Pr(y=4|x):           0.0560   [ 0.0490,   0.0629]
     Pr(y=5|x):           0.0180   [ 0.0149,   0.0212]
     Pr(y=6|x):           0.0048   [ 0.0037,   0.0059]
     Pr(y=7|x):           0.0011   [ 0.0008,   0.0014]
     Pr(y=8|x):           0.0002   [ 0.0001,   0.0003]
     Pr(y=9|x):           0.0000   [ 0.0000,   0.0001]

           fem        mar       kid5        phd       ment
x=   .46010929  .66229508  .49508197  3.1031093  8.7672131
```

and for the NBRM, the output looks like the following:

```
. quietly nbreg art fem mar kid5 phd ment
. prvalue
nbreg: Predictions for art
Confidence intervals by delta method
                                   95% Conf. Interval
     Rate:                 1.602   [ 1.4936,   1.7104]
     Pr(y=0|x):           0.2978   [ 0.2788,   0.3167]
     Pr(y=1|x):           0.2794   [ 0.2727,   0.2860]
     Pr(y=2|x):           0.1889   [ 0.1859,   0.1919]
     Pr(y=3|x):           0.1113   [ 0.1051,   0.1174]
     Pr(y=4|x):           0.0607   [ 0.0549,   0.0664]
     Pr(y=5|x):           0.0315   [ 0.0273,   0.0357]
     Pr(y=6|x):           0.0158   [ 0.0130,   0.0186]
     Pr(y=7|x):           0.0077   [ 0.0061,   0.0094]
     Pr(y=8|x):           0.0037   [ 0.0028,   0.0046]
     Pr(y=9|x):           0.0018   [ 0.0012,   0.0023]

           fem        mar       kid5        phd       ment
x=   .46010929  .66229508  .49508197  3.1031093  8.7672131
```

The predicted rate is nearly identical for both models: 1.610 versus 1.602. This illustrates that even with overdispersion (which there is in this example), the estimates
from the PRM are consistent. But substantial differences emerge when we examine pre-
dicted probabilities: $\widehat{\Pr}_{PRM}(y = 0 \mid \overline{x}) = 0.200$ compared with $\widehat{\Pr}_{NBRM}(y = 0 \mid \overline{x}) =$
0.298. We also find higher probabilities in the NBRM for larger counts. For exam-
ple, $\widehat{\Pr}_{NBRM}(y = 5 \mid \overline{x}) = 0.0315$ compared with $\widehat{\Pr}_{PRM}(y = 5 \mid \overline{x}) = 0.0180$. These
probabilities reflect the greater dispersion in the NBRM compared with the PRM.

Another way to see the greater probability for 0 counts in the NBRM is to plot the probability of 0s as values of an independent variable change. This is done with `prgen`:

```
. quietly nbreg art fem mar kid5 phd ment, nolog
. prgen ment, rest(mean) f(0) t(50) gen(nb) n(20)
nbreg: Predicted values as ment varies from 0 to 50.
          fem         mar        kid5         phd        ment
x=  .46010929   .66229508   .49508197   3.1031093   8.7672131
. quietly poisson art fem mar kid5 phd ment
. prgen ment, rest(mean) f(0) t(50) gen(psn) n(20)
poisson: Predicted values as ment varies from 0 to 50.
          fem         mar        kid5         phd        ment
x=  .46010929   .66229508   .49508197   3.1031093   8.7672131
. label var psnp0 "Pr(0) for PRM"
. label var nbp0 "Pr(0) for NBRM"
. graph twoway connected psnp0 nbp0 nbx,      ///
> ylabel(0(.1).4) yline(.1 .2 .3)             ///
> ytitle("Probability of a Zero Count")
```

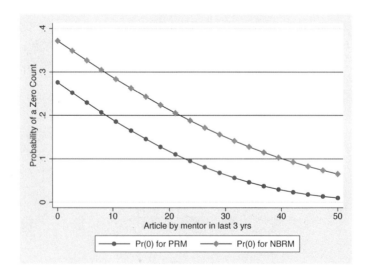

The probability of having zero publications is computed when each variable except the mentor's number of articles is held at its mean. For both models, the probability of a zero decreases as the mentor's articles increase. But the proportion of predicted zeros is significantly higher for the NBRM. As both models have the same expected number of publications, the higher proportion of predicted zeros for the NBRM is offset by the higher proportion of larger counts that are also predicted by this model.

Although the prior graph clearly shows the greater proportion of zeros predicted by the NBRM, we should also consider whether these differences are significant. One way to approach this question is to add confidence intervals around our predictions, as shown in this graph:

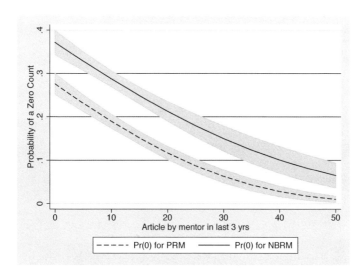

To create this graph, we used the command:

```
. graph twoway                                        ///
>       (rarea psnp0lb psnp0ub nbx, color(gs14))      ///
>       (rarea nbp0lb nbp0ub nbx, color(gs14))        ///
>       (connected psnp0 nbx, lpattern(dash) msize(zero))  ///
>       (connected nbp0 nbx, lpattern(solid) msize(zero)), ///
>       legend(on order(3 4))                         ///
>       ylabel(0(.1).4) yline(.1 .2 .3)               ///
>       ytitle("Probability of a Zero Count")
```

The two `rarea` graphs add shaded regions for the estimated 95% confidence intervals.
The two `connected` graphs plot the predicted probability of no publications as `ment`
increases. Normally, `graph` includes a legend for each of the overlaid graphs, which
would produce a legend for the two `rarea` graphs as well as the two `connected` graphs.
The `order` subcommand of `legend` indicates that we want to order the lines in the
legend to start with the third graph, followed by the fourth. Since 1 and 2 are not
included, those legends are dropped, which is what we wanted.

8.4   Models for truncated counts

A second type of problem with zeros occurs when observations with outcomes equal to
zero are missing from the sample because of the way the data were collected. Suppose
that we are gathering data to study scientific productivity among chemists but that
we do not have a sampling roster of all chemists with Ph.D.s. One solution is to take
a sample from those scientists who published at least one article that was listed in
Chemical Abstracts. But this approach excludes all nonpublishers. Other examples of
such truncation are easy to find. A study of how often people visit national parks could
be based on a survey given to people entering the park. Or, when you fill out a warranty
card for buying a TV, you might be asked how many TVs are in your household, leading

to a dataset in which everyone has at least one TV. Zero-truncated count models are designed for data in which observations with an outcome of zero have been excluded from the sample.

The zero-truncated Poisson model begins with the Poisson regression model from section 8.2:

$$\Pr(y_i \mid \mathbf{x}) = \frac{\exp(-\mu_i)\,\mu_i^{y_i}}{y_i!} \tag{8.4}$$

where $\mu = \exp(\mathbf{x}\boldsymbol{\beta})$. For a given set of x's, the probability of observing a zero is $\Pr(y = 0 \mid \mathbf{x}) = \exp(-\mu)$, so the probability of a nonzero count is $\Pr(y > 0 \mid \mathbf{x}) = 1 - \exp(-\mu)$. Since our counts are truncated at zero, we want to compute the probability for each positive outcome given that we know that the outcome is greater than zero. By the law of conditional probability, $\Pr(A \mid B) = \Pr(A \text{ and } B) / \Pr(B)$, the probability of observing a specific value of y given that we know the count is not zero is

$$\Pr(y_i \mid y_i > 0, \mathbf{x}_i) = \frac{\Pr(y_i \mid \mathbf{x}_i)}{\Pr(y_i > 0 \mid \mathbf{x}_i)} = \frac{\Pr(y_i \mid \mathbf{x}_i)}{1 - \exp(-\mu_i)} \quad \text{for } y > 0 \tag{8.5}$$

This formula simply increases each unconditional probability by the factor $\{1 - \exp(-\mu)\}^{-1}$, forcing the probability mass of the truncated distribution to sum to 1.

With a truncated model, there are two types of expected counts that might be of interest. First, we can compute the expectation without truncation. For example, we might want to estimate the expected number of publications for scientists in the entire population, not just among those who have at least one article. In the ZTP, the expected rate is the same as for the standard PRM:

$$E(y_i \mid \mathbf{x}_i) = \exp(\mathbf{x}_i\boldsymbol{\beta})$$

Second, we might be interested in the expected rate given that the count is positive, written as $E(y \mid y > 0, \mathbf{x})$. For example, among those who publish, what is the expected number of publications? The conditional rate must be larger than the unconditional rate since we are excluding zeros. As with the conditional probability in (8.5), the rate increases proportionally to the probability of a positive count:

$$E(y_i \mid y_i > 0, \mathbf{x}_i) = \frac{\mu_i}{\Pr(y_i > 0 \mid \mathbf{x}_i)} = \frac{\mu_i}{1 - \exp(-\mu_i)} \tag{8.6}$$

The same ideas applies to the NBRM where

$$\Pr(y_i \mid \mathbf{x}_i) = \frac{\Gamma\!\left(y_i + \alpha^{-1}\right)}{y_i!\,\Gamma(\alpha^{-1})} \left(\frac{\alpha^{-1}}{\alpha^{-1} + \mu_i}\right)^{\alpha^{-1}} \left(\frac{\mu_i}{\alpha^{-1} + \mu_i}\right)^{y_i} \tag{8.7}$$

so that $\Pr(y = 0 \mid \mathbf{x}) = (1 + \alpha\mu)^{-1/\alpha}$ and $\Pr(y > 0 \mid \mathbf{x}) = 1 - (1 + \alpha\mu)^{-1/\alpha}$. Accordingly, the conditional probability in the ZTNB is

$$\Pr(y_i \mid y_i > 0, \mathbf{x}_i) = \frac{\Pr(y_i \mid \mathbf{x}_i)}{1 - (1 + \alpha\mu_i)^{-1/\alpha}} \tag{8.8}$$

and the conditional mean equals

$$E(y_i \mid y_i > 0, \mathbf{x}_i) = \frac{\mu_i}{1 - (1 + \alpha\mu_i)^{-1/\alpha}} \qquad (8.9)$$

When dealing with truncated data, realize that the adverse effects of overdispersion are worse with truncated models. When the sample is not truncated, using the PRM in the presence of overdispersion does not bias the estimated βs. But with the ZTP model, overdispersion results in biased and inconsistent estimates of the βs and, consequently, in biased estimates of probabilities (Grogger and Carson 1991, 229). Accordingly, before using the zero-truncated Poisson model, you must check for overdispersion by first estimating the zero-truncated negative binomial model. As with the NBRM, overdispersion in the ZTNB is based on an LR test of $\alpha = 0$, which is included in the output for the ZTNB model (see page 376).

8.4.1   Fitting zero-truncated models

The zero-truncated Poisson model is fitted with the command

ztp *depvar* [*indepvars*] [*if*] [*in*] [*weight*] [**,** *options*]

where *options* are the same as those for **poisson** (see page 357). For the zero-truncated negative binomial model, the command is **ztnb**, with the same options.

8.4.2   Example of fitting zero-truncated models

To illustrate truncated count models, we continue our example of scientific productivity but begin by artificially truncating the sample to exclude scientists who have no publications.

```
. use http://www.stata-press.com/data/lf2/couart2, clear
(Academic Biochemist / S Long)
. drop if art==0                        // to artificially truncate the sample
(275 observations deleted)
```

(*Continued on next page*)

We then fit the model using the truncated sample:

```
. ztp art fem mar kid5 phd ment, nolog
Zero-truncated Poisson regression              Number of obs   =        640
                                               LR chi2(5)      =      98.52
                                               Prob > chi2     =     0.0000
Log likelihood = -1080.0336                    Pseudo R2       =     0.0436
```

art	Coef.	Std. Err.	z	P>|z|	[95% Conf. Interval]	
fem	-.2285826	.0652157	-3.51	0.000	-.3564031	-.1007621
mar	.096485	.0728252	1.32	0.185	-.0462497	.2392197
kid5	-.1421872	.0484538	-2.93	0.003	-.237155	-.0472195
phd	-.0127266	.0313043	-0.41	0.684	-.0740818	.0486287
ment	.0187455	.0022805	8.22	0.000	.0142758	.0232152
_cons	.6711393	.122456	5.48	0.000	.43113	.9111487

The output looks like that from `poisson`, except that the title now says "Zero-truncated Poisson regression". We then fit the ZTNB model:

```
. ztnb art fem mar kid5 phd ment, nolog
Zero-truncated negative binomial regression    Number of obs   =        640
                                               LR chi2(5)      =      44.58
Dispersion       = mean                        Prob > chi2     =     0.0000
Log likelihood = -1027.3185                    Pseudo R2       =     0.0212
```

art	Coef.	Std. Err.	z	P>|z|	[95% Conf. Interval]	
fem	-.2446712	.0972181	-2.52	0.012	-.4352153	-.0541272
mar	.1034172	.1094297	0.95	0.345	-.1110611	.3178955
kid5	-.1532593	.0722291	-2.12	0.034	-.2948257	-.011693
phd	-.0029336	.0480673	-0.06	0.951	-.0971437	.0912766
ment	.0237382	.0042868	5.54	0.000	.0153362	.0321402
_cons	.355125	.1968307	1.80	0.071	-.0306562	.7409062
/lnalpha	-.6034753	.2249915			-1.044451	-.1625001
alpha	.5469076	.1230496			.3518851	.850016

```
Likelihood-ratio test of alpha=0:   chibar2(01) =  105.43 Prob>=chibar2 = 0.000
```

The LR test provides clear evidence of overdispersion, favoring the ZTNB. Also note how the z-values in the ZTNB have decreased compared with the ZTP.

8.4.3   Interpretation of parameters

If zero counts are missing from your sample because of the way the data were collected, the parameters can be can be interpreted the same way as those for the PRM and the NBRM. Essentially, the model fills in the data that were lost by the way data were collected. If, however, you believe that the zeros are generated by a different process from that of positive counts, you need to interpret the parameters differently, as discussed in the section 8.5 on the hurdle model. Assuming that the zeros were generated by the same process as positive counts, $\exp\widehat{\beta}_k$ indicates the factor increase in the rate for a unit increase in x_k, holding all other variables constant. Keep in mind that we are referring to the factor change in the unconditional rate $E(y \mid \mathbf{x})$, not the conditional rate $E(y \mid y > 0, \mathbf{x})$.

There are several ways to compute the exponential of the coefficients. First, you can use the `irr` option (`irr` stands for incidence-rate ratios) with `ztp` or `ztnb`. Second, `listcoef` will compute factor change coefficients. Or, we could use the `eform` option with `estimates table`. For example,

```
. ztp art fem mar kid5 phd ment, nolog
(output omitted)
. est store ztp
. ztnb art fem mar kid5 phd ment, nolog
(output omitted)
. est store ztnb
. estimates table ztp ztnb, stats(N chi2 aic bic) eform b(%9.3f) t(%6.2f)
```

Variable	ztp	ztnb
art		
fem	0.796	0.783
	-3.51	-2.52
mar	1.101	1.109
	1.32	0.95
kid5	0.867	0.858
	-2.93	-2.12
phd	0.987	0.997
	-0.41	-0.06
ment	1.019	1.024
	8.22	5.54
_cons	1.956	1.426
	5.48	1.80
lnalpha		
_cons		0.547
		-2.68
Statistics		
N	640.000	640.000
chi2	98.519	44.578
aic	2172.067	2068.637
bic	2198.836	2099.867

legend: b/t

Do not be confused by the legend. Although the output reports `b/t`, it would be more accurate if it said `exp(b)/z`. We interpret the estimates as we did for the PRM or the NBRM. For example, the coefficient for `fem` in the ZTNB can be interpreted as follows:

> Being a female scientist decreases the expected number of papers by a factor of .78, holding all other variables constant.

8.4.4   Interpretation using predicted probabilities and rates

Estimates of predicted probabilities for y are computed using (8.4) and (8.7), with predicted probabilities for the truncated distribution computed using (8.5) and (8.8). As with other count models, probabilities and rates can be computed at specified values of the independent variables. prvalue and prgen compute both conditional and unconditional predictions. prchange and prtab compute unconditional rates and probabilities by default, but they compute conditional predictions when the conditional option is specified.[3]

For example, we can compute predictions for a married, female scientist without young children who is average on all other characteristics:

```
. prvalue, x(fem=1 mar=1 kid5=0)

ztnb: Predictions for art
                            Uncond      Cond
        Rate:               1.5611    2.3076
        Pr(y=0|x):          0.3235         .
        Pr(y=1|x):          0.2724    0.4027
        Pr(y=2|x):          0.1774    0.2623
        Pr(y=3|x):          0.1043    0.1542
        Pr(y=4|x):          0.0580    0.0857
        Pr(y=5|x):          0.0311    0.0460
        Pr(y=6|x):          0.0163    0.0241
        Pr(y=7|x):          0.0084    0.0124
        Pr(y=8|x):          0.0043    0.0063
        Pr(y=9|x):          0.0021    0.0032

                 fem         mar        kid5         phd        ment
        x=         1           1           0   3.1539765    10.14375
```

There are several things to note. First, the unconditional rate is lower than the conditional rate, which must occur since the conditional rate considers only those with at least one paper. Second, the unconditional probability of a zero count is .32, indicating that we expect that 32% of the scientists with these characteristics would not publish. The conditional probability of zero is missing since we are conditioning on having at least one publication (i.e., you cannot have zero publications if you know you have at least one publication). Finally, the conditional probabilities are greater than the unconditional probabilities since the 32% zeros in the unconditional distribution are distributed across these other values. To see the effect of overdispersion in the estimates, we encourage you to compare these results with those that you would get from the ztp model.

---

3. Confidence intervals are not available for the ztp and ztnb commands.

8.4.5   Computing predicted rates and probabilities in the estimation sample

We can also compute predictions for the entire sample using `prcounts`. For example,

```
. ztnb art fem mar kid5 phd ment, nolog
  (output omitted)
. prcounts mztnb
```

`prcounts` generates the variables `mztnbrate` and `mztnbCrate`, which contain unconditional and conditional rates for each observation in the estimation sample (i.e., those observations that were used when fitting the model). The values in `mztnbrate` and `mztnbCrate` are identical to those you would obtain from `predict` *name*, `ir`, where `ir` stands for incidence rate and `predict` *name*, `cm`, where `cm` stands for conditional mean. Unconditional probabilities are computed with the names `mztnbpr0`, `mztnbpr1`, etc., and conditional probabilities with names `mztnbCpr0`, `mztnbCpr1`, etc. These quantities can be used for computing mean predictions, as we show with the PRM and NBRM (page 388).

8.5   The hurdle regression model[*]

The hurdle regression model (HRM) combines a binary model to predict zeros and a zero-truncated Poisson or zero-truncated negative binomial model to predict nonzero counts (Mullahy 1986; Cameron and Trivedi 1998). Let y be a count outcome that ranges from zero to some maximum value. Suppose that zero counts are generated by a binary process. Here we use a logit to model the binary outcome $y = 0$ versus $y > 0$, but other binary models could be used:

$$\Pr(y_i = 0 \mid \mathbf{x}_i) = \frac{\exp(\mathbf{x}_i \boldsymbol{\gamma})}{1 + \exp(\mathbf{x}_i \boldsymbol{\gamma})} = \pi_i$$

Positive counts are generated by either the ZTP or ZTNB models from section 8.4. In this two-equation model, zero is viewed as a hurdle that you have to get past before reaching positive counts. Since there are separate equations to predict zero counts and positive counts, this allows the process predicting zeros to be different from the process that determines positive counts. The predicted rates and probabilities for the HRM are computed by mixing the results of the binary model and the zero-truncated model. The probability of a zero is

$$\Pr(y_i = 0 \mid \mathbf{x}_i) = \pi_i$$

as fitted by our binary model. Since positive counts can occur only if you get past the zero hurdle (which occurs with probability $1 - \pi$), we weight the conditional probability from the zero-truncated model:

$$\Pr(y_i \mid \mathbf{x}_i) = (1 - \pi_i) \Pr(y_i \mid y_i > 0, \mathbf{x}_i) \quad \text{for } y > 0 \tag{8.10}$$

The unconditional rate is computed by combining the mean rates for those with $y = 0$ (which, of course, is 0) and the mean rate for those with positive counts:

$$E\left(y_i \mid \mathbf{x}_i\right) = \left[\pi_i \times 0\right] + \left\{(1 - \pi_i) \times E(y_i \mid y_i > 0, \mathbf{x}_i)\right\}$$
$$= (1 - \pi_i) \times E(y_i \mid y_i > 0, \mathbf{x}_i)$$

where $E(y_i \mid y_i > 0, \mathbf{x}_i)$ is defined by (8.6) for the ZTP and (8.9) for the ZTNB.

Although Stata does not include the hurdle model as a command, it can be fitted by combining results from `logit` with those from either `ztnb` or `ztp`. Using these commands along with `prvalue` and `prcounts`, we can compute everything that we need. This process involves many more steps than the other examples in this book. In the process, however, this section reveals much about how to accomplish postestimation analyses "by hand".

8.5.1 In-sample predictions for the hurdle model

We begin by fitting the logit model and storing our results:

```
. use http://www.stata-press.com/data/lf2/couart2, clear
(Academic Biochemists / S Long)

. gen iszero = (art==0)

. label var iszero "Does scientists have 0 articles?"

. label def iszero 1 NoArticles 0 HasArticles

. label val iszero iszero

. logit iszero fem-ment, nolog or
(output omitted )

. est store hlogit
```

Next we fit a ZTNB model using `if art > 0` to restrict our sample to those with at least one paper:

```
. ztnb art fem-ment if art>0, nolog irr
(output omitted )

. est store hztnb
```

It would be handy to print the model with the coefficients in two columns. Although this is not possible with `estimates table` (try `estimates table hlogit hztnb` to see what happens), it is possible to generate a nice table using `estout`:

```
. estout hlogit hztnb, eq(1) style(fixed)         ///
> cells(b(star fmt(%9.3f)) se(par fmt(%9.3f)))    ///
> collabels("") mlabels("h_logit" "h_ztnb")       ///
> eqlabels(,none) stats(N ll, fmt(%9.0f %9.3f))   ///
> varlabels(lnalpha:_cons _alpha) legend
                        h_logit             h_ztnb

fem                      0.251             -0.245*
                        (0.159)            (0.097)
mar                     -0.326              0.103
                        (0.181)            (0.109)
kid5                     0.285*            -0.153*
                        (0.111)            (0.072)
phd                     -0.022             -0.003
                        (0.080)            (0.048)
ment                    -0.080***           0.024***
                        (0.013)            (0.004)
_cons                   -0.237              0.355
                        (0.296)            (0.197)
_alpha                                     -0.603**
                                           (0.225)
N                          915                640
ll                    -525.278          -1027.319
* p<0.05, ** p<0.01, *** p<0.001
```

If a coefficient for x_k is greater than 1 in the logit portion of the model, it means that a unit increase in x_k increases the odds of no publications by a factor of $\exp\widehat{\beta}_k$, holding other variables constant. For example, consider the effect of young children:

> For an additional young child, the odds of not publishing increase by a factor of 1.3 [=exp(.285)], holding all other variables constant.

In the ZTNB part of the model, positive coefficients show that the variable increases the rate of publication. A variable can be significant in one part of the model, but not in the other part. For example, women are not significantly different from men in terms of "getting over the hurdle", but they do have a significantly lower rate of publication once over the hurdle.

We can compute the probability of a zero from the logit portion of the model using `predict`:

```
. estimates restore hlogit
(results hlogit are active now)
. predict Hlgt_pry0, p
. label var Hlgt_pry0 "blm: Pr(y=0)=Pr(iszero=1)"
```

The average predicted probability of a zero is computed using `summarize`:

```
. summarize Hlgt_pry0
```

Variable	Obs	Mean	Std. Dev.	Min	Max
Hlgt_pry0	915	.3005464	.1180336	.0015213	.5612054

To compute predictions of positive counts, we restore the results from the ZTNB and use `prcounts`. Here is the shortened output, where we look only at the conditional predictions that will be used later:

```
. estimates restore hztnb
(results hztnb are active now)
. prcounts Hztnb_
. describe Hztnb_C*
```

variable name	storage type	display format	value label	variable label
Hztnb_Crate	float	%9.0g		Predicted conditional rate from ztnb
Hztnb_Cpr1	float	%9.0g		Pr(y=1|y>0) from ztnb
Hztnb_Cpr2	float	%9.0g		Pr(y=2|y>0) from ztnb
(output omitted)				
Hztnb_Cpr9	float	%9.0g		Pr(y=9|y>0) from ztnb
(output omitted)				

```
. summarize Hztnb_C*
```

Variable	Obs	Mean	Std. Dev.	Min	Max
Hztnb_Crate	915	2.356661	.52643	1.73671	8.774148
Hztnb_Cpr1	915	.4076762	.0722457	.0673948	.558071
Hztnb_Cpr2	915	.2573494	.0191666	.078269	.2678436
(output omitted)					
Hztnb_Cpr9	915	.0043409	.0057484	.0003102	.0542898
(output omitted)					

With these quantities, we can compute the predicted rate for the hurdle model:

```
. gen Hmu = (1-Hlgt_pry0) * Hztnb_Crate
. label var Hmu "Hurdle: mu"
. summarize Hmu
```

Variable	Obs	Mean	Std. Dev.	Min	Max
Hmu	915	1.697686	.7016245	.7705519	8.7608

The predicted probabilities are computed using a `foreach` loop:

```
. gen Hpr0 = Hlgt_pry0
. label var Hpr0 "Hurdle: pr(y=0)"
. foreach k of numlist 1(1)9 {
  2.     gen Hpr`k' = (1-Hlgt_pry0) * Hztnb_Cpr`k'
  3.     label var Hpr`k' "Hurdle: pr(y=`k')"
  4. }
```

```
. summarize Hpr*
    Variable |      Obs        Mean    Std. Dev.        Min         Max
-------------+--------------------------------------------------------
        Hpr0 |      915    .3005464    .1180336    .0015213    .5612054
        Hpr1 |      915    .2772253    .0264024    .0672922    .3398655
        Hpr2 |      915    .1783374    .0225025      .07815    .2233984
        Hpr3 |      915    .1053164    .0246465    .0489526    .1597355
        Hpr4 |      915    .0599405     .022171    .0196098    .1201168
-------------+--------------------------------------------------------
        Hpr5 |      915    .0336258    .0178631    .0075859    .0961457
        Hpr6 |      915    .0188354    .0136543     .002865    .0806172
        Hpr7 |      915    .0106266    .0102113    .0010633    .0676752
        Hpr8 |      915    .0060773    .0075988    .0003894    .0604464
        Hpr9 |      915    .0035407    .0056788    .0001406    .0542072
```

The mean predicted probabilities can be compared with the observed predictions, as illustrated for the PRM and NBRM. The mean predicted probability of a zero (.3005464) corresponds exactly to the observed percentage of zeros calculated by `tab art` (30.05%). This will always be the case, since the mean prediction from a binary logit model equals the observed probability.

8.5.2 Predictions for user-specified values

We can use `prvalue` to compute predictions at specific values of the independent variables. First, we restore the logit model and compute predictions for men (`fem=0`) from weak Ph.D. programs (`phd=1`) who have mentors that have not published (`ment=0`):

```
. est restore hlogit
(results hlogit are active now)
. prvalue, x(fem=0 phd=1 ment=0) rest(mean)

logit: Predictions for iszero
Confidence intervals by delta method
                          95% Conf. Interval
    Pr(y=1|x):    0.4173   [ 0.3241,   0.5105]
    Pr(y=0|x):    0.5827   [ 0.4895,   0.6759]

           fem        mar        kid5        phd       ment
x=           0  .66229508   .49508197          1          0
```

`prvalue` saves its results in the matrix `pepred`, which looks like this:

```
. matrix list pepred
pepred[7,2]
                   c1          c2
   1values          0           1
     2prob   .58269751   .41730249
     3misc  -.33385682           .
  4sav_prob          .           .
  5sav_misc          .           .
  6dif_prob          .           .
  7dif_misc          .           .
```

Rows four and five are missing but would hold the predicted values if the save option had been used with prvalue. Rows six and seven are filled in when the diff option is used. Here we need only the information in the second row. The probability of a zero count, which corresponds to a one in the outcome for the logit, is located in the second row and second column. We save this value in a scalar variable for use in later computations:

```
. scalar hrdlp0 = pepred[2,2]
```

Now we restore the zero-truncated part of the model, compute prvalue, and grab the values that we want from peCpred. We use peCpred, not pepred, since we want the conditional predictions:

```
. est restore hztnb
(results hztnb are active now)
. prvalue, x(fem=0 phd=1 ment=0) rest(mean) all

ztnb: Predictions for art
                     Uncond      Cond
    Rate:            1.4117    2.1762
    Pr(y=0|x):       0.3513         .
    Pr(y=1|x):       0.2798    0.4314
    Pr(y=2|x):       0.1724    0.2658
    Pr(y=3|x):       0.0959    0.1478
    Pr(y=4|x):       0.0504    0.0777
    Pr(y=5|x):       0.0256    0.0395
    Pr(y=6|x):       0.0127    0.0196
    Pr(y=7|x):       0.0062    0.0095
    Pr(y=8|x):       0.0030    0.0046
    Pr(y=9|x):       0.0014    0.0022

             fem        mar        kid5        phd       ment
    x=         0  .66229508   .49508197          1          0
```

The all option in the last prvalue specifies that all observations in memory should be used when computing mean values, not just those in the estimation sample. This is necessary since ztnb was estimated using only cases with positive counts, but we want the means to be based on the sample used for logit. More details are discussed in the next section.

We now want to compute the probabilities for the positive counts. We do this by looping through the count values 1–9, retrieving the needed quantities from the peCpred matrix, and applying the formula from (8.10):

```
. forvalues i = 1(1)9 {
  2.        scalar ztnbp'i' = peCpred[2,'i'+1]
  3.        scalar hrdlp'i' = (1-hrdlp0) * ztnbp'i'
  4.        display "Prob(art='i'|X) = " hrdlp'i'
  5. }
Prob(art=1|X) = .25136843
Prob(art=2|X) = .15488527
Prob(art=3|X) = .08611766
Prob(art=4|X) = .04529185
Prob(art=5|X) = .02300292
Prob(art=6|X) = .01140601
Prob(art=7|X) = .00555765
Prob(art=8|X) = .00267218
Prob(art=9|X) = .00127141

. display "Prob(art=0|X) = " hrdlp0
Prob(art=0|X) = .41730249
```

These predictions are for a male scientist from a low-prestige department who is married with an average number of children. Given these characteristics, we predict the probability of not publishing to be .25, with an expected rate of publication equal to .42.

Warning regarding sample specification

Remember: the two parts of the hurdle model are functionally independent, which allows us to fit the model by first fitting a binary model on the entire sample and then fitting a truncated model using only cases with positive counts. Accordingly, the two parts of the model are fitted with different samples. When computing predictions at specified values of the independent variables—say, the mean—you need to compute the mean with the entire sample used in the binary portion of the model, not the smaller sample used for the truncated model. By default, the **prvalue** command computes the values specified by rest() based on the estimation sample. For example, after fitting the ztnb model, **prvalue, rest(mean)** computes the means using only those cases with positive outcomes. That is, the cases with zero counts will not be included. What we want, however, is to compute the means using the entire sample used for the binary portion of the model.

First, before fitting the binary portion of the model, drop any cases with missing data or that you would otherwise drop using an **if** or **in** condition. (This was not necessary in our example above since we wanted to use all the cases.) Second, fit the truncated model with the **if** condition used to drop cases that are zero on the outcome (e.g., ztnb art fem-ment if art>0, nolog irr). Finally, when using **prvalue**, include the **all** option, which specifies that all cases in memory should be used to compute the values specified by rest(), rather than using only the estimation sample.

8.6   Zero-inflated count models

The NBRM improves upon the underprediction of zeros in the PRM by increasing the conditional variance without changing the conditional mean. The hurdle model addresses the underprediction of zeros by using two equations, a binary model to predict zeros and a zero-truncated model for the remaining counts. Zero-inflated count models, introduced by Lambert (1992), change the mean structure to allow zeros to be generated by two distinct processes, compared with one process generating zeros in the hurdle model. Consider our example of scientific productivity. The PRM, NBRM, and hurdle models assume that *every* scientist has a positive probability of publishing any given number of papers. The probability differs across individuals according to their characteristics, but *all* scientists have some probability of publishing. Substantively, this would be unrealistic if some scientists are not potential publishers. For example, they could hold positions, perhaps in industry, where publishing is not allowed. Zero-inflated models allow for this possibility, thereby increasing the conditional variance and the probability of zero counts.

The zero-inflated model assumes that there are two *latent* (i.e., unobserved) groups. An individual in the *Always Zero group* (Group A) has an outcome of 0 with a probability of 1, whereas an individual in the *Not Always Zero group* (Group ~A) might have a zero count, but there is a nonzero probability that she has a positive count. This process is developed in three steps: (1) model membership into the latent groups; (2) Model counts for those in Group ~A; and (3) compute observed probabilities as a mixture of the probabilities for the two groups.

Step 1: Membership in Group A

Let $A = 1$ if someone is in Group A, else $A = 0$. Group membership is a binary outcome that can be modeled using the logit or probit model of chapter 4

$$\psi_i = \Pr\left(A_i = 1 \mid z_i\right) = F\left(z_i\gamma\right) \tag{8.11}$$

where ψ_i is the probability of being in Group A for individual i. The z-variables are referred to as *inflation* variables because they inflate the number of 0s as shown below. To illustrate (8.11), assume that two variables affect the probability of an individual being in Group A and that we model this with a logit equation:

$$\psi_i = \frac{\exp\left(\gamma_0 + \gamma_1 z_1 + \gamma_2 z_2\right)}{1 + \exp\left(\gamma_0 + \gamma_1 z_1 + \gamma_2 z_2\right)}$$

If we had an observed variable indicating group membership, this would be a standard, binary regression model. But because group membership is a latent variable, we do not know whether an individual is in Group A or Group ~A.

Step 2: Counts for those in Group ~A

Among those who are *not* always zero, the probability of each count (including zeros) is determined by either a Poisson or a negative binomial regression. In the equations that follow, we are conditioning both on the x_k's and on $A = 0$. Also the x_k's are not necessarily the same as the inflation variables z_k in the first step (although the two sets of variables can be the same). For the *zero-inflated Poisson* (ZIP) *model*, we have

$$\Pr(y_i \mid \mathbf{x}_i, \ A_i = 0) = \frac{e^{-\mu_i} \mu_i^{y_i}}{y_i!}$$

or, for the *zero-inflated negative binomial* (ZINB) *model*,

$$\Pr\left(y_i \mid \mathbf{x}_i, \ A_i = 0\right) = \frac{\Gamma(y_i + \alpha^{-1})}{y_i! \Gamma(\alpha^{-1})} \left(\frac{\alpha^{-1}}{\alpha^{-1} + \mu_i}\right)^{\alpha^{-1}} \left(\frac{\mu_i}{\alpha^{-1} + \mu_i}\right)^{y_i}$$

In both equations, $\mu_i = \exp(\mathbf{x}_i \boldsymbol{\beta})$. If we knew which observations were in Group ~A, these equations would define the PRM and the NBRM. But here the equations apply only to those observations in Group ~A, and we do not have an observed variable indicating group membership.

Step 3: Mixing Groups A and ~A

The simplest way to understand the mixing is to start with an example. Suppose that retirement status is indicated by $r = 1$ for retired folks and $r = 0$ for those not retired, where

$$\Pr(r = 1) = .2$$
$$\Pr(r = 0) = 1 - .2 = .8$$

Let y indicate living in a warm climate, with $y = 1$ for yes and $y = 0$ for no. Suppose that the conditional probabilities are

$$\Pr(y = 1 \mid r = 1) = .5$$
$$\Pr(y = 1 \mid r = 0) = .3$$

so that people are more likely to live in a warm climate if they are retired. What is the probability of living in a warm climate for the population as a whole? The answer is a mixture of the probabilities for the two groups weighted by the proportion in each group:

$$\Pr(y = 1) = \{\Pr(r = 1) \times \Pr(y = 1 \mid r = 1)\}$$
$$+ \{\Pr(r = 0) \times \Pr(y = 1 \mid r = 0)\}$$
$$= [.2 \times .5] + [.8 \times .3] = .34$$

In other words, the two groups are mixed according to their proportions in the population to determine the overall rate. The same thing is done for the zero-inflated models.

The proportion in each group is defined by

$$\Pr(A_i = 1) = \psi_i$$
$$\Pr(A_i = 0) = 1 - \psi_i$$

and the probabilities of a zero within each group are

$$\Pr(y_i = 0 \mid A_i = 1, \mathbf{x}_i, \mathbf{z}_i) = 1 \quad \text{by definition of the } A \text{ group}$$
$$\Pr(y_i = 0 \mid A_i = 0, \mathbf{x}_i, \mathbf{z}_i) = \text{outcome of PRM or NBRM.}$$

Then the overall probability of a zero count is

$$\Pr(y_i = 0 \mid \mathbf{x}_i, \mathbf{z}_i) = [\psi_i \times 1] + \{(1 - \psi_i) \times \Pr(y_i = 0 \mid \mathbf{x}_i, A_i = 0)\}$$
$$= \psi_i + \{(1 - \psi_i) \times \Pr(y_i = 0 \mid \mathbf{x}_i, A_i = 0)\}$$

For outcomes other than 0,

$$\Pr(y_i = k \mid \mathbf{x}_i, \mathbf{z}_i) = [\psi_i \times 0] + \{(1 - \psi_i) \times \Pr(y_i = k \mid \mathbf{x}_i, A_i = 0)\}$$
$$= (1 - \psi_i) \times \Pr(y_i = k \mid \mathbf{x}_i, A_i = 0)$$

where we use the assumption that the probability of a positive count in Group A is 0.

Expected counts are computed similarly :

$$E(y \mid \mathbf{x}, \mathbf{z}) = [0 \times \psi] + \{\mu \times (1 - \psi)\}$$
$$= \mu(1 - \psi)$$

Because $0 \leq \psi \leq 1$, the expected value will be smaller than μ, which shows that the mean structure in zero-inflated models differs from that in the PRM or NBRM.

8.6.1   Fitting zero-inflated models with zinb and zip

The ZIP and ZINB models are fitted with the zip and zinb commands and their basic options:

zip *depvar* [*indepvars*] [*if*] [*in*] [*weight*], <u>inf</u>late(*indepvars2*) [<u>noco</u>nstant

exposure(*varname*) <u>constr</u>aints(*constraints*) probit <u>r</u>obust

<u>cl</u>uster(*varname*) <u>l</u>evel(*#*) irr vuong <u>nolog</u>]

zinb *depvar* [*indepvars*] [*if*] [*in*] [*weight*], <u>inf</u>late(*indepvars2*)

[<u>noco</u>nstant exposure(*varname*) <u>constr</u>aints(*constraints*) probit <u>r</u>obust

<u>cl</u>uster(*varname*) <u>l</u>evel(*#*) irr vuong <u>nolog</u>]

Variable lists

depvar is the dependent variable, which must be a count variable.

indepvars is a list of independent variables that determine counts among those who are not always zeros. If *indepvars* is not included, a model with only an intercept is fitted.

indepvars2 is a list of inflation variables that determine if you are in the Always Zero group or the Not Always Zero group.

indepvars and *indepvars2* can be the same variables but do not have to be.

Options

Here we consider only options that differ from those in earlier models for this chapter.

probit specifies that the model determining the probability of being in the Always Zero group versus the Not Always Zero group is to be a binary probit model. By default, a binary logit model is used.

vuong requests a Vuong (1989) test of the ZIP model versus the PRM, or of the ZINB versus the NBRM. Details are given in section 8.7.2.

8.6.2   Example of fitting the ZIP and ZINB models

The output from zip and zinb is similar, so here we show only the output for zinb:

(*Continued on next page*)

```
. zinb art fem mar kid5 phd ment, inf(fem mar kid5 phd ment) nolog
Zero-inflated negative binomial regression        Number of obs   =        915
                                                  Nonzero obs     =        640
                                                  Zero obs        =        275
Inflation model = logit                           LR chi2(5)      =      67.97
Log likelihood  = -1549.991                       Prob > chi2     =     0.0000
```

	Coef.	Std. Err.	z	P>|z|	[95% Conf.	Interval]
art						
fem	-.1955068	.0755926	-2.59	0.010	-.3436655	-.0473481
mar	.0975826	.084452	1.16	0.248	-.0679402	.2631054
kid5	-.1517325	.054206	-2.80	0.005	-.2579744	-.0454906
phd	-.0007001	.0362696	-0.02	0.985	-.0717872	.0703869
ment	.0247862	.0034924	7.10	0.000	.0179412	.0316312
_cons	.4167466	.1435962	2.90	0.004	.1353032	.69819
inflate						
fem	.6359328	.8489175	0.75	0.454	-1.027915	2.299781
mar	-1.499469	.9386701	-1.60	0.110	-3.339228	.3402909
kid5	.6284274	.4427825	1.42	0.156	-.2394105	1.496265
phd	-.0377153	.3080086	-0.12	0.903	-.641401	.5659705
ment	-.8822932	.3162276	-2.79	0.005	-1.502088	-.2624984
_cons	-.1916865	1.322821	-0.14	0.885	-2.784368	2.400995
/lnalpha	-.9763565	.1354679	-7.21	0.000	-1.241869	-.7108443
alpha	.3766811	.0510282			.288844	.4912293

The top set of coefficients, labeled `art` at the left margin, corresponds to the NBRM for those in the Not Always Zero group. The lower set of coefficients, labeled `inflate`, corresponds to the binary model predicting group membership.

8.6.3 Interpretation of coefficients

When interpreting zero-inflated models, it is easy to be confused by the direction of the coefficients. `listcoef` makes interpretation simpler. For example, consider the results for the ZINB:

```
. listcoef, help
zinb (N=915): Factor Change in Expected Count
 Observed SD: 1.926069
Count Equation: Factor Change in Expected Count for Those Not Always 0
```

art	b	z	P>|z|	e^b	e^bStdX	SDofX
fem	-0.19551	-2.586	0.010	0.8224	0.9071	0.4987
mar	0.09758	1.155	0.248	1.1025	1.0473	0.4732
kid5	-0.15173	-2.799	0.005	0.8592	0.8904	0.7649
phd	-0.00070	-0.019	0.985	0.9993	0.9993	0.9842
ment	0.02479	7.097	0.000	1.0251	1.2650	9.4839
ln alpha	-0.97636					
alpha	0.37668	SE(alpha) = 0.05103				

```
       b = raw coefficient
       z = z-score for test of b=0
   P>|z| = p-value for z-test
     e^b = exp(b) = factor change in expected count for unit increase in X
e^bStdX = exp(b*SD of X) = change in expected count for SD increase in X
   SDofX = standard deviation of X
Binary Equation: Factor Change in Odds of Always 0
```

Always0	b	z	P>|z|	e^b	e^bStdX	SDofX
fem	0.63593	0.749	0.454	1.8888	1.3732	0.4987
mar	-1.49947	-1.597	0.110	0.2232	0.4919	0.4732
kid5	0.62843	1.419	0.156	1.8747	1.6172	0.7649
phd	-0.03772	-0.122	0.903	0.9630	0.9636	0.9842
ment	-0.88229	-2.790	0.005	0.4138	0.0002	9.4839

```
       b = raw coefficient
       z = z-score for test of b=0
   P>|z| = p-value for z-test
     e^b = exp(b) = factor change in odds for unit increase in X
e^bStdX = exp(b*SD of X) = change in odds for SD increase in X
   SDofX = standard deviation of X
```

The top half of the output, labeled `Count Equation`, contains coefficients for the factor change in the expected count for those in the Not Always Zero group. This group comprises those scientists who have the opportunity to publish. The coefficients can be interpreted in the same way as coefficients from the PRM or the NBRM. For example, among those who have the opportunity to publish, being a woman decreases the expected rate of publication by a factor of .82, holding all other factors constant.

The bottom half, labeled `Binary Equation`, contains coefficients for the factor change in the odds of being in the Always Zero group compared with the Not Always Zero group. These can be interpreted just as the coefficients for a binary logit model. For example, being a woman increases the odds of not having the opportunity to publish by a factor of 1.89, holding all other variables constant.

As we found in this example, when the same variables are included in both equations, the signs of the corresponding coefficients from the binary equation are often in the opposite direction of those from the count equation. This often makes substantive sense because the binary process is predicting membership in the group that always has zero counts, so a positive coefficient implies lower productivity. The count process predicts number of publications so that a negative coefficient would indicate lower productivity.

8.6.4 Interpretation of predicted probabilities

For the ZIP model,

$$\widehat{\Pr}\left(y = 0 \mid \mathbf{x}, \mathbf{z}\right) = \widehat{\psi} + \left(1 - \widehat{\psi}\right) e^{-\widehat{\mu}}$$

where $\widehat{\mu} = \exp\left(\mathbf{x}\boldsymbol{\beta}\right)$ and $\widehat{\psi} = F\left(\mathbf{z}\widehat{\boldsymbol{\gamma}}\right)$. The predicted probability of a positive count applies only to the $1 - \widehat{\psi}$ observations in the Not Always Zero group:

$$\widehat{\Pr}(y \mid \mathbf{x}) = \left(1 - \widehat{\psi}\right) \frac{e^{-\widehat{\mu}_i} \widehat{\mu}^y}{y!}$$

Similarly, for the ZINB model,

$$\widehat{\Pr}\left(y = 0 \mid \mathbf{x}, \mathbf{z}\right) = \widehat{\psi} + \left(1 - \widehat{\psi}\right) \left(\frac{\widehat{\alpha}^{-1}}{\widehat{\alpha}^{-1} + \widehat{\mu}_i}\right)^{\widehat{\alpha}^{-1}}$$

and the predicted probability for a positive count is

$$\widehat{\Pr}(y \mid \mathbf{x}) = \left(1 - \widehat{\psi}\right) \frac{\Gamma\left(y + \widehat{\alpha}^{-1}\right)}{y!\,\Gamma(\widehat{\alpha}^{-1})} \left(\frac{\widehat{\alpha}^{-1}}{\widehat{\alpha}^{-1} + \widehat{\mu}}\right)^{\widehat{\alpha}^{-1}} \left(\frac{\widehat{\mu}}{\widehat{\alpha}^{-1} + \widehat{\mu}}\right)^{y}$$

The probabilities can be computed with `prvalue`, `prcounts`, and `prgen`.

Predicted probabilities with prvalue

`prvalue` works in the same way for `zip` and `zinb` as it did for earlier count models, although the output is slightly different. Suppose that we want to compare the predicted probabilities for a married female scientist with young children who came from a weak graduate program with those for a married male from a strong department who had a productive mentor:

```
. quietly prvalue, x(fem=0 mar=1 kid5=3 phd=3 ment=10) save
. prvalue, x(fem=1 mar=1 kid5=3 phd=1 ment=0) diff
zinb: Change in Predictions for art
                    Current      Saved     Change
     Expected y:     0.2717     1.3563    -1.0845
  Pr(Always0|z):     0.6883     0.0002     0.6882
   Pr(y=0|x,z):      0.8350     0.3344     0.5006
    Pr(y=1|x):       0.0962     0.3001    -0.2038
    Pr(y=2|x):       0.0435     0.1854    -0.1419
    Pr(y=3|x):       0.0167     0.0973    -0.0806
    Pr(y=4|x):       0.0058     0.0465    -0.0407
    Pr(y=5|x):       0.0019     0.0209    -0.0190
    Pr(y=6|x):       0.0006     0.0090    -0.0084
    Pr(y=7|x):       0.0002     0.0038    -0.0036
    Pr(y=8|x):       0.0001     0.0015    -0.0015
    Pr(y=9|x):       0.0000     0.0006    -0.0006

x values for count equation
            fem   mar   kid5   phd   ment
Current=     1     1     3      1     0
  Saved=     0     1     3      3     10
   Diff=     1     0     0     -2    -10

z values for binary equation
            fem   mar   kid5   phd   ment
Current=     1     1     3      1     0
  Saved=     0     1     3      3     10
   Diff=     1     0     0     -2    -10
```

There are two major differences in the output of `prvalue` for `zip` and `zinb` compared with other count models. First, the output lists levels of both the x variables from the count equation and the z variables from the binary equation. Here they are the same variables, but they could be different. Second, there are two probabilities of zero counts. For example, for our female scientists, `prvalue` lists `Pr(y=0|x,z): 0.8350`, which is the probability of having no publications, either because a scientist does not have the opportunity to publish or because a scientist is a potential publisher who by chance did not publish. The quantity `Pr(Always0|z): 0.6883` is the probability of not having the opportunity to publish. Thus most of the 0s for women are due to being in the group that never publishes. The remaining probabilities listed are the probabilities of observing each count of publications for the specified set of characteristics.

Confidence intervals with prvalue

By specifying the `bootstrap` option, we can add confidence intervals to `prvalue` for the `zip` and `zinb` models. Suppose that we want to compare the predicted probabilities and rates for men from a strong graduate program with women from an adequate program. For both groups, we assume that the mentor had 10 publications (`ment=10`) and that they are married (`mar=1`) with three young children (`k5=3`). Since we are using the bootstrap, which draws repeated, random samples, we begin by specifying the seed for the random-number generator so that when we run our do-file again, we will get the same answer.

```
. set seed 14972132
```

In general, the `zip` and `zinb` models fail to converge more often than other count models that we consider in this book. When you are taking 1,000 bootstrap samples, it is not uncommon to draw at least one sample for which `zip` and `zinb` will not converge, even after 16,000 iterations (the default number). If you wait for 16,000 iterations to be computed before concluding that the model will not converge, you are going to wait a long time. We find that if the model has not converged by iteration 50, it will not converge with 16,000 iterations either. Accordingly, we set the maximum number of iterations to 50.

```
. set maxiter 50
```

Another way is to fit your model for the full sample and count how many iterations were required. Quadruple this number and use it for the maximum number of iterations to compute before concluding that the model is not going to converge.

Next we compute the predicted values for men from strong programs:

```
. prvalue, x(fem=0 mar=1 kid5=3 phd=3 ment=10) save boot dots
```

We save the results so that we can compute discrete changes. The `dots` option prints a period after each iteration and periodically updates you on what percentage of the iterations has been completed. If the dots "get stuck", it probably means that you forgot to `set maxiter` and you have encountered a sample where the model will not converge. Our results, editing out the dots, look like this:

```
zinb: Predictions for art
Bootstrapped confidence intervals using percentile method
(949 of 1000 replications completed)
                                   95% Conf. Interval
    Expected y:            1.3563   [ 1.0135,   1.7999]
    Pr(Always0|z):         0.0002   [ 0.0000,   0.0471]
    Pr(y=0|x,z):           0.3344   [ 0.2508,   0.4217]
    Pr(y=1|x):             0.3001   [ 0.2634,   0.3190]
    Pr(y=2|x):             0.1854   [ 0.1571,   0.2075]
    Pr(y=3|x):             0.0973   [ 0.0675,   0.1279]
    Pr(y=4|x):             0.0465   [ 0.0256,   0.0717]
    Pr(y=5|x):             0.0209   [ 0.0092,   0.0383]
    Pr(y=6|x):             0.0090   [ 0.0031,   0.0202]
    Pr(y=7|x):             0.0038   [ 0.0010,   0.0104]
    Pr(y=8|x):             0.0015   [ 0.0003,   0.0051]
    Pr(y=9|x):             0.0006   [ 0.0001,   0.0025]

x values for count equation
       fem   mar   kid5   phd   ment
  x=     0     1     3      3     10

z values for binary equation
       fem   mar   kid5   phd   ment
  z=     0     1     3      3     10
```

Next we compute predictions for women from adequate programs whose mentors publish infrequently and use `diff` to compute the difference in predictions:

```
. prvalue, x(fem=1 mar=1 kid5=3 phd=1 ment=2) diff boot dots

zinb: Change in Predictions for art

Bootstrapped confidence intervals using percentile method
(961 of 1000 replications completed)
                        Current     Saved    Change    95% CI for Change
    Expected y:          .66478    1.3563   -.69149   [-1.4494,  -0.3482]
    Pr(Always0|z):       0.2744    0.0002    0.2743   [ 0.0000,   0.9391]
    Pr(y=0|x,z):         0.6047    0.3344    0.2703   [ 0.1091,   0.6564]
    Pr(y=1|x):           0.2249    0.3001   -0.0751   [-0.2757,   0.0134]
    Pr(y=2|x):           0.1055    0.1854   -0.0800   [-0.1796,  -0.0359]
    Pr(y=3|x):           0.0420    0.0973   -0.0553   [-0.1028,  -0.0316]
    Pr(y=4|x):           0.0152    0.0465   -0.0313   [-0.0576,  -0.0177]
    Pr(y=5|x):           0.0052    0.0209   -0.0157   [-0.0306,  -0.0075]
    Pr(y=6|x):           0.0017    0.0090   -0.0073   [-0.0164,  -0.0029]
    Pr(y=7|x):           0.0005    0.0038   -0.0032   [-0.0084,  -0.0011]
    Pr(y=8|x):           0.0002    0.0015   -0.0014   [-0.0041,  -0.0003]
    Pr(y=9|x):           0.0001    0.0006   -0.0006   [-0.0020,  -0.0001]

x values for count equation
              fem   mar   kid5   phd   ment
    Current=    1     1      3     1      2
      Saved=    0     1      3     3     10
       Diff=    1     0      0    -2     -8

z values for binary equation
              fem   mar   kid5   phd   ment
    Current=    1     1      3     1      2
      Saved=    0     1      3     3     10
       Diff=    1     0      0    -2     -8
```

We find that women with weaker graduate training are expected to have .69 fewer publications than men with better training, with a 95% confidence interval from .35 to 1.45.

There are two major differences in the output of `prvalue` for `zip` and `zinb` compared with other count models. First, the output lists levels of both the x variables from the count equation and the z variables from the binary equation. Here they are the same variables, but they could be different. Second, there are two probabilities of zero counts. For example, for female scientists, `prvalue` lists `Pr(y=0|x,z): 0.6047`, which is the probability of having no publications, either because a scientist does not have the opportunity to publish or because a scientist is a potential publisher who by chance did not publish. The quantity `Pr(Always0|z): 0.2744` is the probability of not being a potential publisher. Thus most of the zeros for women are due to being in the group that never publishes. The remaining probabilities listed are the probabilities of observing each count of publications for the specified set of characteristics.

Even with setting `maxiter` to 50, the pair of `prvalue` commands used above took 29 minutes with Stata 9 under Windows XP using an Intel® Pentium® 4 running at 3.4 GHz (without setting `maxiter`, it would have taking roughly 1,500,000 more iterations and several more days). Although it is tempting to use fewer replications in the bootstrap (e.g., reps(100)), we are convinced that 1,000 is the minimum needed to obtain stable confidence intervals.

How do you figure out how long it takes commands to run? If you include the line display "* $S_DATE - $S_TIME" in a do-file, it will print the date and time. Putting this line before and after a series of commands will allow you to figure out how long the commands took to estimate.

Predicted probabilities with prgen

prgen is used to plot predictions. Here we examine the two sources of 0s. First, we call prgen to compute the predicted values to be plotted:

```
. prgen ment, rest(mean) f(0) t(20) gen(zinb) n(21)
zinb: Predicted values as ment varies from 0 to 20.

base x values for count equation:
         fem        mar       kid5        phd       ment
x=  .46010929  .66229508  .49508197  3.1031093  8.7672131

base z values for binary equation:
         fem        mar       kid5        phd       ment
z=  .46010929  .66229508  .49508197  3.1031093  8.7672131
```

prgen created two probabilities for zero counts: zinbp0 contains the probability of a zero count from both the count and the binary equation. zinball0 is the probability due to observations being in the Always Zero group. We use generate zinbnb0 = zinbp0 - zinball0 to compute the probability of 0s from the count portion of the model:

```
. generate zinbnb0 = zinbp0 - zinball0
(894 missing values generated)
. label var zinbp0 "0s from Both Equations"
. label var zinball0 "0s from Binary Equation"
. label var zinbnb0  "0s from Count Equation"
. label var zinbx "Mentor's Publications"
```

These are plotted with the command

```
. graph twoway connected zinball0 zinbnb0 zinbp0 zinbx, ///
> xlabel(0(5)20) ylabel(0(.1).7)                        ///
> ytitle(Probability of zero) msymbol(Oh Sh O)
```

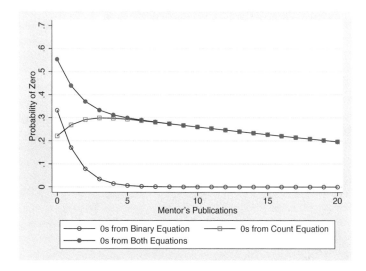

The curve marked with $\bigcirc$s is a probability curve just like those shown in chapter 4 for binary models. The curve marked with $\square$s shows the probability of 0s from a series of negative binomial distributions, each with different rate parameters, μ, determined by the level of mentor's publications. The overall probability of a zero count is the sum of the two curves, which is shown by the curve with $\bullet$s.

8.7 Comparisons among count models

There are two methods that can be used to compare the results of PRM, NBRM, ZIP, and ZINB. First, we can compute the mean predicted probabilities and compare these with the observed probabilities for each count. Second, we can use various test and fit statistics to compare models. We begin by showing you how to make these computations using Stata commands. Then we demonstrate the use of the SPost command `countfit`, which automates this process. Although `countfit` is the simplest way to compare models, it is useful to understand how these computations are made to more fully understand the output of `countfit`.

8.7.1 Comparing mean probabilities

One way to compare these models is to compare predicted probabilities across models. First, we compute the mean predicted probability. For example, in the PRM,

$$\overline{\Pr}_{\text{PRM}}(y = m) = \frac{1}{N} \sum_{i=1}^{N} \widehat{\Pr}_{\text{PRM}}(y_i = m \mid \mathbf{x}_i)$$

This is simply the average across all observations of the probability of each count. The difference between the observed probabilities and the mean prediction can be computed as

$$\Delta \overline{\Pr}_{\mathrm{PRM}}(y = m) = \widehat{\Pr}_{\mathrm{Observed}}(y = m) - \overline{\Pr}_{\mathrm{PRM}}(y = m)$$

This can be done for each model and then plotted. The commands are as follows:

```
. use http://www.stata-press.com/data/lf2/couart2, clear
(Academic Biochemists / S Long)
. quietly poisson art fem mar kid5 phd ment, nolog
. prcounts prm, plot max(9)
. label var prmpreq "Predicted: PRM"
. label var prmobeq "Observed"
. quietly nbreg art fem mar kid5 phd ment, nolog
. prcounts nbrm, plot max(9)
. label var nbrmpreq "Predicted: NBRM"
. quietly zip art fem mar kid5 phd ment,
      inf(fem mar kid5 phd ment) vuong nolog
. prcounts zip, plot max(9)
. label var zippreq "Predicted: ZIP"
. quietly zinb art fem mar kid5 phd ment, inf(fem mar kid5 phd ment) vuong nolog
. prcounts zinb, plot max(9)
. label var zinbpreq "Predicted: ZINB"
. * create deviations
. gen obs = prmobeq
(905 missing values generated)
. gen dprm = obs - prmpreq
(905 missing values generated)
. label var dprm "PRM"
. gen dnbrm = obs - nbrmpreq
(905 missing values generated)
. label var dnbrm "NBRM"
. gen dzip = obs - zippreq
(905 missing values generated)
. label var dzip "ZIP"
. gen dzinb = obs - zinbpreq
(905 missing values generated)
. label var dzinb "ZINB"
. * plot deviations
. graph twoway connected dprm dnbrm dzip dzinb prmval, ///
> ytitle(Observed-Predicted) ylabel(-.10(.05).10)      ///
> xlabel(0(1)9) msymbol(Oh Sh O S)
```

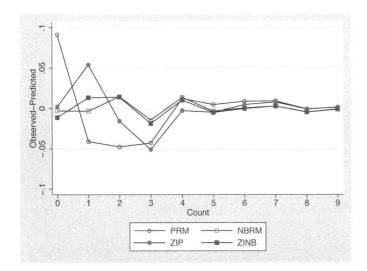

Points above 0 on the y-axis indicate more observed counts than predicted; those below 0 indicate more predicted counts than observed. The graph shows that only the PRM has a problem predicting the average number of 0s. The ZIP does less well, predicting too many 1s and too few 2s and 3s. The NBRM and ZINB do about equally well. From these results, we might prefer the NBRM because it is simpler.

8.7.2   Tests to compare count models

Plotting predictions is only an informal method of assessing the fit of a count model. More formal testing can be done with an LR test of overdispersion and a Vuong test to compare two models.

LR tests of α

Because the NBRM reduces to the PRM when $\alpha=0$, the PRM and NBRM can be compared by testing H_0: $\alpha = 0$. As shown in section 8.3.3, we find that

```
Likelihood-ratio test of alpha=0:  chibar2(01) =  180.20 Prob>=chibar2 = 0.000
```

which provides strong evidence for preferring the NBRM over the PRM.

Because the ZIP and ZINB models are also nested, the same LR test can be applied to compare them. Although Stata does not compute this for you, it is simple to do. First, we fit the ZIP model:

```
. quietly zip art fem mar kid5 phd ment, inf(fem mar kid5 phd ment) vuong nolog
. scalar llzip = e(ll)
```

The command `scalar llzip = e(ll)` saves the log likelihood that was left in memory by `zip`. Next we do the same thing for `zinb` and compute the difference between the two log likelihoods:

```
. quietly zinb art fem mar kid5 phd ment, inf(fem mar kid5 phd ment) nolog
. scalar llzinb = e(ll)
. scalar lr = -2*(llzip-llzinb)
```

The following commands can be used to compute the *p*-value. If you do this with your own model, you need to substitute the value of `lnalpha`, which is listed as part of the output for `zinb`:

```
. scalar pvalue = chiprob(1,lr)/2
. scalar lnalpha = -.9763565
. if (lnalpha < -20)  scalar pvalue= 1
. di as text "Likelihood-ratio test comparing ZIP to ZINB: " as res %8.3f
     lr as text " Prob>=" as res %5.3f pvalue
```

The first line is the standard way to compute the *p*-value for a chi-squared test with 1 degree of freedom, except that we divide by 2. This is because α cannot be negative, as we discussed earlier with regard to comparing the `poisson` and `nbreg` models (Gutierrez et al. 2001). The next line assigns the estimated value of $\ln \alpha$ to a scalar. If this value is very close to 0, we conclude that the *p*-value is 1. The last line simply prints the result:

```
Likelihood-ratio test comparing ZIP to ZINB:  109.564 Prob>=0.000
```

We conclude that the ZINB significantly improves the fit over the ZIP model.

Vuong test of nonnested models

Greene (1994) points out that the PRM and the ZIP are not nested. For the ZIP model to reduce to the PRM, ψ must equal zero. This does *not* occur when $\gamma = \mathbf{0}$ because $\psi = F(\mathbf{z0}) = .5$. Similarly, the NBRM and the ZINB are not nested. Consequently, Greene proposes using a test by Vuong (1989, 319) for nonnested models. This test considers two models, where $\widehat{\Pr}_1(y_i \mid \mathbf{x}_i)$ is the predicted probability of observing y in the first model and $\widehat{\Pr}_2(y_i \mid \mathbf{x}_i)$ is the predicted probability for the second model. Defining

$$m_i = \ln \left\{ \frac{\widehat{\Pr}_1(y_i \mid \mathbf{x}_i)}{\widehat{\Pr}_2(y_i \mid \mathbf{x}_i)} \right\}$$

let $\overline{m}$ be the mean and let s_m be the standard deviation of m_i. The Vuong statistic to test the hypothesis that $E(m) = 0$ equals

$$V = \frac{\sqrt{N}\,\overline{m}}{s_m}$$

V has an asymptotic normal distribution. If $V > 1.96$, the first model is favored; if $V < -1.96$, the second model is favored.

For `zip`, the `vuong` option computes the Vuong statistic, comparing the ZIP model with the PRM; for `zinb` it compares ZINB with NBRM. For example,

```
. zip art fem mar kid5 phd ment, inf(fem mar kid5 phd ment) vuong nolog
  (output omitted)
Vuong test of zip vs. standard Poisson:            z =    4.18  Pr>z = 0.0000
```

The significant, positive value of V supports the ZIP model over the PRM. If you use
`listcoef`, you get more guidance in interpreting the result:

```
. listcoef, help
zip (N=915): Factor Change in Expected Count
  (output omitted)
  Vuong Test =   4.18 (p=0.000) favoring ZIP over PRM.
```

For the ZINB,

```
. listcoef, help
zinb (N=915): Factor Change in Expected Count
  (output omitted)
  Vuong Test =   2.24 (p=0.012) favoring ZINB over NBRM.
```

Although it is possible to compute a Vuong statistic to compare other pairs of models,
such as ZIP and NBRM, these are currently not available in Stata.

Overall, these tests provide evidence that the ZINB model fits the data best. However,
when fitting a series of models with no theoretical rationale, it is easy to overfit the
data. Here the most compelling evidence for the ZINB is that it makes substantive
sense. Within the realm of science, there are some scientists who for structural reasons
cannot publish, but for other scientists, the failure to publish in a given period is a
matter of chance. This is the basis of the zero-inflated models. The negative binomial
version of the model seems preferable to the Poisson version, since there are probably
unobserved sources of heterogeneity that differentiate the scientists. In sum, the ZINB
makes substantive sense and fits the data well.

8.8   Using countfit to compare count models

`countfit` automates the analyses described in the last two sections. The command
compares the fit of PRM, NBRM, ZIP, and ZINB, optionally generating a table of estimates,
a table of differences between observed and average estimated probabilities, a graph of
these differences, and various tests and measures of fit used to compare count models.
The syntax is

countfit *varlist* $\big[$*if*$\big]$ $\big[$*in*$\big]$ $\big[$, <u>inf</u>late(*varlist2*) <u>noc</u>onstant <u>prm</u> <u>n</u>breg zip
   <u>zin</u>b <u>g</u>enerate(*prefix*) replace note(*string*) <u>graph</u>export(*filename*$\big[$,
   replace$\big]$) <u>nog</u>raph <u>nod</u>ifferences <u>nop</u>rtable <u>noe</u>stimates <u>nof</u>it nodash
   <u>max</u>count(#) <u>noi</u>sily$\big]$

Options for specifying the model

varlist is the variable list for the model, beginning with the count outcome variable.

if *exp* and in *range* specify the sample used for estimating the models.

inflate(*varlist2*) specifies the inflation variables for zip and zinb.

noconstant specifies that there is no constant included in the model.

Options to select the models to fit

By default, poisson, nbreg, zip, and zinb are all fitted. If you want only some of these models, specify the models you want:

prm fits the poisson model.

nbreg fits the nbreg model.

zip fits the zip model.

zinb fits the zinb model.

Options to label and save results

generate(*prefix*) is up to five letters to prefix the variables that are created and to label the models in the output. This name is placed in front of the type of model (e.g., *name*PRM). These labels help keep track of results from multiple specifications of models.

replace replaces variables created by generate() if they already exist.

note(*string*) adds a label to the graph.

graphexport(*filename*$\big[$, replace$\big]$) saves the graph to *filename*.

Options to control what is printed

nograph suppresses the graph of differences from observed counts.

nodifferences suppresses the table of differences from observed counts.

noprtable suppresses the table of predictions for each model.

noestimates suppresses the table of estimated coefficients.

`nofit` suppresses the table of fit statistics and tests of fit.

`nodash` suppresses dashed lines between measures of fit.

`maxcount(#)` is the number of counts to evaluate.

`noisily` includes output from Stata estimation commands; without this option the
    results are shown only in the `estimates table` output.

To explain further, we use `countfit` with our publication example and discuss the
output that is generated. We specify the variables in our model but do not limit the
output that is generated:

```
. countfit art fem mar kid5 phd ment, inf(ment fem)
```

First, `countfit` presents estimates of the exponentiated parameters for the four models.
As we would expect, for all models the general direction of coefficients for a given variable
is the same.

Variable	PRM	NBRM	ZIP	ZINB
art				
Gender: 1=female 0=male	0.799	0.805	0.812	0.836
	-4.11	-2.98	-3.31	-2.40
Married: 1=yes 0=no	1.168	1.162	1.142	1.150
	2.53	1.83	2.01	1.72
Number of children < 6	0.831	0.838	0.849	0.845
	-4.61	-3.32	-3.77	-3.22
PhD prestige	1.013	1.015	0.993	1.001
	0.49	0.42	-0.24	0.04
Article by mentor in last 3 yrs	1.026	1.030	1.018	1.025
	12.73	8.38	8.09	7.07
Constant	1.356	1.292	1.874	1.465
	2.96	1.85	5.54	2.69
lnalpha				
Constant		0.442		0.375
		-6.81		-7.06
inflate				
Article by mentor in last 3 yrs			0.876	0.470
			-3.23	-2.55
Gender: 1=female 0=male			1.120	2.868
			0.42	1.40
Constant			0.484	0.275
			-3.15	-2.14
Statistics				
alpha		0.442		
N	915.000	915.000	915.000	915.000
ll	-1651.056	-1560.958	-1605.644	-1552.034
bic	3343.026	3169.649	3272.659	3172.257
aic	3314.113	3135.917	3229.288	3124.068

legend: b/t

Second, `countfit` lists the count for which the deviation between the observed and average expected count is the greatest. For the PRM, the biggest problem is the prediction of zero counts, with a difference that is much larger than the maximum for the other models. Also the average difference between observed and predicted is largest for the PRM (.026) and smallest for the NBRM (.006) and ZINB (.008).

```
Comparison of Mean Observed and Predicted Count

           Maximum      At      Mean
Model     Difference   Value   |Diff|
-------------------------------------------
PRM         0.091        0      0.026
NBRM       -0.015        3      0.006
ZIP         0.052        1      0.014
ZINB       -0.019        3      0.008
```

Third, `countfit` provides an expanded comparison of observed and predicted counts for each of the four models.

```
PRM: Predicted and actual probabilities

Count   Actual    Predicted    |Diff|    Pearson
-----------------------------------------------------
0        0.301      0.209       0.091     36.489
1        0.269      0.310       0.041      4.962
2        0.195      0.242       0.048      8.549
3        0.092      0.135       0.043     12.483
4        0.073      0.061       0.012      2.174
5        0.030      0.025       0.005      0.760
6        0.019      0.010       0.009      6.883
7        0.013      0.004       0.009     17.815
8        0.001      0.002       0.001      0.300
9        0.002      0.001       0.001      1.550
-----------------------------------------------------
Sum      0.993      0.999       0.259     91.964

NBRM: Predicted and actual probabilities

Count   Actual    Predicted    |Diff|    Pearson
-----------------------------------------------------
0        0.301      0.304       0.003      0.028
1        0.269      0.272       0.003      0.039
2        0.195      0.180       0.014      1.066
3        0.092      0.106       0.015      1.818
4        0.073      0.060       0.013      2.753
5        0.030      0.033       0.004      0.348
6        0.019      0.018       0.000      0.004
7        0.013      0.010       0.003      0.719
8        0.001      0.006       0.005      3.593
9        0.002      0.004       0.001      0.456
-----------------------------------------------------
Sum      0.993      0.993       0.062     10.824
```

```
ZIP: Predicted and actual probabilities
Count   Actual    Predicted     |Diff|    Pearson
--------------------------------------------------
0       0.301       0.298        0.003      0.022
1       0.269       0.217        0.052     11.526
2       0.195       0.210        0.016      1.095
3       0.092       0.142        0.050     16.281
4       0.073       0.076        0.002      0.071
5       0.030       0.034        0.005      0.612
6       0.019       0.014        0.005      1.346
7       0.013       0.005        0.008      9.840
8       0.001       0.002        0.001      0.447
9       0.002       0.001        0.001      1.985
--------------------------------------------------
Sum     0.993       0.999        0.143     43.225
ZINB: Predicted and actual probabilities
Count   Actual    Predicted     |Diff|    Pearson
--------------------------------------------------
0       0.301       0.312        0.012      0.396
1       0.269       0.256        0.013      0.611
2       0.195       0.181        0.014      0.981
3       0.092       0.110        0.019      2.889
4       0.073       0.063        0.010      1.524
5       0.030       0.035        0.005      0.709
6       0.019       0.019        0.000      0.004
7       0.013       0.010        0.003      0.710
8       0.001       0.006        0.005      3.397
9       0.002       0.003        0.001      0.302
--------------------------------------------------
Sum     0.993       0.995        0.081     11.522
```

The |Diff| columns of these tables are then plotted:

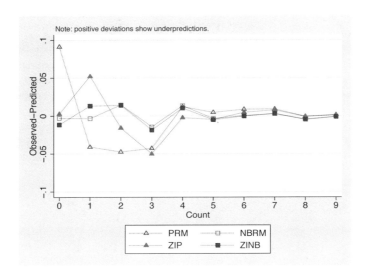

Finally, `countfit` compares the fit of the four models by several standard criteria and tests, including BIC and AIC. For each statistic comparing models, the last three columns indicate which model is preferred. Both the NBRM and ZINB consistently fit better than either the PRM or the ZIP. This example also provides a good example of how BIC penalizes extra parameters more severely than does AIC, as the more parsimonious NBRM model is preferred over the ZINB according to the BIC statistic, but not according to the AIC statistic.

```
Tests and Fit Statistics
PRM             BIC= -2896.289  AIC=      3.622  Prefer  Over  Evidence
-------------------------------------------------------------------------
   vs NBRM      BIC= -3069.666  dif=    173.377  NBRM    PRM   Very strong
                AIC=      3.427  dif=      0.195  NBRM    PRM
                LRX2=  180.196  prob=     0.000  NBRM    PRM   p=0.000
-------------------------------------------------------------------------
   vs ZIP       BIC= -2966.657  dif=     70.367  ZIP     PRM   Very strong
                AIC=      3.529  dif=      0.093  ZIP     PRM
                Vuong=    4.133  prob=     0.000  ZIP     PRM   p=0.000
-------------------------------------------------------------------------
   vs ZINB      BIC= -3067.059  dif=    170.769  ZINB    PRM   Very strong
                AIC=      3.414  dif=      0.208  ZINB    PRM
-------------------------------------------------------------------------
NBRM            BIC= -3069.666  AIC=      3.427  Prefer  Over  Evidence
-------------------------------------------------------------------------
   vs ZIP       BIC= -2966.657  dif=   -103.010  NBRM    ZIP   Very strong
                AIC=      3.529  dif=     -0.102  NBRM    ZIP
-------------------------------------------------------------------------
   vs ZINB      BIC= -3067.059  dif=     -2.608  NBRM    ZINB  Positive
                AIC=      3.414  dif=      0.013  ZINB    NBRM
                Vuong=    2.069  prob=     0.019  ZINB    NBRM  p=0.019
-------------------------------------------------------------------------
ZIP             BIC= -2966.657  AIC=      3.529  Prefer  Over  Evidence
-------------------------------------------------------------------------
   vs ZINB      BIC= -3067.059  dif=    100.402  ZINB    ZIP   Very strong
                AIC=      3.414  dif=      0.115  ZINB    ZIP
                LRX2=  107.221  prob=     0.000  ZINB    ZIP   p=0.000
-------------------------------------------------------------------------
```

9 More topics

In this final chapter, we discuss some disparate topics that were not covered in the preceding chapters. We begin by considering complications on the right-hand side of the model: nonlinearities, interactions, and nominal or ordinal variables coded as a set of dummy variables. Although the same principles of interpretation apply in these cases, several tricks are necessary for computing the appropriate quantities. We then examine ways you can use basic Stata programming to simplify data analysis. Next we discuss briefly what is required if you want to modify SPost to work with other estimation commands. The final section discusses a menagerie of Stata "tricks" that we find useful for working more efficiently in Stata.

9.1 Ordinal and nominal independent variables

When an independent variable is categorical, it should be entered into the model as a set of binary indicator variables. Although our example uses an ordinal variable, the discussion applies equally to nominal independent variables, with one exception.

9.1.1 Coding a categorical independent variable as a set of dummy variables

A categorical independent variable with J categories can be included in a regression model as a set of $J - 1$ dummy variables. Here we use a binary logit model to analyze factors affecting whether a scientist has published. The outcome is a dummy variable, `hasarts`, that is equal to 1 if the scientist has one or more publications and equals 0 otherwise. In our analysis in the last chapter, we included the independent variable `ment`, which we treated as continuous. But suppose instead that the data were from a survey in which the mentor was asked to indicate whether he or she had 0 articles (none), 1–3 articles (few), 4–9 (some), 10–20 (many), or more than 20 articles (lots). The resulting variable, which we call `mentord`, has the following frequency distribution:[1]

---

1. Details on creating `mentord` from the data in `couart2.dta` are located in `st9ch9other.do`, which is part of the `spost9_do` package. For details, when you are in Stata and online, type `search spost9_do, net`.

```
. tab mentord, missing
```

Ordinal measure of mentor's articles	Freq.	Percent	Cum.
None	90	9.84	9.84
Few	201	21.97	31.80
Some	324	35.41	67.21
Many	213	23.28	90.49
Lots	87	9.51	100.00
Total	915	100.00	

We can convert `mentord` into a set of dummy variables using a series of `generate` commands. Because the dummy variables are used to indicate in which category an observation belongs, they are often referred to as *indicator variables*. First, we construct `none` to indicate that the mentor had no publications:

```
. gen none = (mentord == 0) if mentord < .
```

Expressions in Stata equal 1 if true and 0 if false. Accordingly, `gen none = (mentord==0)` creates `none` equal to 1 for scientists whose mentors had no publications and equal to 0 otherwise. Although we have no missing values for `mentord`, it is a good habit to always add an `if` condition so that missing values continue to be missing. This is done by adding `if mentord < .` to the command (remember that missing values are treated by Stata as larger than all nonmissing values when evaluating expressions). We use `tab` to verify that `none` was constructed correctly:

```
. tab none mentord, missing
```

none	Ordinal measure of mentor's articles					Total
	None	Few	Some	Many	Lots	
0	0	201	324	213	87	825
1	90	0	0	0	0	90
Total	90	201	324	213	87	915

Likewise, we create indicator variables for the other categories of `mentord`:

```
. gen few = (mentord == 1) if mentord < .
. gen some = (mentord == 2) if mentord < .
. gen many = (mentord == 3) if mentord < .
. gen lots = (mentord == 4) if mentord < .
```

Note You can also construct indicator variables using `xi` or `tabulate`'s `gen()` option. For more information, type `help xi` or `help tabulate`.

9.1.2 Estimation and interpretation with categorical independent variables

Since `mentord` has $J = 5$ categories, we must include $J - 1 = 4$ indicator variables as independent variables in our model. To see why one of the indicators must be dropped, consider our example. If you know that `none`, `few`, `some`, and `many` are all 0, it must be that `lots` equals 1 because a person has to be in one of the five categories. Another way to think of this is to note that `none + few + some + many + lots = 1` so that including all J categories would lead to perfect collinearity. If you include all five indicator variables, Stata automatically drops one of them. For example,

```
. logit hasarts fem mar kid5 phd none few some many lots, nolog
note: lots dropped due to collinearity
Logistic regression                       Number of obs   =        915
    (output omitted)
```

The category that is excluded is the *reference category*, as the coefficients for the included indicators are interpreted relative to the excluded category, which serves as a point of reference. Which category you exclude is arbitrary, but with an ordinal independent variable, it is generally easier to interpret the results when you exclude an extreme category. For nominal categories, it is often useful to exclude the most important category. For example, we fit a binary logit, excluding the indicator variable `none`:

```
. logit hasarts fem mar kid5 phd few some many lots, nolog
Logistic regression                       Number of obs   =        915
                                          LR chi2(8)      =      73.80
                                          Prob > chi2     =     0.0000
Log likelihood = -522.46467               Pseudo R2       =     0.0660
```

hasarts	Coef.	Std. Err.	z	P>|z|	[95% Conf. Interval]	
fem	-.2579293	.1601187	-1.61	0.107	-.5717562	.0558976
mar	.3300817	.1822141	1.81	0.070	-.0270514	.6872147
kid5	-.2795751	.1118578	-2.50	0.012	-.4988123	-.0603379
phd	.0121703	.0802726	0.15	0.879	-.145161	.1695017
few	.3859147	.2586461	1.49	0.136	-.1210223	.8928517
some	.9602176	.2490498	3.86	0.000	.4720889	1.448346
many	1.463606	.2829625	5.17	0.000	.9090099	2.018203
lots	2.335227	.4368715	5.35	0.000	1.478975	3.19148
_cons	-.0521187	.3361977	-0.16	0.877	-.7110542	.6068167

Logit models can be interpreted in terms of factor changes in the odds, which we compute using `listcoef`:

```
. listcoef

logit (N=915): Factor Change in Odds
  Odds of: Arts vs NoArts
```

hasarts	b	z	P>|z|	e^b	e^bStdX	SDofX
fem	-0.25793	-1.611	0.107	0.7726	0.8793	0.4987
mar	0.33008	1.812	0.070	1.3911	1.1690	0.4732
kid5	-0.27958	-2.499	0.012	0.7561	0.8075	0.7649
phd	0.01217	0.152	0.879	1.0122	1.0121	0.9842
few	0.38591	1.492	0.136	1.4710	1.1734	0.4143
some	0.96022	3.856	0.000	2.6123	1.5832	0.4785
many	1.46361	5.172	0.000	4.3215	1.8568	0.4228
lots	2.33523	5.345	0.000	10.3318	1.9845	0.2935

The effect of an indicator variable can be interpreted just as we interpreted dummy variables in chapter 4, but with comparisons being relative to the reference category. For example, the odds ratio of 10.33 for lots can be interpreted as

> The odds of a scientist publishing are 10.3 times larger if his or her mentor had lots of publications compared with no publications, holding other variables constant.

Or, you can say equivalently:

> If a scientist's mentor has lots of publications as opposed to no publications, the odds of a scientist publishing are 10.3 times larger, holding other variables constant.

The odds ratios for the other indicators can be interpreted in the same way.

9.1.3   Tests with categorical independent variables

The basic ideas and commands for tests that involve categorical independent variables are the same as those used in prior chapters. But, because the tests involve some special considerations, we will review them here.

Testing the effect of membership in one category versus the reference category

When a set of indicator variables is included in a regression, a test of the significance of the coefficient for any indicator variable is a test of whether being in that category compared with being in the reference category affects the outcome. For example, the coefficient for few can be used to test whether having a mentor with few publications compared with having a mentor with no publications significantly affects the scientist's publishing. Here $z = 1.492$ and $p = 0.136$, so we conclude that the effect of having a mentor with a few publications compared with none is not significant using a two-tailed test ($z = 1.492$, $p = 0.14$).

Often the significance of an indicator variable is reported without mentioning the reference category. For example, the test of **many** could be reported as follows:

Having a mentor with many publications significantly affects a scientist's productivity ($z = 5.17$, $p < .01$).

Here the comparison is implicitly being made to mentors with no publications. Such interpretations should be used only if you are confident that the implicit comparison will be apparent to the reader.

Testing the effect of membership in two nonreference categories

What if neither of the categories that we wish to compare is the reference category? A simple solution is to refit the model with a different reference category. For example, to test the effect of having a mentor with some articles compared with a mentor with many publications, we can refit the model using **some** as the reference category:

```
. logit hasarts fem mar kid5 phd none few many lots, nolog
Logistic regression                          Number of obs   =        915
                                             LR chi2(8)      =      73.80
                                             Prob > chi2     =     0.0000
Log likelihood = -522.46467                  Pseudo R2       =     0.0660
```

hasarts	Coef.	Std. Err.	z	P>|z|	[95% Conf. Interval]	
fem	-.2579293	.1601187	-1.61	0.107	-.5717562	.0558976
mar	.3300817	.1822141	1.81	0.070	-.0270514	.6872147
kid5	-.2795751	.1118578	-2.50	0.012	-.4988123	-.0603379
phd	.0121703	.0802726	0.15	0.879	-.145161	.1695017
none	-.9602176	.2490498	-3.86	0.000	-1.448346	-.4720889
few	-.5743029	.1897376	-3.03	0.002	-.9461818	-.2024241
many	.5033886	.2143001	2.35	0.019	.0833682	.9234091
lots	1.37501	.3945447	3.49	0.000	.6017161	2.148303
_cons	.9080989	.3182603	2.85	0.004	.2843202	1.531878

The z-statistics for the mentor indicator variables are now tests comparing a given category with that of the mentor having some publications.

(Continued on next page)

Advanced: lincom For the model that excludes some, the estimated coefficient for many equals the difference between the coefficients for many and some in the earlier model that excluded none. This suggests that instead of refitting the model, we could have used lincom to fit $\beta_{\text{many}} - \beta_{\text{some}}$:

```
. lincom many-some
 ( 1) - some + many = 0
```

hasarts	Coef.	Std. Err.	z	P>|z|	[95% Conf. Interval]
(1)	.5033886	.2143001	2.35	0.019	.0833682 .9234091

The result is identical to that obtained by refitting the model with a different base outcome.

Testing that a categorical independent variable has no effect

For an omnibus test of a categorical variable, our null hypothesis is that the coefficients for all the indicator variables are zero. In our model where none is the excluded variable, the hypothesis to test is

$$H_0\colon \beta_{\text{few}} = \beta_{\text{some}} = \beta_{\text{lots}} = \beta_{\text{many}} = 0$$

This hypothesis can be tested with an LR test by comparing the model with the four indicators with the model that drops the four indicator variables:

```
. logit hasarts fem mar kid5 phd few some many lots, nolog
 (output omitted )
. estimates store fmodel
. logit hasarts fem mar kid5 phd, nolog
 (output omitted )
. lrtest fmodel
Likelihood-ratio test                            LR chi2(4)  =     58.32
(Assumption: . nested in fmodel)                 Prob > chi2 =    0.0000
```

The key result is the change as few changes from 0->1 (which, because it is a dummy variable, is also the change from min->max). A Wald test can also be used, although the LR test is generally preferred:

```
. logit hasarts fem mar kid5 phd few some many lots, nolog
 (output omitted )
. test few some many lots
 ( 1)  few = 0
 ( 2)  some = 0
 ( 3)  many = 0
 ( 4)  lots = 0
           chi2(  4) =     51.60
         Prob > chi2 =     0.0000
```

which leads to the same conclusion as the LR test.

Exactly the same results would be obtained for either test if we had used a different reference category and tested, for example,

$$H_0: \beta_{\text{none}} = \beta_{\text{few}} = \beta_{\text{lots}} = \beta_{\text{many}} = 0$$

Testing whether treating an ordinal variable as interval loses information

Ordinal independent variables are often treated as interval in regression models. For example, rather than include the four indicator variables that were created from `mentord`, we might simply include only `mentord` in our model:

```
. logit hasarts fem mar kid5 phd mentord, nolog
Logistic regression                             Number of obs   =         915
                                                LR chi2(5)      =       72.73
                                                Prob > chi2     =      0.0000
Log likelihood = -522.99932                     Pseudo R2       =      0.0650
```

hasarts	Coef.	Std. Err.	z	P>|z|	[95% Conf. Interval]	
fem	-.266308	.1598617	-1.67	0.096	-.5796312	.0470153
mar	.3329119	.1823256	1.83	0.068	-.0244397	.6902635
kid5	-.2812119	.1118409	-2.51	0.012	-.500416	-.0620078
phd	.0100783	.0802174	0.13	0.900	-.147145	.1673016
mentord	.5429222	.0747143	7.27	0.000	.3964848	.6893595
_cons	-.1553251	.3050814	-0.51	0.611	-.7532736	.4426234

The advantage of this approach is that interpretation is simpler, but to take advantage of this simplicity you must make the strong assumption that successive categories of the ordinal independent variable are equally spaced. For example, it implies that an increase from no publications by the mentor to a few publications involves an increase of the same amount of productivity as an increase from a few to some, from some to many, and from many to lots of publications.

Accordingly, before treating an ordinal independent variable as if it were interval, you should test whether this leads to a loss of information about the association between the independent and dependent variable. A likelihood-ratio test can be computed by comparing the model with only `mentord` to the model that includes both the ordinal variable (`mentord`) and all but two of the indicator variables. Below, we add `some`, `many`, and `lots`, but including any three of the indicators leads to the same results. If the categories of the ordinal variables are equally spaced, the coefficients of the $J - 2$ indicator variables should all be 0. For example,

```
. logit hasarts fem mar kid5 phd mentord some many lots, nolog
  (output omitted)
. estimates store fmodel
. logit hasarts fem mar kid5 phd mentord, nolog
  (output omitted)
. lrtest fmodel
Likelihood-ratio test                           LR chi2(3)   =        1.07
(Assumption: . nested in fmodel)                Prob > chi2  =      0.7845
```

We conclude that the indicator variables do not add more information to the model ($LRX^2 = 1.07$, $df = 3$, $p = .78$). If the test were significant, we would have evidence that the categories of mentord are not evenly spaced, and so we should not treat mentord as interval. A Wald test can also be computed, leading to the same conclusion:

```
. logit hasarts fem mar kid5 phd mentord some many lots, nolog
  (output omitted)
. test some many lots

 ( 1)   some = 0
 ( 2)   many = 0
 ( 3)   lots = 0

          chi2(  3) =     1.03
        Prob > chi2 =    0.7950
```

9.1.4 Discrete change for categorical independent variables

There are a few tricks that you must be aware of when computing discrete change for categorical independent variables. To show how this is done, we will compute the change in the probability of publishing for those with a mentor with few publications compared with a mentor with no publications. There are two ways to compute this discrete change. The first way is easier, but the second is more flexible.

Computing discrete change with prchange

The easy way is to use prchange, where we set all the indicator variables to 0:

```
. logit hasarts fem mar kid5 phd few some many lots, nolog
  (output omitted)
. prchange few, x(some=0 many=0 lots=0)

logit: Changes in Probabilities for hasarts
          min->max        0->1       -+1/2      -+sd/2   MargEfct
few        0.0957      0.0957      0.0962      0.0399     0.0965

           NoArts        Arts
Pr(y|x)    0.4920      0.5080

              fem        mar       kid5        phd        few       some       many       lots
   x=     .460109    .662295    .495082    3.10311    .219672          0          0          0
sd(x)=    .498679    .473186     .76488    .984249    .414251    .478501    .422839    .293489
```

We conclude that having a mentor with a few publications compared with none increases a scientist's probability of publishing by .10, holding all other variables at their mean.

The effect of the mentor's productivity is significant at the .01 level ($LRX^2 = 58.32$, $df = 4$, $p < .01$).

Even though we say "holding all other variables at their mean", which is clear within the context of reporting substantive results, the key to getting the right answer from prchange is holding all the indicator variables at 0, not at their mean. It does not make sense to change few from 0 to 1 when some, many, and lots are at their means.

Computing discrete change with prvalue

A second approach to computing discrete change is to use two calls to `prvalue`. The advantage of this approach is that it works in situations where `prchange` does not. For example, how does the predicted probability change if we compare a mentor with a few publications to a mentor with some publications, holding all other variables constant? This involves computing probabilities as we move from `few=1` and `some=0` to `few=0` and `some=1`. We cannot compute this with `prchange` because two variables are changing at the same time. Instead, we use two calls to `prvalue`:[2]

```
. quietly prvalue, x(few=1 some=0 many=0 lots=0) save
. prvalue, x(few=0 some=1 many=0 lots=0) diff

logit: Change in Predictions for hasarts
Confidence intervals by delta method
                    Current      Saved    Change    95% CI for Change
    Pr(y=Arts|x):    0.7125     0.5825    0.1300   [ 0.0452,   0.2147]
  Pr(y=NoArts|x):    0.2875     0.4175   -0.1300   [-0.2147,  -0.0452]

                fem         mar        kid5        phd        few      some
  Current= .46010929  .66229508  .49508197  3.1031093          0         1
    Saved= .46010929  .66229508  .49508197  3.1031093          1         0
     Diff=         0          0          0          0         -1         1

                many        lots
  Current=         0          0
    Saved=         0          0
     Diff=         0          0
```

Because we have used the `save` and `diff` options, the difference in the predicted probability (i.e., the discrete change) is reported. When we use the `save` and `diff` options, we usually add `quietly` to the first `prvalue` because all the information is listed by the second `prvalue`.

9.2 Interactions

Interaction terms are commonly included in regression models when the effect of an independent variable is thought to vary depending on the value of another independent variable. To illustrate how interactions are used, we extend the example from chapter 5, where the dependent variable is a respondent's level of agreement that a working mother can establish as warm a relationship with her children as mothers who do not work.

The effect of education on attitudes toward working mothers might vary by sex. To allow this possibility, we add the interaction of education (`ed`) and sex (`male`) by adding the variable `maleXed = male*ed`. In fitting this model, we find that

---

2. We could also have refitted the model adding `none` and excluding either `few` or `some`, and then used `prchange`.

```
. use http://www.stata-press.com/data/lf2/ordwarm2, clear
(77 & 89 General Social Survey)

. gen maleXed = male*ed

. ologit warm age prst yr89 white male ed maleXed, nolog
```

```
Ordered logistic regression                       Number of obs   =        2293
                                                  LR chi2(7)      =      305.30
                                                  Prob > chi2     =      0.0000
Log likelihood = -2843.1198                       Pseudo R2       =      0.0510
```

warm	Coef.	Std. Err.	z	P>|z|	[95% Conf. Interval]	
age	-.0212523	.0024775	-8.58	0.000	-.0261082	-.0163965
prst	.0052597	.0033198	1.58	0.113	-.001247	.0117664
yr89	.5238686	.0799287	6.55	0.000	.3672111	.680526
white	-.3908743	.1184189	-3.30	0.001	-.622971	-.1587776
male	-.1505216	.3176105	-0.47	0.636	-.7730268	.4719836
ed	.0976341	.0226886	4.30	0.000	.0531651	.142103
maleXed	-.047534	.0251183	-1.89	0.058	-.0967649	.001697
/cut1	-2.107903	.3043008			-2.704322	-1.511484
/cut2	-.2761098	.2992857			-.862699	.3104793
/cut3	1.621787	.3018749			1.030123	2.213451

The interaction is marginally significant ($p = .06$) for a two-tailed Wald test. We can also compute an LR test

```
. ologit warm age prst yr89 white male ed maleXed, nolog
  (output omitted )

. estimates store fmodel

. ologit warm age prst yr89 white male ed, nolog
  (output omitted )

. lrtest fmodel

Likelihood-ratio test                             LR chi2(1)  =        3.59
(Assumption: . nested in fmodel)                  Prob > chi2 =      0.0583
```

which leads to the same conclusion.

9.2.1 Computing sex differences in predictions with interactions

What if we want to compute the difference between men and women in the predicted probabilities for the outcome categories? Sex differences are reflected in two ways in the model. First, we want to change male from 0 to 1 to indicate women versus men. If this were the only variable affected by changing the value of male, we could use prchange. But when the value of male changes, this necessarily changes the value of maleXed (except in the case when ed is 0). For women, maleXed = male * ed = 0 * ed = 0, whereas for men, maleXed = male * ed = 1 * ed = ed. Accordingly, we must examine the change in the outcome probabilities when two variables change, so prvalue must be used. We start by computing the predicted values for women, which requires fixing male=0 and maleXed=0:

```
. prvalue, x(male=0 maleXed=0) rest(mean) save
ologit: Predictions for warm
Confidence intervals by delta method
                         95% Conf. Interval
    Pr(y=SD|x):     0.0816   [ 0.0701,    0.0932]
    Pr(y=D|x):      0.2754   [ 0.2539,    0.2968]
    Pr(y=A|x):      0.4304   [ 0.4074,    0.4534]
    Pr(y=SA|x):     0.2126   [ 0.1914,    0.2337]
          age       prst        yr89       white      male          ed
x=   44.935456   39.585259   .39860445   .8765809       0   12.218055
        maleXed
x=          0
```

Next we compute the predicted probability for men, where male=1 and maleXed equals the average value of education (because for men, maleXed=male*ed= 1*ed=ed). The value for maleXed can be obtained by computing the mean of ed:

```
. summarize ed
    Variable |       Obs        Mean    Std. Dev.       Min        Max
-------------+--------------------------------------------------------
          ed |      2293    12.21805    3.160827         0         20
. global meaned = r(mean)
```

summarize returns the mean to r(mean). The command global meaned = r(mean) assigns the mean of ed to the global macro meaned. In the prvalue command, we specify x(male=1 maleXed=$meaned), where $meaned tells Stata to substitute the value contained in the global macro:

```
. prvalue, x(male=1 maleXed=$meaned) diff

ologit: Change in Predictions for warm
Confidence intervals by delta method
                        Current      Saved    Change   95% CI for Change
        Pr(y=SD|x):      0.1559     0.0816    0.0743   [ 0.0574,   0.0911]
        Pr(y=D|x):       0.3797     0.2754    0.1044   [ 0.0818,   0.1270]
        Pr(y=A|x):       0.3494     0.4304   -0.0811   [-0.1001,  -0.0621]
        Pr(y=SA|x):      0.1150     0.2126   -0.0976   [-0.1185,  -0.0767]

                    age        prst        yr89       white      male          ed
Current=      44.935456   39.585259   .39860445   .8765809         1   12.218055
  Saved=      44.935456   39.585259   .39860445   .8765809         0   12.218055
   Diff=              0           0           0           0         1           0
                 maleXed
Current=       12.218055
  Saved=               0
   Diff=       12.218055
```

Warning The mean of maleXed does not equal the mean of ed. That is why we could not use the option x(male=1 maleXed=mean) and instead had to compute the mean with summarize.

Although the trick of using maleXed=$meaned may seem like a lot of trouble to avoid having to type maleXed=12.21805, it can help you avoid errors, and in some cases (illustrated below), it saves a lot of time.

Substantively, we conclude that the probability of strongly agreeing that working mothers can be good mothers is .10 higher for woman than men, taking the interaction with education into account and holding other variables constant at their means. The probability of strongly disagreeing is .07 higher for men than women.

9.2.2 Computing sex differences in discrete change with interactions

We might also be interested in how the predicted outcomes are affected by a change in education from having a high school diploma (12 years of education) to having a college degree (16 years). The interaction term suggests that the effect of education varies by sex, so we must look at the discrete change separately for men and women. Again repeated calls to prvalue using the save and diff options allow us to do this. For women, we hold both male and maleXed to 0 and allow ed to vary. For men, we hold male to 1 and allow both ed and maleXed to vary. For women, we find that

```
. quietly prvalue, x(male=0 maleXed=0 ed=12) rest(mean) save
. prvalue, x(male=0 maleXed=0 ed=16) rest(mean) diff
ologit: Change in Predictions for warm
Confidence intervals by delta method
                    Current      Saved     Change     95% CI for Change
     Pr(y=SD|x):     0.0579     0.0833    -0.0254   [-0.0357,   -0.0150]
     Pr(y=D|x):      0.2194     0.2786    -0.0592   [-0.0855,   -0.0329]
     Pr(y=A|x):      0.4418     0.4291     0.0127   [ 0.0055,    0.0200]
     Pr(y=SA|x):     0.2809     0.2090     0.0718   [ 0.0361,    0.1075]
                    age       prst      yr89      white       male       ed
    Current= 44.935456  39.585259  .39860445  .8765809          0       16
      Saved= 44.935456  39.585259  .39860445  .8765809          0       12
      Diff=          0          0          0          0          0        4
                  maleXed
    Current=          0
      Saved=          0
      Diff=           0
```

For men,

```
. quietly prvalue, x(male=1 maleXed=12 ed=12) rest(mean) save
. prvalue, x(male=1 maleXed=16 ed=16) rest(mean) diff
ologit: Change in Predictions for warm
Confidence intervals by delta method
                    Current      Saved     Change     95% CI for Change
     Pr(y=SD|x):     0.1326     0.1574    -0.0248   [-0.0416,   -0.0080]
     Pr(y=D|x):      0.3558     0.3810    -0.0252   [-0.0447,   -0.0058]
     Pr(y=A|x):      0.3759     0.3477     0.0282   [ 0.0088,    0.0476]
     Pr(y=SA|x):     0.1357     0.1139     0.0218   [ 0.0051,    0.0386]
                    age       prst      yr89      white       male       ed
    Current= 44.935456  39.585259  .39860445  .8765809          1       16
      Saved= 44.935456  39.585259  .39860445  .8765809          1       12
      Diff=          0          0          0          0          0        4
                  maleXed
    Current=         16
      Saved=         12
      Diff=           4
```

The largest difference in the discrete change between the sexes is for the probability of answering "strongly agree". For both men and women, an increase in education from 12 years to 16 years increases the probability of strong agreement, but the increase is .07 for women and only .02 for men.

9.3 Nonlinear nonlinear models

The models that we consider in this book are nonlinear models in that the effect of a change in an independent variable on the predicted probability or predicted count depends on the values of all the independent variables. However, the right-hand side of the model includes a linear combination of variables just like the linear regression model. For example,

Linear regression: $$y = \beta_0 + \beta_1 x_1 + \beta_2 x_2 + \varepsilon$$

Binary logit: $$\Pr\left(y = 1 \mid \mathbf{x}\right) = \frac{\exp\left(\beta_0 + \beta_1 x_1 + \beta_2 x_2 + \varepsilon\right)}{1 + \exp\left(\beta_0 + \beta_1 x_1 + \beta_2 x_2 + \varepsilon\right)}$$

In the terminology of the generalized linear model, we would say that both models have the same *linear predictor*: $\beta_0 + \beta_1 x_1 + \beta_2 x_2$. In the linear regression model, this leads to predictions that are linear surfaces. For example, with one independent variable the predictions are a line, with two a plane, and so on. In the binary logit model, the prediction is a curved surface, as illustrated in chapter 4.

9.3.1 Adding nonlinearities to linear predictors

Nonlinearities in the LRM can be introduced by adding transformations on the right-hand side. For example, in the model

$$y = \alpha + \beta_1 x + \beta_2 x^2 + \varepsilon$$

we include x and x^2 to allow predictions that are a quadratic form. For example, if the fitted model is $\widehat{y} = 1 + -.1x + .1x^2$, the plot is far from linear:

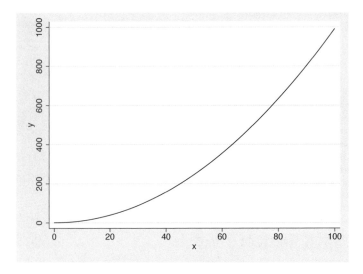

Likewise, nonlinearities can be added to the right-hand side of the models for categorical outcomes that we have been considering. What may seem odd is that adding nonlinearities to a nonlinear model can sometimes make the predictions *more* linear.

9.3.2   Discrete change in nonlinear models

In the model of labor force participation from chapter 4, we included a woman's age as an independent variable. Often when age is used in a model, terms for both age and age-squared are included to allow for diminishing (or increasing) effects of an additional year of age. First, we fit the model *without* age-squared and compute the effect of a change in age from 30 to 50 for an average respondent:

```
. use http://www.stata-press.com/data/lf2/binlfp2, clear
(Data from 1976 PSID-T Mroz)
. logit lfp k5 k618 wc hc lwg inc age, nolog
  (output omitted)
. prchange age, x(age=30) delta(20) uncentered
logit: Changes in Probabilities for lfp
(Note: delta = 20)
        min->max      0->1    +delta      +sd  MargEfct
age     -0.4372   -0.0030   -0.2894  -0.1062   -0.0118

           NotInLF      inLF
Pr(y|x)     0.2494    0.7506
                k5       k618        wc        hc       lwg       inc       age
     x=  .237716   1.35325   .281541   .391766   1.09711    20.129        30
sd(x)=  .523959   1.31987   .450049   .488469   .587556   11.6348   8.07257
```

We have taken advantage of the `delta()` and `uncentered` options (see chapter 3). We find that the predicted probability of a woman working decreases by .29 as age increases from 30 to 50, with all other variables at the mean. Now we add age-squared to the model:

```
. gen age2 = age*age
. logit lfp k5 k618 wc hc lwg inc age age2, nolog
Logistic regression                              Number of obs   =        753
                                                 LR chi2(8)      =     125.67
                                                 Prob > chi2     =     0.0000
Log likelihood = -452.03836                      Pseudo R2       =     0.1220
```

lfp	Coef.	Std. Err.	z	P>|z|	[95% Conf. Interval]	
k5	-1.411597	.2001829	-7.05	0.000	-1.803948	-1.019246
k618	-.0815087	.0696247	-1.17	0.242	-.2179706	.0549531
wc	.8098626	.2299065	3.52	0.000	.3592542	1.260471
hc	.1340998	.207023	0.65	0.517	-.2716579	.5398575
lwg	.5925741	.1507807	3.93	0.000	.2970495	.8880988
inc	-.0355964	.0083188	-4.28	0.000	-.0519009	-.0192919
age	.0659135	.1188199	0.55	0.579	-.1669693	.2987962
age2	-.0014784	.0013584	-1.09	0.276	-.0041408	.001184
_cons	.511489	2.527194	0.20	0.840	-4.44172	5.464698

To test for the joint significance of `age` and `age2`, we use a likelihood-ratio test:

```
. quietly logit lfp k5 k618 wc hc lwg inc age age2, nolog
. estimates store fmodel
```

```
. quietly logit lfp k5 k618 wc hc lwg inc, nolog

. lrtest fmodel

Likelihood-ratio test                              LR chi2(2)  =     26.79
(Assumption: . nested in fmodel)                   Prob > chi2 =    0.0000
```

We can no longer use `prchange` to compute the discrete change because we need to change two variables at the same time. Once again, we use a pair of `prvalue` commands, where we change `age` from 30 to 50 and change `age2` from 30^2 (=900) to 50^2 (=2,500). First, we compute the prediction with `age` at 30:

```
. global age30 = 30

. global age30sq = $age30*$age30

. quietly prvalue, x(age=$age30 age2=$age30sq) rest(mean) save
```

Then we let `age` equal 50 and compute the difference:

```
. global age50 = 50

. global age50sq = $age50*$age50

. prvalue, x(age=$age50 age2=$age50sq) rest(mean) diff

logit: Change in Predictions for lfp

Confidence intervals by delta method
                        Current    Saved     Change    95% CI for Change
   Pr(y=inLF|x):         0.4699    0.7164    -0.2465   [-0.3810,  -0.1120]
   Pr(y=NotInLF|x):      0.5301    0.2836     0.2465   [ 0.1120,   0.3810]

               k5        k618        wc         hc        lwg        inc
Current=  .2377158   1.3532537   .2815405   .39176627   1.0971148   20.128965
  Saved=  .2377158   1.3532537   .2815405   .39176627   1.0971148   20.128965
   Diff=         0           0          0           0           0           0

              age        age2
Current=       50        2500
  Saved=       30         900
   Diff=       20        1600
```

We conclude that an increase in age from 30 to 50 years decreases the probability of being in the labor force by .25, holding other variables at their mean.

By adding the squared term, we have decreased our estimate of the change. Although here the difference is not large, the example illustrates the general point of how to add nonlinearities to the model.

9.4 Using praccum and forvalues to plot predictions

In previous chapters, we used `prgen` to generate predicted probabilities over the range of one variable, while holding other variables constant. Although `prgen` is a relatively simple way of generating predictions for graphs, it can be used only when the specification of the right-hand side of the model is straightforward. When interactions or polynomials are included in the model, graphing the effects of a change in an independent variable often requires computing changes in the probabilities as more than one of the variables in the model changes (e.g., `age` and `age2`). We created `praccum` to handle such situations. The user calculates each of the points to be plotted through several calls to `prvalue`. Executing `praccum` immediately after `prvalue` accumulates these predictions.

The first time `praccum` is run, the predicted values are saved in a new matrix. Each subsequent call to `praccum` adds new predictions to this matrix. When all the calls to `prvalue` have been completed, the accumulated predictions in the matrix can be added as new variables to the dataset in an arrangement ideal for plotting, just like with `prgen`. The syntax of `praccum` is

praccum , { <u>us</u>ing(*matrixname*) | <u>s</u>aving(*matrixname*) } [xis(*value*)

   <u>gen</u>erate(*prefix*)]

where either `using()` or `saving()` is required.

Options

using(*matrixname*) specifies the name of the matrix where the predictions from the previous call to `prvalue` should be added. An error is generated if the matrix does not have the correct number of columns. This can happen if you try to append values to a matrix generated from calls to `praccum` based on a different model. Matrix *matrixname* will be created if it does not already exist.

saving(*matrixname*) specifies that a new matrix should be generated to contain the predicted values from the previous call to `prvalue`. You use this option only when you initially create the matrix. After the matrix is created, you add to it with `using()`. The difference between `saving()` and `using()` is that `saving()` will overwrite *matrixname* if it exists, whereas `using()` will append results to it.

xis(*value*) indicates the value of the *x* variable associated with the predicted values that are accumulated. For example, this could be the value of age if you wish to plot changes in predicted values as age changes. You do *not* need to include the values of variables created as transformations of this variable. To continue the example, you would not include the value of age-squared.

generate(*prefix*) indicates that new variables are to be added to the current dataset. These variables begin with *prefix* and contain the values accumulated in the matrix in prior calls to `praccum`.

The generality of `praccum` requires it to be more complicated to use than `prgen`.

9.4.1   Example using age and age-squared

To illustrate the command, we use `praccum` to plot the effects of age on labor force participation for a model in which both age and age-squared are included. First, we compute the predictions from the model without `age2`:

```
. use http://www.stata-press.com/data/lf2/binlfp2,clear
(Data from 1976 PSID-T Mroz)
. quietly logit lfp k5 k618 age wc hc lwg inc
. prgen age, from(20) to(60) gen(prage) ncases(9)
logit: Predicted values as age varies from 20 to 60.
           k5       k618        age         wc         hc        lwg
x=   .2377158  1.3532537  42.537849   .2815405  .39176627  1.0971148

          inc
x=   20.128965
. label var pragep1 "Pr(lpf | age)"
```

This is the same thing we did using `prgen` in earlier chapters. Next we fit the model with `age2` added:

```
. logit lfp k5 k618 age age2 wc hc lwg inc
  (output omitted)
```

To compute the predictions from this model, we use several calls to `prvalue`. For these predictions, we let `age` change by 5-year increments from 20 to 60 and `age2` increase from 20^2 (= 400) to 60^2 (= 3,600). In the first call of `praccum`, we use the `saving()` option to declare that `mat_age` is the matrix that will hold the results. The `xis()` option is required because it specifies the value for the x-axis of the graph that will plot these probabilities:

```
. quietly prvalue, x(age=20 age2=400) rest(mean)
. praccum, saving(mat_age) xis(20)
```

We execute `prvalue` quietly to suppress the output, because we are generating these predictions only to save them with `praccum`. The next set of calls adds new predictions to `mat_age`, as indicated by the option `using()`:

```
. quietly prvalue, x(age=25 age2=625) rest(mean)
. praccum, using(mat_age) xis(25)
. quietly prvalue, x(age=30 age2=900) rest(mean)
. praccum, using(mat_age) xis(30)
  (and so on)
. quietly prvalue, x(age=55 age2=3025) rest(mean)
. praccum, using(mat_age) xis(55)
```

The last call includes not only the `using()` option but also `gen()`, which tells `praccum` to save the predicted values from the matrix to variables that begin with the specified root, in this case `agesq`:

```
. quietly prvalue, x(age 60 age2 3600) rest(mean)
. * the last call of praccum generates agesqp1 and agesqx variables
. praccum, using(mat_age) xis(60) gen(agesq)
New variables created by praccum:
```

Variable	Obs	Mean	Std. Dev.	Min	Max
agesqx	9	40	13.69306	20	60
agesqp0	9	.4282142	.1752595	.2676314	.7479599
agesqp1	9	.5717858	.1752595	.2520402	.7323686

To understand what has been done, it helps to look at the new variables that were created:

```
. list agesqx agesqp0 agesqp1 in 1/10
```

	agesqx	agesqp0	agesqp1
1.	20	.2676314	.7323686
2.	25	.2682353	.7317647
3.	30	.2836163	.7163837
4.	35	.3152536	.6847464
5.	40	.3656723	.6343277
6.	45	.4373158	.5626842
7.	50	.5301194	.4698806
8.	55	.6381241	.3618759
9.	60	.7479599	.2520402
10.	.	.	.

The 10th observation is all missing values because we made only nine calls to `praccum`. Each value of `agesqx` reproduces the value specified in `xis()`. The values of `agesqp0` and `agesqp1` are the probabilities of $y = 0$ and $y = 1$ that were computed by `prvalue`. We see that the probability of observing a 1, that is, being in the labor force, was .73 the first time we executed `prvalue` with `age` at 20; the probability was .25 the last time we executed `prvalue` with `age` at 60. Now that these predictions have been added to the dataset, we can use `graph` to show how the predicted probability of being in the labor force changes with age:

```
. label var agesqp1 "Pr(lpf | age,age2)"
. label var agesqx  "Age"
. graph twoway connected pragep1 agesqp1 agesqx,      ///
> msymbol(Sh Dh) xlabel(20(5)60)                      ///
> ytitle("Pr(Being in the Labor Force)")  ylabel(0(.2)1)
```

We are also plotting `pragep1`, which was computed earlier in this section using `prgen`. The `graph` command leads to the following plot:

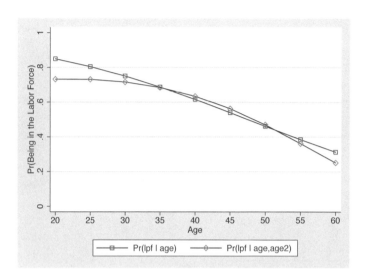

The graph shows that, as age increases from 20 to 60, a woman's probability of being in the labor force declines. In the model with only age, the decline is from .85 to .31, whereas in the model with age-squared, the decrease is from .73 to .25. Overall, the changes are smaller during younger years and larger after age 50.

9.4.2 Using forvalues with praccum

The use of praccum is often greatly simplified by Stata's forvalues command. The forvalues command allows you to repeat a set of commands where the only thing that you vary between successive repetitions is the value of some key number. As a trivial example, we can use forvalues to have Stata count from 0 to 100 by fives. Enter the following three lines either interactively or in a do-file:

```
forvalues count = 0(5)100 {
    display 'count'
}
```

In the forvalues statement, count is the name of a local macro that will contain the successive values of interest (see chapter 2 if you are unfamiliar with local macros). The combination 0(5)100 indicates that Stata should begin by setting the value of count at 0 and should increase its value by 5 with each repetition until it reaches 100. The braces enclose the commands that will be repeated for each value of count. Here all we want to do is to display the value of count. This is done with the command display 'count'. To indicate that count is a local macro, we use the pair of single quote marks (i.e., 'count'). The output produced is

```
0
5
10
  (and so on)
95
100
```

In our earlier example, we graphed the effect of age as it increased from 20 to 60 by 5-year increments. If we specify `forvalues count = 20(5)60`, Stata will repeatedly execute the code we enclose in braces with the value of `count` updated from 20 to 60 by increments of 5. The following lines reproduce the results we obtained earlier:

```
capture matrix drop mage
forvalues count = 20(5)60 {
    local countsq = 'count'^2
    prvalue, x(age='count' age2='countsq') rest(mean) brief
    praccum, using(mage) xis('count')
}
praccum, using(mage) gen(agsq)
```

The command `capture matrix drop mage` at the beginning will drop the matrix `mage` if it exists, but the do-file will not halt with an error if the matrix does not exist. Within the `forvalues` loop, `count` is set to the appropriate value of `age`, and we use the `local` command to create the local macro `countsq`, which contains the square of `count`. After all the predictions have been computed and accumulated to matrix `mage`, we make a last call to `praccum` in which we use the `generate()` option to specify the stem of names of the new variables to be generated.

9.4.3   Using praccum for graphing a transformed variable

`praccum` can also be used when an independent variable is a transformation of the original variable. For example, you might want to include the natural log of `age` as independent variable rather than `age`. Such a model can be easily fitted:

```
. gen ageln = ln(age)
. logit lfp k5 k618 ageln wc hc lwg inc
  (output omitted)
```

As before, we use `forvalues` to execute several calls to `prvalue` and `praccum` to generate predictions:

```
capture matrix drop mat_ln
forvalues count = 20(5)60 {
    local countln = ln('count')
    prvalue, x(ageln='countln') rest(mean) brief
    praccum, using(mat_ln) xis('count')
}
praccum, using(mat_ln) gen(ageln)
```

We use a local value to compute the log of `age`, the value of which is passed to `prvalue` with the option `x(ageln='countln')`. But in `praccum` we specify `xis('count')`, not

xis(`countln`) because we want to plot the probability against age in its original
units. The saved values can then be plotted:

```
. label var agelnp1 "Pr(lpf | log of age)"
. graph twoway connected pragep1 agesqp1 agelnp1 agesqx,   ///
> xlabel(20(5)60) msymbol(Sh Dh Th)                        ///
> ytitle("Pr(Being in the Labor Force)")                   ///
> ylabel(0(.2)1)
```

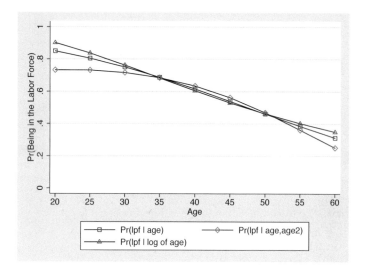

9.4.4 Using praccum to graph interactions

Earlier in this chapter, we examined an ordinal regression model of support for working
mothers that included an interaction between a respondent's sex and education. Another
way to examine the effects of the interaction is to plot the effect of education on the
predicted probability of strongly agreeing for men and women separately. First, we fit
the model:

```
. use http://www.stata-press.com/data/lf2/ordwarm2, clear
(77 & 89 General Social Survey)
. gen maleXed = male*ed
. ologit warm age prst yr89 white male ed maleXed
(output omitted)
```

Next we compute the predicted values of strongly agreeing as education increases for
women who are average on all other characteristics. This is done using forvalues to
make several calls to prvalue and praccum. For women, maleXed is always 0 because
male is 0:

```
forvalues count = 8(2)20 {
    quietly prvalue, x(male=0 ed='count' maleXed=0) rest(mean)
    praccum, using(mat_f) xis('count')
}
praccum, using(mat_f) gen(pfem)
```

In the successive calls to `prvalue`, only the variable `ed` is changing. Accordingly, we could have used `prgen`. For the men, however, we must use `praccum` because both `ed` and `maleXed` change together:

```
. forvalues count = 8(2)20 {
  2.     quietly prvalue, x(male=1 ed='count' maleXed='count') rest(mean)
  3.     praccum, using(mat_m) xis('count')
  4. }
. praccum, using(mat_m) gen(pmal)
New variables created by praccum:
```

Variable	Obs	Mean	Std. Dev.	Min	Max
pmalx	7	14	4.320494	8	20
pmalp1	7	.1462868	.0268927	.1111754	.1857918
pmalp2	7	.3669779	.0269781	.3273872	.4018448
pmalp3	7	.3607055	.0301248	.317195	.40045
pmalp4	7	.1260299	.0237202	.0951684	.1609874
pmals1	7	.1462868	.0268927	.1111754	.1857918
pmals2	7	.5132647	.0537622	.4385626	.5876365
pmals3	7	.8739701	.0237202	.8390126	.9048315
pmals4	7	1	2.25e-08	.9999999	1

Years of education, as specified with `xis()`, are stored in `pfemx` and `pmalx`. These variables are identical because we used the same levels for both men and women. The probabilities for women are contained in the variables `pfemp`k, where k is the category value; for models for ordered or count data, the variables `pfems`k store the cumulative probabilities $\Pr(y \leq k)$. The corresponding predictions for men are contained in `pmalp`k and `pmals`k. All that remains is to clean up the variable labels and plot the predictions:

```
. label var pfemp4 "Pr(SA | female)"
. label var pmalp4 "Pr(SA | male)"
. label var pfemx "Education in Years"
. graph twoway connected pfemp4 pmalp4 pfemx,   ///
> msymbol(Sh Dh) xlabel(8(2)20) ylabel(0(.1).4) ///
> ytitle("Pr(Strongly Agreeing)")
```

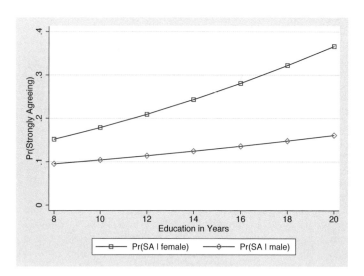

For all levels of education, women are more likely than men to strongly agree that working mothers can be good mothers, holding other variables to their mean. This difference between men and women is much larger at higher levels of education than at lower levels.

9.4.5   Using forvalues with prvalue to create tables

You can also use a `forvalues` loop to create a table of statistics. To illustrate this, we start by rerunning the model of labor force participation:

```
. use http://www.stata-press.com/data/lf2/binlfp2, clear
(Data from 1976 PSID-T Mroz)
. logit lfp-inc, nolog
  (output omitted)
```

Suppose that we wanted a table of predicted probabilities of labor force participation for all combinations of whether a woman attended college (`wc`) and the number of young children (`k5`) in the family. The easiest approach is to use `prtab`:

```
. prtab wc k5, novarlbl

logit: Predicted probabilities of positive outcome for lfp
```

wc	k5 0	1	2	3
NoCol	0.6069	0.2633	0.0764	0.0188
College	0.7758	0.4449	0.1565	0.0412

	k5	k618	age	wc	hc	lwg
x=	.2377158	1.3532537	42.537849	.2815405	.39176627	1.0971148

	inc
x=	20.128965

Although this is simple, there are no upper and lower bounds for the predictions. To compute bounds, we could use eight `prvalue` commands:

```
. prvalue, x(wc=0 k5=0)
. prvalue, x(wc=0 k5=1)
. prvalue, x(wc=0 k5=2)
. prvalue, x(wc=0 k5=3)
. prvalue, x(wc=1 k5=0)
. prvalue, x(wc=1 k5=1)
. prvalue, x(wc=1 k5=2)
. prvalue, x(wc=1 k5=3)
```

Although this method would work, it is tedious, error prone, and leads to output that is not convenient. If we use a bit of Stata programming, we can compute the same information more efficiently and present the results more effectively. As you read what follows, you might think that we are going to a lot of work to avoid entering eight commands, but our main purpose is to show a powerful tool that can be used to simplify many routine tasks. First, we use a loop to compute `prvalue` when `wc=0` and `k5` varies from 0 to 3:

```
. forvalues k = 0/3 {
  2.      prvalue, x(k5='k' wc=0)
  3. }

logit: Predictions for lfp
Confidence intervals by delta method
                              95% Conf. Interval
  Pr(y=inLF|x):        0.6069   [ 0.5567,   0.6570]
  Pr(y=NotInLF|x):     0.3931   [ 0.3430,   0.4433]

           k5       k618        age      wc         hc         lwg         inc
x=          0  1.3532537  42.537849       0  .39176627  1.0971148   20.128965

logit: Predictions for lfp
Confidence intervals by delta method
                              95% Conf. Interval
  Pr(y=inLF|x):        0.2633   [ 0.1932,   0.3335]
  Pr(y=NotInLF|x):     0.7367   [ 0.6665,   0.8068]

           k5       k618        age      wc         hc         lwg         inc
x=          1  1.3532537  42.537849       0  .39176627  1.0971148   20.128965
  (output omitted )
```

We do this again setting `wc=1` but do not show the output:

```
. forvalues k = 0/3 {
  2.      prvalue, x(k5='k' wc=1)
  3. }
```

We can combine the last two steps by using two loops (again, we do not show the output):

```
. forvalues w = 0/1 {
  2.      forvalues k = 0/3 {
  3.          prvalue, x(k5='k' wc='w')
  4.      }
  5. }
```

This program first sets `wc` to 0 with `k5` equal to 0. We increase `k5` to 1, then 2, and finally 3. Then we increment `wc` to 1 and proceed through the values of `k5` again. This program produces the same output as we obtained above.

Next we want to make the output more convenient by saving the results to matrices. To do this, we use the matrices left in memory each time `prvalue` is run. For example, we can `quietly` run `prvalue`:

```
. quietly prvalue, x(wc=0 k5=0)
```

Even though `prvalue` did not print results since it was run `quietly`, it still created the matrix `pepred` with predictions and placed the upper and lower bounds in the matrices `peupper` and `pelower`. To see what these contain, we use the `matrix list` command (we have edited the output to delete rows 4–7, which are not needed):

```
. matrix list pepred
pepred[7,2]
                    c1          c2
  1values            0           1
   2prob    .39312333   .60687667
   3misc     .4342021           .
. matrix list peupper
peupper[7,2]
                     c1          c2
   1values            0           1
    2up_pr    .44328529   .65703863
   3up_misc   .64445657           .
. matrix list pelower
pelower[7,2]
                     c1          c2
   1values            0           1
    2lo_pr    .34296137   .55671471
   3lo_misc   .22394763           .
```

If you compare the numbers in these matrices with those in the `prvalue` output above, you will see that they match.

Now we are ready to modify our program. First, we create a matrix `prob` to hold the predicted probabilities:

```
. matrix prob = J(4,2,.)
```

`J(4,2,.)` means that we want a matrix with four rows (for the four values of `k5`) and two columns (for the two values of `wc`) that is filled with missing values, indicated by ".". Next we add labels to the rows and columns so that the output is clearer when we print the matrix.

```
. matrix rownames prob = "k5=0" "k5=1" "k5=2" "k5=3"
. matrix colnames prob = "wc=0" "wc=1"
```

Now that `prob` is how we want it, we create matrices just like it to hold upper and lower bounds:

```
. matrix LBprob = prob
. matrix UBprob = prob
```

In our `forvalues` loop we suppress the output from `prvalue` and move the values from `pepred`, `pelower`, and `peupper` into the matrices we just created:

```
. forvalues w = 0/1 {
  2.      forvalues k = 0/3 {
  3.          qui prvalue, x(k5='k' wc='w')
  4.          matrix prob['k'+1,'w'+1] = pepred[2,2]
  5.          matrix LBprob['k'+1,'w'+1] = pelower[2,2]
  6.          matrix UBprob['k'+1,'w'+1] = peupper[2,2]
  7.      }
  8. }
```

Listing the results conveniently summarizes the information:

```
. matrix list prob, format(%6.3f)

prob[4,2]
        wc=0    wc=1
k5=0   0.607   0.776
k5=1   0.263   0.445
k5=2   0.076   0.157
k5=3   0.019   0.041

. matrix list LBprob, format(%6.3f)

LBprob[4,2]
         wc=0     wc=1
k5=0    0.557    0.708
k5=1    0.193    0.333
k5=2    0.026    0.058
k5=3   -0.001   -0.002

. matrix list UBprob, format(%6.3f)

UBprob[4,2]
        wc=0    wc=1
k5=0   0.657   0.844
k5=1   0.333   0.557
k5=2   0.127   0.255
k5=3   0.039   0.085
```

9.4.6   A more advanced example*

Here we briefly include a more elaborate program that places the predicted values and the bounds into a matrix that will look like this:

	wc=0	lb	ub	wc=1	lb	ub
k5=0	0.607	0.557	0.657	0.776	0.708	0.844
k5=1	0.263	0.193	0.333	0.445	0.333	0.557
k5=2	0.076	0.026	0.127	0.157	0.058	0.255
k5=3	0.019	-0.001	0.039	0.041	-0.002	0.085

The key to understanding the program below is to keep track of the row and column indicators that are used to place the information into the correct position.

```
. matrix out = J(4,6,.)
. local colnm ""
. forvalues w = 0/1 {
  2.        local colnm "'colnm' wc='w' lb ub"
  3.        local rownm ""
  4.        forvalues k = 0/3 {
  5.                local rownm "'rownm' k5='k'"
  6.                qui prvalue, x(k5='k' wc='w')  .
  7.                scalar p = pepred[2,2] // get information from matrices
  8.                scalar lb = pelower[2,2]
  9.                scalar ub = peupper[2,2]
 10.                local colnum = ('w'*3) + 1 // colnum=1 if wc=0, 4 if wc=1
 11.                matrix out['k'+1,'colnum'] = p
 12.                local ++colnum   // same as local colnum = 'colnum' + 1
 13.                matrix out['k'+1,'colnum'] = lb
 14.                local ++colnum
 15.                matrix out['k'+1,'colnum'] = ub
 16.        }
 17. }
```

We add row and column labels to our matrix and print the results:

```
. matrix colnames out = 'colnm'

. matrix rownames out = 'rownm'

. matrix list out, noheader format(%6.3f)
         wc=0      lb      ub    wc=1      lb      ub
k5=0    0.607   0.557   0.657   0.776   0.708   0.844
k5=1    0.263   0.193   0.333   0.445   0.333   0.557
k5=2    0.076   0.026   0.127   0.157   0.058   0.255
k5=3    0.019  -0.001   0.039   0.041  -0.002   0.085
```

9.4.7 Using forvalues to create tables with other commands

The basic idea of looping through values and storing predictions can be used with commands other than prvalue. Suppose that we want to know the mean age for those with various combinations of wc and k5. The summarize command prints results,

```
. summarize age
    Variable |      Obs        Mean    Std. Dev.       Min        Max
-------------+--------------------------------------------------------
         age |      753    42.53785    8.072574         30         60
```

The summarize command also returns results that we can use in our forvalues loop:

```
. return list
scalars:
                  r(N) =  753
              r(sum_w) =  753
               r(mean) =  42.53784860557769
                r(Var) =  65.16645121641095
                 r(sd) =  8.072574014303674
                r(min) =  30
                r(max) =  60
                r(sum) =  32031
```

The following program uses summarize with if conditions and retrieves information
from the returned results. Again we begin by setting up matrices to hold the results:

```
. matrix mn = J(4,2,.)                    // matrix for means
. matrix colnames mn = wc=0 wc=1
. matrix rownames mn = k5=0 k5=1 k5=2 k5=3
. matrix sd = mn                          // matrix for SDs
. matrix n = mn                           // matrix for Ns
```

Then we use a forvalues loop:

```
. forvalues w = 0/1 {
  2.      forvalues k = 0/3 {
  3.          summarize age if wc=='w' & k5=='k'
  4.          matrix n['k'+1,'w'+1] = r(N)
  5.          matrix mn['k'+1,'w'+1] = r(mean)
  6.          matrix sd['k'+1,'w'+1] = r(sd)
  7.      }
  8. }
```

The results are

```
. matrix list n, format(%6.0f) noheader

        wc=0  wc=1
k5=0    444   162
k5=1     85    33
k5=2     12    14
k5=3      0     3
. matrix list mn, format(%6.3f) noheader

         wc=0    wc=1
k5=0   44.450  43.951
k5=1   35.953  35.121
k5=2   32.917  33.286
k5=3        .  33.000
. matrix list sd, format(%6.3f) noheader

        wc=0   wc=1
k5=0   7.565  7.874
k5=1   5.661  4.379
k5=2   2.875  4.122
k5=3       .  3.464
```

9.5   Extending SPost to other estimation commands

The commands in SPost work only with some of the many estimation commands available in Stata. If you try to use our commands after fitting other types of models, you will be told that the SPost command does not work for the last model fitted. Over the past year as we developed these commands, we have received many inquiries about whether we can modify SPost to work with more estimation commands. Although we would like to accommodate such requests, extensions are likely to be made mainly to estimation commands that we are using in our own work. Our time is limited, and we want to be sure that we fully understand the specifics of each model before we incorporate it into SPost. Still, users who know how to program in Stata are welcome to extend our programs to work with other models. However, we can provide only limited support. Although we have attempted to write each command to make it as simple as possible to expand, some of the programs are complex, and you will need to be adept at programming in Stata.[3]

Here are some brief points that may be useful for a programmer wishing to modify our commands. First, our commands use ancillary programs that we have also written, all which begin with _pe (e.g., _pebase). As will be apparent when you trace through the logic of one of our ado-files, extending a command to a new model might require modifications to these ancillary programs as well. As the _pe*.ado files are used by many different commands, be careful that you do not make changes that break other commands. Second, our programs use information returned in e() by the estimation command. Some user-written estimation commands, especially older ones, do not return the appropriate information in e(), and extending programs to work after these estimation commands will be extremely difficult.

9.6   Using Stata more efficiently

Our introduction to Stata in chapter 2 focused on the basics. But as you use Stata, you will discover various tricks that make your use of Stata more enjoyable and efficient. While what constitutes a "good trick" depends on the needs and skills of the particular users, here we describe some things that we have found useful.

9.6.1   profile.do

When Stata is launched, it looks for a do-file called `profile.do` in the directories listed when you type `sysdir`.[4] If `profile.do` is found in one of these directories, Stata runs it. Accordingly, you can customize Stata by including commands in `profile.do`. Although you should consult the *Getting Started with Stata* manual for full details or enter the command `help profile`, the following examples show you some things that

---

3. StataCorp offers both introductory and advanced NetCourses in programming; more information on this can be obtained from *http://www.stata.com*.

4. The preferred place for the file is in your default data directory (e.g., `c:\data`).

we find useful. We have added detailed comments after the //s. The comments do not need to be included in `profile.do`.

```
// In Stata, all data are kept in memory.  You can set the amount of memory
// Stata reserves for data.  If you get memory errors when loading a dataset
// or while fitting a model, you need more memory.

set memory 30m

// Many programs in official Stata and many of our commands use matrices.
// Some of our commands, such as -prchange-, use a lot of memory. So, we
// suggest setting the amount of space for matrices to the largest value
// allowed. Type -help matsize- for details.
// The limit is 11,000 for Stata/SE and 800 for Intercooled.

set matsize 800

// Output in log files can be written either as text (as with earlier
// versions of Stata) or in SMCL. We find it easier to save logs as text
// since they can be more easily printed, copied to a word processor,
// and so on. Type -help log- for details.

set logtype text

// You can assign commands to function keys F2 through F9. After assigning
// a text string to a key, when you press that key, the string is
// inserted into the Command window.

global F8 "set trace on"
global F9 "set trace off"

// You can tell Stata what you want your default working directory
// to be.

cd d:\statastart

// You can also add notes to yourself. Here we post a reminder that
// the command -spost- will change the working directory to the directory
// where we have the files for this book.

noisily di "spost == cd d:\spost\examples"

// Alternatively, you can also use global macros to assign names to
// frequently used directories; then you can make that the working
// directory just by typing -cd "$macroname"-

global spost "d:\spost\examples"
```

Some tasks that can be done using the `profile.do` file can also be done using the permanently option of the `set` command. Specifying this option means that the new setting will remain on subsequent occasions when Stata is launched. The `set memory`, `set matsize`, and `set logtype` commands can be specified using the permanently option.

9.6.2   Changing screen fonts and window preferences

If you use a higher resolution for your monitor (e.g., 1280×1024), you might prefer a larger font size than Stata's default. To change the font, right-click in the Results window; select Font...; and choose a font you like. You do not need to select one of the fonts that are named "Stata ..." as any fixed-width font will work.

You can also change the size and position of the windows using the usual methods of clicking and dragging. After the font is selected and any new placement of windows is done, you can save your new options to be the defaults with the Preference menu and the Save Windowing Preferences option. Your windowing preferences are automatically saved when you exit Stata.

9.6.3   Using ado-files for changing directories

One of the things we like best about Stata is that you can create your own commands using ado-files. These commands work just like the commands that are part of official Stata, and indeed many commands in Stata are written as ado-files. If you are like us, at any one time you are working on several different projects. We like to keep each project in a different directory. For example, `d:\nas` includes research for the National Academy of Sciences, `d:\kinsey` is a project associated with the Kinsey Institute, and `d:\spost\examples` is (you guessed it) for this book. While you can change to these directories with the `cd` command, one of us keeps forgetting the names of directories. So, he writes a simple ado-file

```
program define spost
    cd d:\spost\examples
end
```

and saves this in his PERSONAL directory as `spost.ado`. Type `sysdir` to see what directory is assigned as the PERSONAL directory. Then whenever he types `spost`, his working directory is immediately changed:

```
. spost
d:\spost\examples
```

9.6.4   me.hlp file

Help files in Stata are plain text or SMCL files that end with the `.hlp` extension. When you type `help` *command*, Stata searches in the same directories used for ado-files until it finds a file called `command.hlp`. We have a file called `me.hlp` that contains information on things we often use but seldom remember. For example,

```
. -
help for ^me^
. -

Reset everything          ^clear^
----------------          ^discard^

List installed packages   ^ado dir^
-----------------------

Axes options              ^x/ylabel()^
------------              ^x/yline()^

. -
Author: Scott Long
```

This file is saved in your PERSONAL directory; typing sysdir will tell you what your PERSONAL directory is. Then whenever we are stuck and want to recall this information, we just need to type help me, and it is displayed on our screen.

9.7 Conclusions

Our goal in writing this book was to make it routine to carry out the complex calculations necessary for the interpretation of regression models for categorical outcomes. We hope that the tools we provide not only help you analyze your data more effectively but also make the process more enjoyable.

A Syntax for SPost commands

This appendix is a quick reference for all commands in SPost. Details on how each statistic is computed and interpreted are provided in the text. The following commands are described:

asprvalue   Compute predicted probabilities for models with alternative-specific variables.

brant   Perform a Brant test of the parallel regression assumption for the ordered logit model.

case2alt   Convert data from case-specific to alternative-specific form.

countfit   Generate plots and measures to assess the fit of count models.

fitstat   Fit statistics for regression models.

leastlikely Identify least-likely observations.

listcoef   List transformed regression coefficients with guidelines for interpretation.

misschk   Document patterns of missing data in a set of variables.

mlogplot   Create odds-ratio and discrete-change plots for multinomial logit.

mlogtest   Test for the multinomial logit model.

mlogview   Access a dialog box for using mlogplot interactively.

praccum   Accumulate results from prvalue to construct plots.

prchange   Compute discrete and marginal change for regression models.

prcounts   Compute predicted probabilities and rates for count models.

prgen   Generate variables with predicted values over a range of a continuous independent variable for use in plots.

prtab   Construct tables of predicted values.

prvalue   Compute predicted values for specified values of the independent variables.

spex   Load example datasets and run example models to illustrate SPost.

A.1   asprvalue

asprvalue   Compute predicted probabilities for models with alternative-specific variables for `clogit`, `rologit`, and `asmprobit`. For a detailed discussion, see chapter 7.

Syntax

asprvalue $\big[$, x(*variables_and_values*) <u>r</u>est(*stat*) <u>b</u>ase(*refcatname*)

<u>c</u>at(*catnames*) <u>s</u>ave <u>d</u>iff <u>br</u>ief$\big]$

Description

asprvalue computes predicted probabilities for logit or probit models that can combine case- and alternative-specific variables. asprvalue allows you to specify the values of the independent variables and presents predicted probabilities for the alternatives. The command works with `clogit`, `rologit`, and `asmprobit`.

Important: If you use asprvalue with `clogit` or `rologit` with case-specific variables, the interaction variables must be named following the conventions used by `case2alt` (although you do not have to use `case2alt` to generate the variables). Namely, the interactions must be named *altname*X*csvname*, where *altname* is the name of an alternative (and the name of the corresponding dummy variable used to indicate the row for that alternative) and *csvname* is the name of the case-specific variable. For example, all the interactions of case-specific variables with the alternative `bus` must begin `busX` and all the interactions with case-specific variable `hinc` must end with `Xhinc`. If there are any alternative by alternative-specific variable interactions, these must be similarly named (e.g., `busXtime`). A capital X cannot be used in any names except to denote these interactions.

Options

x(*variables_and_values*) assigns values to the independent variables for computing the predictions. For case-specific variables, only one value should be provided. For alternative-specific variables, one value can be provided, which will be assigned to all the alternatives, or J values can be provided to assign sequentially to all J alternatives. For `clogit` and `rologit`, values for alternative-specific variables should be ordered the same as they appear in the model (or as specified using `cat()`), with the value to be assigned to the reference category last. For `asmprobit`, the different alternatives are specified using one variable rather than a series of dummy variables, and values for alternative-specific variables should be ordered to correspond to the ascending values of the variable.

rest(*stat*) assigns values to the independent variables not specified in `x()`. The default is `mean`, which assigns the mean value to all variables. `asmean` assigns the mean to case-specific variables but assigns the alternative-specific means to alternative-specific variables. For example, with an alternative-specific variable `time`, `mean` would compute the mean over all individuals and alternatives and assign this same value to each alternative, whereas `asmean` would compute the mean for `time` within each alternative (i.e., separate means for bus, train, and car) and assign each alternative with the mean for that alternative.

`base`(*refcatname*) specifies the name of the base category to be used in the output. If `base()` is not included, the name `base` is used. This option should not be used with `asmprobit`.

`cat`(*catnames*) provides the names of the dummy variables for the alternatives in the model. This option does not need to be specified if case-specific variables are in the model, as then `asprvalue` determines the names of these dummy variables from the names of the interactions. If no case-specific variables were included in the model, however, the alternatives would need to be indicated by, for example, `cat(train bus)`. The reference category should not be listed among the alternatives in `cat()`. This option should not be used after models fitted using `asmprobit`.

`save` and `diff` save predicted probabilities and compares current probabilities with saved probabilities, respectively.

`brief` provides only minimal output.

Examples

We first fit a conditional logit model:

```
. use http://www.stata-press.com/data/lf2/travel2
. gen trainXhinc = hinc*train
. gen busXhinc = hinc*bus
. clogit choice time trainXhinc busXhinc train bus, group(id) nolog
```

Then we use `asprvalue` to compute the predicted probabilities of a trip that takes 4 hours by train, 3 hours by bus, and 5 hours by car (the reference category):

```
. asprvalue, x(time=240 180 300) base(car)
clogit: Predictions for choice
            prob
train   .30936006
  bus   .66694695
  car   .02369296
case-specific variables
          hinc
x=   31.809211
alternative-specific variables
        train    bus    car
time      240    180    300
```

We can examine how these predictions change, for example, as the value of `hinc` changes from 40 to 50, with the values for `time` held at their alternative-specific means:

```
. quietly asprvalue, x(hinc=40) rest(asmean) base(car) save
. asprvalue, x(hinc=50) rest(asmean) base(car) diff
clogit: Predictions for choice
            Current      Saved       Diff
train    .26957342   .36417952   -.0946061
  bus    .18044217   .17694971   .00349246
  car     .5499844   .45887077   .09111363

case-specific variables

              hinc
Current        50
  Saved        40
   Diff        10

alternative-specific variables

                    train        bus        car
Current:time    643.44079  674.61842  578.26974
  Saved:time    643.44079  674.61842  578.26974
   Dif:time             0          0          0
```

A.2 brant

brant   Brant test of parallel regression assumption for the ordered logit model fitted by `ologit`. For a detailed discussion, see section 5.6.

Syntax

brant $\left[\text{, detail}\right]$

Description

`brant` performs a Brant test (Brant 1990) of the parallel regression assumption (also called the proportional odds assumption) for the ordered logit model fitted by `ologit`. The test compares the slope coefficients from the $J-1$ binary logits implied by the ordered regression model. An omnibus test of all variables and tests for each independent variable are reported. To compute the tests, `brant` fits the $J-1$ binary logits defined by whether the outcome y is greater than or equal to j. *Note*: If there is perfect prediction in one of these binary logits, the Brant test cannot be computed.

Option

`detail` specifies that the coefficients for each of the binary logits should be presented.

Examples

`brant` requires the most recent model to have been fitted with `ologit`. For example,

```
. use http://www.stata-press.com/data/lf2/ordwarm2,clear
(77 & 89 General Social Survey)
. ologit warm yr89 male white age ed prst, nolog
  (output omitted )
. brant

Brant Test of Parallel Regression Assumption

    Variable |    chi2   p>chi2    df
    ---------+-------------------------
         All |   49.18    0.000     12
    ---------+-------------------------
        yr89 |   13.01    0.001      2
        male |   22.24    0.000      2
       white |    1.27    0.531      2
         age |    7.38    0.025      2
          ed |    4.31    0.116      2
        prst |    4.33    0.115      2
    ---------+-------------------------

A significant test statistic provides evidence that the parallel
regression assumption has been violated.
```

Here we find that the omnibus test rejects the hypothesis of parallel regressions, whereas the tests for individual coefficients show that the assumption is violated only for the variables `yr89`, `male`, and `age`. To see the results from the binary logits used to compute the Brant test, we add the `detail` option:

```
. brant, detail

Estimated coefficients from j-1 binary regressions

              y>1          y>2          y>3
 yr89     .9647422    .56540626    .31907316
 male   -.30536425   -.69054232   -1.0837888
white   -.55265759   -.31427081   -.39299842
  age    -.0164704   -.02533448   -.01859051
   ed    .10479624    .05285265    .05755466
 prst   -.00141118    .00953216    .00553043
_cons    1.8584045    .73032873   -1.0245168

Brant Test of Parallel Regression Assumption

    Variable |    chi2   p>chi2    df
    ---------+-------------------------
         All |   49.18    0.000     12
    ---------+-------------------------
        yr89 |   13.01    0.001      2
        male |   22.24    0.000      2
       white |    1.27    0.531      2
         age |    7.38    0.025      2
          ed |    4.31    0.116      2
        prst |    4.33    0.115      2
    ---------+-------------------------

A significant test statistic provides evidence that the parallel
regression assumption has been violated.
```

Saved results

Scalars
  r(chi2)   chi-squared statistic of omnibus test
  r(df)   degrees of freedom of omnibus test
  r(p) p-value of omnibus test

Matrices
  r(ivtests) contains the test statistics in the first column and the associated p-values
         in the second column for the tests of each independent variable. The
         row labels are the names of the independent variables.

A.3 case2alt

case2alt Convert data from case-specific to alternative-specific form.

Syntax

case2alt , {choice(*varname*) | rank(*stubname*)} [alt(*stubnames*)

 casevars(*varlist*) case(*varname*) generate(*newvar*) replace

 altnum(*varname*) nonames]

 case2alt assumes that the values of the outcome variable are positive integers.

Description

case2alt reshapes data arranged as one row per case to data arranged as one row per
alternative (per case). This is useful when fitting the models for alternative-specific
data discussed in chapter 7, which require data to be arranged in the latter form.

Options

choice(*varname*) or rank(*stubname*) is required.

 choice(*varname*) specifies the outcomes that are not ranks; *varname* is the variable
  that indicates the value of the selected alternative.

 rank(*stubname*) specifies the ranked outcomes, *stubname* is the prefix of the variable that contains the information about rank; i.e., if *stubname* is outcome,
  outcome1 would contain the rank information for alternative 1, outcome2 the
  rank information for alternative 2, and so on.

alt(*stubnames*) contains the *stubnames* for alternative-specific variables. This requires
 that variables *stubname#* exist for each value of an alternative.

casevars(*varlist*) contains the names of the case-specific variables (not including the
 ID or outcome variable).

case(*varname*) indicates the variable, either existing or to be created, that identifies individual cases. If *varname* is unspecified, a new variable named _id will be created.

generate(*newvar*) and replace are used to name the variable that contains 1 for the selected alternative and 0 for nonselected alternatives. The variable will be named *newvar* if *newvar* is specified, will be the name of the variable specified in choice() if replace is specified, and will be named choice otherwise.

altnum(*varname*) indicates the name of the new variable used to indicate the alternatives. _altnum will be used if altnum() is not specified.

nonames indicates that the case-specific interactions should be named y# instead of using the value labels of the outcome variable.

Example

We want to convert the wlsrnk dataset from one case per row to one alternative per row. wlsrnk.dta contains the case-specific variables fem and hn, the alternative-specific variables hashi1–hashi4 and haslo1–haslo4, the outcome variables value1–value4 (containing the ranks of the four alternatives), and the identifying variable id.

```
. use http://www.stata-press.com/data/lf2/wlsrnk, clear
. label variable value1 "est"
. label variable value2 "var"
. label variable value3 "aut"
. label variable value4 "sec"
. case2alt, case(id) rank(value) alt(hashi haslo) casevars(fem hn) gen(rank) ///
> altnum(alt)
ranks indicated by: rank
case identifier: id
case-specific interactions: est* var* aut* sec*
alternative-specific variables: hashi haslo
```

A.4   countfit

countfit   Compare the fit of the Poisson, negative binomial, zero-inflated Poisson, and zero-inflated negative binomial models.

(*Continued on next page*)

Syntax

>   countfit *varlist* [*if*] [*in*] [, inflate(*varlist2*) noconstant prm nbreg zip
>   zinb generate(*prefix*) replace note(*string*)
>   graphexport(*filename*[, replace]) nograph nodifferences noprtable
>   noestimates nofit nodash maxcount(*#*) noisily]

Description

>   countfit automates the process of comparing models fit with poisson, nbreg, zip,
>   and zinb. It can generate a table of estimates, a table of differences between observed
>   and average estimated probabilities, a graph of these differences, and various tests and
>   measures of fit used to compare count models.

Options for specifying the model

>   *varlist* is the variable list for the model, beginning with the outcome variable.
>
>   if *exp* and in *range* specify the sample used for fitting the models.
>
>   inflate(*varlist2*) specifies the inflation variables for zip and zinb.
>
>   noconstant specifies that there is no constant term in the model.

Options to select the models to fit

>   By default, poisson, nbreg, zip, and zinb are all fitted. If you want only some of these
>   models, specify the models you want with the following:
>
>   prm fits the poisson model.
>
>   nbreg fits the nbreg model.
>
>   zip fits the zip model.
>
>   zinb fits the zinb model.

Options to label and save results

>   generate(*prefix*) specifies up to five letters to prefix the new variables and to label
>   the models in the output. This name is placed in front of the type of model (e.g.,
>   *name*PRM). These labels help keep track of results from multiple specifications of
>   models.
>
>   replace replaces variables created by generate() if they already exist.
>
>   note(*string*) adds a label to the graph.
>
>   graphexport(*filename*[, replace]) specifies the name of file with the saved graph.

Options to control what is printed

nograph suppresses the graph of differences from observed counts.

nodifferences suppresses the table of differences from observed counts.

noprtable suppresses the table of predictions for each model.

noestimates suppresses the table of estimated coefficients.

nofit suppresses the table of fit statistics and tests of fit.

nodash suppresses dashed lines between measures of fit.

maxcount(#) specifies the number of counts to evaluate. For example, maxcount(5) indicates that only counts from 0 to 5 should be shown.

noisily includes output from Stata estimation commands; without this option the results are shown only in the estimates table output.

Example

If we wanted only to compare results for nbreg and zip and did not want the table or graph comparing predicted and actual probabilities, we can use countfit as follows:

(Continued on next page)

```
. use http://www.stata-press.com/data/lf2/couart2, clear
(Academic Biochemists / S Long)
. countfit art fem mar kid5 phd ment, inf(ment fem) nbreg zinb nograph noprtable
```

Variable	NBRM	ZINB
art		
Gender: 1=female 0=male	0.805	0.836
	-2.98	-2.40
Married: 1=yes 0=no	1.162	1.150
	1.83	1.72
Number of children < 6	0.838	0.845
	-3.32	-3.22
PhD prestige	1.015	1.001
	0.42	0.04
Article by mentor in last 3 yrs	1.030	1.025
	8.38	7.07
Constant	1.292	1.465
	1.85	2.69
lnalpha		
Constant	0.442	0.375
	-6.81	-7.06
inflate		
Article by mentor in last 3 yrs		0.470
		-2.55
Gender: 1=female 0=male		2.868
		1.40
Constant		0.275
		-2.14
Statistics		
alpha	0.442	
N	915.000	915.000
ll	-1560.958	-1552.034
bic	3169.649	3172.257
aic	3135.917	3124.068

```
                                                    legend: b/t
Comparison of Mean Observed and Predicted Count
          Maximum      At      Mean
Model     Difference   Value   |Diff|
------------------------------------------------
NBRM       -0.015       3      0.006
ZINB       -0.019       3      0.008
Tests and Fit Statistics
------------------------------------------------------------------------
NBRM          BIC= -3069.666  AIC=     3.427  Prefer  Over  Evidence
------------------------------------------------------------------------
    vs ZINB   BIC= -3067.059  dif=    -2.608  NBRM    ZINB  Positive
              AIC=     3.414  dif=     0.013  ZINB    NBRM
              Vuong=   2.069  prob=    0.019  ZINB    NBRM  p=0.019
```

A.5   fitstat

fitstat   Compute scalar measures of fit for regression models. For a detailed discussion, see section 3.5.

Syntax

fitstat [, <u>sav</u>ing(*name*) <u>us</u>ing(*name*) <u>b</u>ic force save <u>d</u>iff]

Description

fitstat is a postestimation command that computes several measures of fit for the following regression models: clogit, cloglog, cnreg, intreg, logit, mlogit, mprobit, nbreg, ologit, oprobit, poisson, probit, regress, rologit, slogit, tobit, zinb, zip, ztnb, and ztp. With the saving and using options, it can be used to compare fit from two models. For all models, fitstat reports the log likelihoods of the full and intercept-only models, the deviance (D), the likelihood-ratio chi-square (G^2), Akaike's Information Criterion (AIC), AIC*N, the Bayesian Information Criterion (BIC), and BIC'. For OLS regression, fitstat reports R^2 and the adjusted R^2. For other models, fitstat reports McFadden's R^2, McFadden's adjusted R^2, the maximum likelihood R^2, and Cragg and Uhler's R^2. For categorical outcomes, fitstat reports the regular and adjusted count R^2. For ordered, binary, and censored outcomes, McKelvey and Zavoina's R^2 is computed. Also Efron's R^2 is computed for binary outcomes. Not all measures are provided for models fitted with pweights or iweights.

Options

saving(*name*) saves the computed measures in a matrix for later comparisons. *name* cannot be longer than four characters.

using(*name*) compares the fit measures for the current model with those of the model saved as *name*.

bic presents only information measures.

force compares two models, even if the number of observations or the type of model differs for the models.

save and diff are equivalents of saving(0) and using(0) and do not require the user to provide model names.

Examples

For a binary logit,

```
. use http://www.stata-press.com/data/lf2/binlfp2, clear
(Data from 1976 PSID-T Mroz)
```

```
. logit lfp k5 k618 age wc hc lwg inc, nolog
  (output omitted)
. fitstat

Measures of Fit for logit of lfp
Log-Lik Intercept Only:     -514.873   Log-Lik Full Model:            -452.633
D(745):                      905.266   LR(7):                          124.480
                                       Prob > LR:                        0.000
McFadden's R2:                 0.121   McFadden's Adj R2:                0.105
ML (Cox-Snell) R2:             0.152   Cragg-Uhler(Nagelkerke) R2:      0.204
McKelvey & Zavoina's R2:       0.217   Efron's R2:                      0.155
Variance of y*:                4.203   Variance of error:               3.290
Count R2:                      0.693   Adj Count R2:                    0.289
AIC:                           1.223   AIC*n:                         921.266
BIC:                       -4029.663   BIC':                          -78.112
BIC used by Stata:           958.258   AIC used by Stata:             921.266
```

To compare the fit statistics for two models:

```
. logit lfp k5 k618 age wc hc lwg inc, nolog
  (output omitted)
. quietly fitstat, saving(mod1)
. gen age2 = age*age
. logit lfp k5 age age2 wc inc, nolog
  (output omitted)
. fitstat, using(mod1)

Measures of Fit for logit of lfp
                              Current       Saved       Difference
Model:                          logit       logit
N:                                753         753                0
Log-Lik Intercept Only       -514.873    -514.873            0.000
Log-Lik Full Model           -461.653    -452.633           -9.020
D                          923.306(747) 905.266(745)     18.040(2)
LR                          106.441(5)   124.480(7)       18.040(2)
Prob > LR                       0.000       0.000            0.000
McFadden's R2                   0.103       0.121           -0.018
McFadden's Adj R2               0.092       0.105           -0.014
ML (Cox-Snell) R2               0.132       0.152           -0.021
Cragg-Uhler(Nagelkerke) R2      0.177       0.204           -0.028
McKelvey & Zavoina's R2         0.182       0.217           -0.035
Efron's R2                      0.135       0.155           -0.020
Variance of y*                  4.023       4.203           -0.180
Variance of error               3.290       3.290            0.000
Count R2                        0.677       0.693           -0.016
Adj Count R2                    0.252       0.289           -0.037
AIC                             1.242       1.223            0.019
AIC*n                         935.306     921.266           14.040
BIC                        -4024.871   -4029.663            4.791
BIC'                          -73.321     -78.112            4.791
BIC used by Stata             963.050     958.258            4.791
AIC used by Stata             935.306     921.266           14.040
Difference of    4.791 in BIC' provides positive support for saved model.
Note: p-value for difference in LR is only valid if models are nested.
```

save and diff could have been used instead of saving(mod1) and using(mod1).

Saved results

The following results are saved. If a statistic is not appropriate for a given model, a missing value is returned.

r(stataaic)	AIC computed by Stata	r(n_parm)	number of parameters
r(statabic)	BIC computed by Stata	r(n_rhs)	number of right-hand-side
r(aic)	AIC		variables
r(aic_n)	AIC$\times N$	r(r2)	R^2 for linear regression
r(bic)	BIC		model
r(bic_p)	BIC$'$	r(r2_adj)	adjusted R^2 for linear
r(dev)	deviance		regression model
r(dev_df)	degrees of freedom for deviance	r(r2_ct)	count R^2
r(ll)	log likelihood for full model	r(r2_ctadj)	adjusted count R^2
r(ll_0)	log likelihood for model with	r(r2_cu)	Cragg and Uhler's (Nagelkerke) R^2
	only intercept	r(r2_ef)	Efron's R^2
r(lrx2)	likelihood-ratio chi-square	r(r2_mz)	McKelvey and Zavoina's R^2
r(lrx2_df)	degrees of freedom for	r(r2_mf)	McFadden's R^2
	likelihood-ratio chi-square	r(r2_mfadj)	McFadden's adjusted R^2
r(lrx2_p)	probability level of chi-squared	r(r2_ml)	maximum likelihood (Cox–Snell) R^2
	test	r(v_error)	variance of error term
r(N)	number of observations	r(v_ystar)	variance of y^*

When `saving(`*name*`)` is specified, the fit statistics are saved in matrix `fs_`*name*. The column names are the names of the measures; the row name is the command used to fit the model. Values of -9999 in the matrix indicate that a measure is not appropriate for the given model.

A.6   leastlikely

leastlikely   Identify the least-likely observations based on predicted probabilities.

Syntax

```
leastlikely [ varlist ] [ if ] [ in ] [ , n(#) generate(varname) [no]display
    nolabel noobs ]
```

Description

`leastlikely` lists the in-sample observations with the lowest predicted probabilities of observing the outcome value that was actually observed. For example, in a model with a binary dependent variable, `leastlikely` lists the observations that have the lowest predicted probability of *depvar* $= 0$ among those cases for which *depvar* $= 0$, and it lists the observations that have the lowest predicted probability of *depvar* $= 1$ among those cases for which *depvar* $= 1$. `leastlikely` works with estimation commands for models of binary outcomes in which option p after `predict` provides the predicted probability of a positive outcome (e.g., `logit`, `probit`), but the dependent variable must be coded as 0 and 1. `leastlikely` also works with estimation commands for models of ordinal or nominal outcomes in which option outcome(#) after predict provides the predicted probability of outcome #.

Options

n(#) specifies the number of observations to be listed for each outcome. The default is 5. For multiple observations with identical predicted probabilities, all will be listed.

generate(*varname*) specifies that the probabilities of observing the outcome value that was observed should be stored in *varname*. If not specified, the variable name Prob will be created but dropped after the output is produced.

Options for listing

leastlikely can also include any of the options available after list. These include the following:

[no]display forces the format into display or tabular (nodisplay) format. If you do not specify one of these two options, Stata chooses one based on its judgment of which would be most readable.

nolabel causes numeric values rather than labels to be displayed.

noobs suppresses printing of the observation numbers.

Examples

```
. use http://www.stata-press.com/data/lf2/binlfp2
(Data from 1976 PSID-T Mroz)
. logit lfp k5 k618 age wc hc lwg inc
  (output omitted )
. leastlikely
Outcome: 0 (NotInLF)
```

	Prob
60.	.1231792
172.	.1490344
221.	.1470691
235.	.1666356
252.	.1088271

```
Outcome: 1 (inLF)
```

	Prob
338.	.1760865
534.	.0910262
568.	.178205
635.	.0916614
662.	.1092709

You can also list the values of variables for the least-likely observations:

```
. use http://www.stata-press.com/data/lf2/nomocc2
(1982 General Social Survey)
. mlogit occ white ed exper
  (output omitted )
. leastlikely white ed exper
Outcome: 1 (Menial)
```

	Prob	white	ed	exper
3.	.0644456	1	12	44
6.	.0564391	1	13	38
8.	.0658833	1	14	19
27.	.0753868	1	14	14
31.	.0249498	1	14	52

Outcome: 2 (BlueCol)

	Prob	white	ed	exper
38.	.0482357	1	16	11
39.	.0386521	1	16	19
56.	.0428928	0	16	22
59.	.0482357	1	16	11
78.	.0592207	0	14	49

Outcome: 3 (Craft)

	Prob	white	ed	exper
117.	.1592302	1	15	11
126.	.1586144	1	15	13
140.	.2177173	1	14	41
164.	.2177173	1	14	41
171.	.1048674	1	16	4
178.	.1586144	1	15	13

Outcome: 4 (WhiteCol)

	Prob	white	ed	exper
192.	.0851777	0	12	35
206.	.085288	1	11	3
207.	.0647436	1	18	49
221.	.0754128	1	10	12
225.	.0665037	1	8	38

```
Outcome: 5 (Prof)

           Prob    white   ed   exper

251.    .0644643      0    12      26
281.    .0746618      1    11      16
295.    .0138355      1     8      41
305.    .0883012      0    12      47
336.    .0061913      1     7      39
```

A.7   listcoef

listcoef   List transformed regression coefficients with guidelines for interpretation. For a detailed discussion, see section 3.1.9.

Syntax

listcoef [*varlist*] [, [factor|percent|std] pvalue(#) reverse lt gt
    adjacent nolabel constant expand matrix help]

If *varlist* is provided, only coefficients for these variables are to be listed. Otherwise, coefficients for all variables are listed.

Description

listcoef lists the estimated coefficients for several regression models, with options that allow you to specify different transformations of the coefficients, such as factor change and percent change. Coefficients can be standardized to a unit variance in the independent and dependent variables. For mlogit, coefficients for all comparisons among outcomes are included. The help option provides a guide for interpreting the coefficients. The listcoef command can be used with clogit, cloglog, cnreg, intreg, logistic, logit, mlogit, mprobit, nbreg, ologit, oprobit, poisson, probit, regress, rologit, slogit, tobit, zinb, zip, ztnb, and ztp. For models with categorical outcomes, the output is easier to understand if you assign value labels to the dependent variable.

Options

factor requests that factor change coefficients (i.e., odds ratios) be listed.

percent requests that percent change coefficients be listed.

std requests that coefficients standardized to a unit variance for the dependent variable be listed.

pvalue(#) specifies that only coefficients that are significant at this level or smaller will be printed.

reverse reverses the order of comparison for factor or percent change coefficients for clogit, logit, ologit, or rologit; that is, it presents results indicating the change in the odds of b versus a instead of a versus b.

constant includes the constants in the output.

matrix returns results in r-class matrices.

help includes details on the meaning of each coefficient.

Options for nominal outcomes

For models with nominal outcomes, listcoef shows the coefficient for each pair of outcome categories. When mlogit is used with ordinal variables, it is often helpful to look at only some of these coefficients. The following options are useful for this purpose:

lt specifies that only comparisons where the first category has a smaller value than the second will be printed (e.g., comparing outcome 1 versus 2, but not 2 versus 1).

gt specifies that only comparisons where the first category has a larger value than the second will be printed (e.g., comparing outcome 2 versus 1, but not 1 versus 2).

adjacent specifies that only comparisons where the two category values are adjacent will be printed (e.g., comparing outcome 1 versus 2 or 2 versus 1, but not 1 versus 3).

nolabel requests that category numbers rather than value labels be used in the output.

expand requests expanded output comparing all pairs of outcome categories for slogit.

Examples

listcoef can be used to list only coefficients significant at a given level:

```
. use http://www.stata-press.com/data/lf2/regjob2, clear
(Academic Biochemists / S Long)

. * regress
. regress job fem phd ment fel art cit
  (output omitted )

. listcoef, pv(.05)
regress (N=408): Unstandardized and Standardized Estimates when P>|t| < 0.05

 Observed SD: .97360294
 SD of Error: .8717482
```

job	b	t	P>|t|	bStdX	bStdY	bStdXY	SDofX
phd	0.27268	5.529	0.000	0.2601	0.2801	0.2671	0.9538
fel	0.23414	2.469	0.014	0.1139	0.2405	0.1170	0.4866
cit	0.00448	2.275	0.023	0.1481	0.0046	0.1521	33.0599

For models with a latent variable, coefficients for the effect on the standardized y^* can be listed. Here we request only the coefficients for `age`:

```
. use http://www.stata-press.com/data/lf2/ordwarm2, clear
(77 & 89 General Social Survey)
. ologit warm age ed prst male yr89 white, nolog
  (output omitted )
. listcoef age, std
ologit (N=2293): Unstandardized and Standardized Estimates
  Observed SD: .9282156
    Latent SD: 1.9410634
```

warm	b	z	P>|z|	bStdX	bStdY	bStdXY	SDofX
age	-0.02167	-8.778	0.000	-0.3635	-0.0112	-0.1873	16.7790

or coefficients for the odds of outcomes $> m$ versus $\le m$ can be listed; with the `reverse` option, the coefficients are listed for the odds of $\le m$ versus $> m$:

```
. listcoef age
ologit (N=2293): Factor Change in Odds
  Odds of: >m vs <=m
```

warm	b	z	P>|z|	e^b	e^bStdX	SDofX
age	-0.02167	-8.778	0.000	0.9786	0.6952	16.7790

```
. listcoef age, rev
ologit (N=2293): Factor Change in Odds
  Odds of: <=m vs >m
```

warm	b	z	P>|z|	e^b	e^bStdX	SDofX
age	-0.02167	-8.778	0.000	1.0219	1.4384	16.7790

For `mlogit`, coefficients for all comparisons of outcomes can be listed. For example, we can request that only the significant percent change coefficients for `age` be listed:

```
. use http://www.stata-press.com/data/lf2/nomocc2, clear
(1982 General Social Survey)
. mlogit occ white ed exp, nolog
  (output omitted )
```

```
. listcoef white, percent pv(.05)

mlogit (N=337): Percentage Change in the Odds of occ when P>|z| < 0.05

Variable: white (sd= .276423)
```

	b	z	P>|z|	%	%StdX
Odds comparing Group 1 vs Group 2					
Menial −Prof	−1.77431	−2.350	0.019	−83.0	−38.8
Craft −Prof	−1.30196	−2.011	0.044	−72.8	−30.2
Prof −Menial	1.77431	2.350	0.019	489.6	63.3
Prof −Craft	1.30196	2.011	0.044	267.7	43.3

Saved results

Scalars
   r(pvalue)   1 or value specified with pvalue(#)

Macros
   r(cmd)   name of most recent estimation command

Matrices

   When the **matrix** option is specified, listcoef saves in r() all the statistics that have been computed for the given model. These are saved only when the **matrix** option is specified because saving these matrices can slow execution for **mlogit**. The row and column labels indicate the variable and type of coefficient that has been saved:

   r(b)   slope or regression coefficients
   r(b_fact)   factor change coefficients (i.e., $\exp\left(\hat{\beta}\right)$)
   r(b_facts) x-standardized factor change coefficients
   r(b_p) p-values for test of regression coefficients
   r(b_pct)   percent change coefficients
   r(b_pcts) x-standardized percent change coefficients
   r(b_sdx)   standard deviations for independent variables used to compute
                   x-standardized and fully standardized coefficients
   r(b_std)   fully standardized coefficients
   r(b_xs) x-standardized coefficients
   r(b_ys) y- or y^*-standardized coefficients
   r(b_z) z-values or t-values for regression coefficients
   r(cons)   constants
   r(cons_p) p-values for constants
   r(cons_z) z-values or t-values for constants
   r(contrast) all contrasts from **mlogit**

For **zip** and **zinb**, the matrices cons2, cons2_p, cons2_z, b2, b2_p, b2_z, b2_fact, b2_facts, b2_pct, and b2_pcts contain corresponding results for the binary equation.

A.8   misschk

misschk   Analyzes patterns of missing data. For a detailed discussion, see chapter 3.

Syntax

misschk *varlist* [*if*] [*in*] [, <u>gen</u>erate(*rootname*) replace dummy <u>nos</u>ort
<u>nonum</u>ber help]

varlist is the list of variables to be examined. By default, it is all the variables in the
dataset. if and in restrict the cases to be examined.

Options

generate(*rootname*) specifies the root for the variables created with information about
missing data. If this option is not used, temporary variables are created and
then deleted when the program is finished. Two variables are generated. *rootname*[*number*] specifies the number of variables from the variable list for which
a given observation has missing data. *rootname*[*pattern*] indicates the pattern of
missing values. This is a string variable with _ indicating valid data for a variable
and a number indicating missing data for that variable.

replace replaces existing variables that begin with *rootname* if they exist.

dummy requests that dummy variables be created for each variable in the variable list.
The dummy variable begins with the root specified with the generate() option,
and then adds the name of the variable. A value of 1 indicates missing data for that
case; 0 indicates data are not missing. For example, with the options generate(m_)
dummy, variables such as m_female and m_income would be generated.

nosort specifies that the list of patterns of missing data should not be sorted to list the
most common patterns first.

nonumber specifies that a variable that has missing cases will be indicated by a dot (.)
rather than by a single digit number corresponding to the position of that variable.
For example, without the nonumber option, a missing-data pattern might look like
_2_4_ 6___ to indicate missing data in the 2nd, 4th, and 6th variables. With the
nonumber option, the pattern would be _._._ .___.

help requests a description of each part of the output.

Example

```
. use http://www.stata-press.com/data/lf2/gsskidvalue2, clear
(1993 and 1994 General Social Survey)
```

```
. misschk female-income91, gen(m_) dummy
Variables examined for missing values
    #  Variable         # Missing   % Missing
--------------------------------------------
    1  female                   0        0.0
    2  black                    0        0.0
    3  othrrace                 0        0.0
    4  degree                  14        0.3
    5  anykids                 14        0.3
    6  kidvalue              1609       35.0
    7  income                 495       10.8
    8  age                      0        0.0
    9  income91                 0        0.0
```

```
Missing for |
      which |
  variables?|      Freq.       Percent        Cum.
------------+-------------------------------------
___45 67__  |          3          0.07        0.07
___4_ 67__  |          1          0.02        0.09
___4_ 6___  |          1          0.02        0.11
___4_ _7__  |          5          0.11        0.22
___4_ ____  |          4          0.09        0.30
____5 67__  |          6          0.13        0.43
____5 6___  |          3          0.07        0.50
____5 _7__  |          1          0.02        0.52
____5 ____  |          1          0.02        0.54
_____ 67__  |        185          4.02        4.57
_____ 6___  |      1,410         30.67       35.23
_____ _7__  |        294          6.39       41.63
_____ ____  |      2,684         58.37      100.00
------------+-------------------------------------
      Total |      4,598        100.00
```

```
Missing for |
   how many |
  variables?|      Freq.       Percent        Cum.
------------+-------------------------------------
          0 |      2,684         58.37       58.37
          1 |      1,709         37.17       95.54
          2 |        195          4.24       99.78
          3 |          7          0.15       99.93
          4 |          3          0.07      100.00
------------+-------------------------------------
      Total |      4,598        100.00
```

```
. describe m_*
              storage   display     value
variable name  type     format      label      variable label
-----------------------------------------------------------------------
m_female       float    %10.0g      lmisschk   Missing value for female?
m_black        float    %10.0g      lmisschk   Missing value for black?
m_othrrace     float    %10.0g      lmisschk   Missing value for othrrace?
m_degree       float    %10.0g      lmisschk   Missing value for degree?
m_anykids      float    %10.0g      lmisschk   Missing value for anykids?
m_kidvalue     float    %10.0g      lmisschk   Missing value for kidvalue?
m_income       float    %10.0g      lmisschk   Missing value for income?
m_age          float    %10.0g      lmisschk   Missing value for age?
m_income91     float    %10.0g      lmisschk   Missing value for income91?
m_pattern      str11    %11s                   Missing for which variables?
m_number       float    %9.0g                  Missing for how many variables?
```

A.9   mlogplot

mlogplot   Odds-ratio and discrete-change plots for multinomial logit. For a detailed discussion, see sections 6.6.7–6.6.10.

Syntax

There are two ways that `mlogplot` can be used. If you have just fitted a model using `mlogit`, you can create a discrete-change or odds-ratio plot for the coefficients in memory using the following syntax:

mlogplot [*varlist*] [, <u>or</u>atio <u>d</u>change <u>std</u>([s|u|0]...[s|u|0]) min(#) max(#)
   packed labels <u>p</u>rob(#) <u>b</u>asecategory(#) <u>nt</u>ics(#) <u>n</u>ote(*string*)
   dcadd(#)]

`prchange` must have been previously run if option `dchange` is used.
Important: you must use value labels for the dependent variable, and these labels must begin with a different letter or number for each outcome; otherwise, the plot will be misleading.

You can also create an odds-ratio plot from coefficients that you have placed into matrices (e.g., coefficients from a published paper or another program). The syntax is

mlogplot [*varlist*] [, matrix <u>v</u>ars(*varlist*) <u>std</u>([s|u|0]...[s|u|0]) min(#)
   max(#) packed <u>b</u>aseoutcome(#) <u>nt</u>ics(#) <u>n</u>ote(*string*)]

Details on the matrices are given below. Discrete change plots cannot be computed from matrices.

Description

`mlogplot` facilitates the interpretation of the multinomial logit model. It can plot the odds-ratios coefficients (i.e., $\exp\left(\widehat{\beta}_{A|B}\right)$) from `mlogit` or coefficients that have been saved in matrices. If `prchange` has been run after `mlogit`, `mlogplot` can create either a plot of discrete-change coefficients or add information about discrete change to the odds-ratio plots. Odds ratios can be plotted for either a unit change or a standard deviation change. Discrete-change coefficients can be plotted for a unit change, a standard deviation change, or a change from 0 to 1 (for dummy variables). In the plots, you can list the same variable more than once if you want to plot the effects of different amounts of change for the same variable. Several options control the way the graph looks. The program `mlogview` provides a convenient dialog box that allows you to use `mlogplot` interactively.

Options

varlist is the list of variables to be plotted. The same variable can be listed more than once if you want to plot its coefficients for different amounts of change.

`oratio` indicates that you want to plot the odds ratios (i.e., $\exp(\beta)$). This option cannot be used if coefficients are entered with matrices.

`dchange` indicates that you want to plot the discrete change. To use this option, you must have first run `prchange`.

- If only `oratio` is specified, an odds-ratio plot is drawn.
- If only `dchange` is specified, a discrete-change plot is drawn.
- If both are specified, an odds-ratio plot is drawn where the size of the letters are proportional to the size of the corresponding discrete-change coefficient.

`std([s|u|0]...[s|u|0])` specifies the type of coefficient to plot for each variable: s indicates standardized coefficients; u indicates unstandardized coefficients; 0 indicates changes from 0 to 1 in discrete-change plots. For example, `std(u0su)` indicates that the first variable is unstandardized, the second is binary, the third is standardized, and the last is unstandardized.

`min(#)` and `max(#)` specify the minimum value and the maximum value on the plotting axis. This is useful if you want to control the labeling of the tick marks or if you want to compare coefficients across plots.

`packed` removes the vertical spacing among the outcome categories in an odds-ratio plot. This allows up to 11 variables on one graph; the maximum otherwise is 6.

`labels` uses variable labels to label each row of the plot. You might need to revise your variable labels to make them fit the graph (i.e., your current labels might be too long). This option cannot be used with the `matrix` option.

`prob(#)` specifies for an odds-ratio plot that if a coefficient contrasting two outcomes is *not* significant at this level, a line is to be drawn connecting the letters. This option cannot be used with the `matrix` option.

`basecategory(#)` is used for an odds-ratio plot to specify which outcome of the outcome measure is to be used as the reference point.

`ntics(#)` sets the number of tick marks on the axes. Used along with `min()` and `max()`, this allows you to determine the numbering on the axes and the location of tick marks.

`note(string)` adds a title at the top of the plot.

`dcadd(#)` is rarely used. In odds-ratio plots with the `dchange` option, the size of the letter corresponds to the square root of the size of the discrete-change coefficient. `dcadd` adds an amount to each discrete change to make the size of all letters larger, making it easier to see the letters for small discrete changes. By default, this quantity

is 0. If your letters are too small (because the discrete change is small), you might
want to increase this by a small amount, say, dcadd(.03).

matrix indicates that the data for the graph will be taken from the following matrices
and global macros, which must be previously created by the user:

  mnlbeta is a matrix that contains estimated βs, where element (i, j) is the jth
    variable for comparison i relative to the reference outcome. That is, columns are
    for variables; rows for different contrasts. Constants are *not* included.

  mnlname is a matrix that contains the names of the variables corresponding to the
    columns of *mnlbeta*.

  mnlsd is a matrix that contains the standard deviations for the variables that cor-
    respond to columns of *mnlbeta*.

  mnlcatnm is a macro string with labels for the outcome categories. The first outcome
    corresponds to the first column of *mnlbeta*, the second to the second, etc. The
    label for the reference outcome should be last.

  mnlrefn is a macro with the number of the outcome that is the reference outcome
    for the contrasts contained in *mnlbeta*.

vars(*varlist*) is required when matrix is specified. *varlist* contains the names of the
  variables listed in *mnlname* whose coefficients you want to plot, in the order that
  you want to plot them.

Examples

Plotting coefficients from mlogit: After fitting a multinomial logit model with mlogit,
you can create a discrete-change plot with

```
. mlogit occ white ed exper
. prchange
. mlogplot white ed exper, dc std(0ss) min(-.5) max(.5)
```

A plot with only the odds ratios is created with

```
. mlogplot white ed exper, or std(0ss) min(-2.75) max(.5)
```

A plot with the size of the letters in the odds ratios being proportional to the amount
of discrete change is created with

```
. mlogplot white ed exper, or  dchange std(0ss) base(4) min(-1.75) max(1.75)
```

Plotting coefficients contained in matrices: To use the matrix options, you must
create matrices and macros that contain the values that you wish to plot. For example,

```
. matrix mnlsd = (2.946427, 13.95936, 2.946427, 13.95936)
. global mnlname = "W_Educ W_Exper NW_Educ NW_Exper"
. global mnlrefn = 5
. global mnlcatnm = "Menial BlueC Craft WhiteC Prof"
```

```
. matrix mnlbeta = (-.83075, -.92255, -.68761, -.41964  -.03380, -.03145,
      -.00026,  .00085  -.70126, -.56070, -.88250, -.53115   -.11084,
      -.02611,  -.15979, -.05209 )
. matrix mnlbeta = mnlbeta'
```

Then the plot can be constructed using `mlogplot`:

```
. mlogplot, vars(W_Educ NW_Educ W_Exper NW_Exper) matrix
      std(ssss) note("Effects of Education")
```

A.10 mlogtest

mlogtest   Statistical tests for the multinomial logit model. For a detailed discussion, see section 6.3.

Syntax

mlogtest $\left[\, varlist \,\right]$ $\big[\,$, <u>a</u>ll <u>lr</u> <u>w</u>ald <u>c</u>ombine <u>lrc</u>omb <u>set</u>(varlist $\left[\backslash\ varlist\right]$) <u>iia</u>

<u>h</u>ausman <u>smh</u>siao <u>d</u>etail <u>b</u>ase $\big]$

varlist indicates that the Wald or LR test for the significance of variables should be computed only for these variables. If no *varlist* is given, tests are run for all independent variables.

Description

`mlogtest` computes a variety of tests for the multinomial logit model. Users select the tests they want by specifying the appropriate options. For each independent variable, `mlogtest` can perform either a likelihood-ratio test or a Wald test of the null hypothesis that the coefficients for the variable equal zero across all equations. `mlogtest` can also perform Wald or likelihood-ratio tests of whether any pair of outcome categories can be combined. Also `mlogtest` computes both the Hausman and Small–Hsiao tests of the assumption of the independence of irrelevance alternatives (IIA) for each possible omitted outcome.

Options

`all` specifies that all available tests should be performed.

`lr` requests LR tests for each independent variable in the *varlist* or for all variables if *varlist* is not specified.

`wald` requests Wald tests for each independent variable in the *varlist* or for all variables if *varlist* is not specified.

`combine` computes Wald tests of whether two outcomes in the `mlogit` model can be combined.

`lrcomb` conducts likelihood-ratio tests of whether two outcomes can be combined.

set(*varlist* [\ *varlist*]) specifies that a set of variables is to be considered together for the Wald test or the likelihood-ratio test. The slash \ is used to specify multiple sets of variables. This option is useful, for example, when a categorical independent variable has been included as a set of dummy variables.

iia specifies that both tests of the IIA assumption should be performed.

hausman computes Hausman tests of the IIA assumption.

smhsiao performs Small–Hsiao tests of the IIA assumption.

detail reports full hausman output from IIA test (the default is to provide only a summary of results).

base also conducts an IIA test omitting the base outcome of the original mlogit estimation. This is done by refitting the model using the largest remaining outcome as the base outcome, although the original estimates are restored to memory afterward.

Examples

If all tests are requested, the following results are obtained:

```
. use http://www.stata-press.com/data/lf2/nomocc2, clear
(1982 General Social Survey)
. mlogit occ white ed exper
  (output omitted)
. set seed 131324
. mlogtest, all
**** Likelihood-ratio tests for independent variables (N=337)
Ho: All coefficients associated with given variable(s) are 0.
```

occ	chi2	df	P>chi2
white	8.095	4	0.088
ed	156.937	4	0.000
exper	8.561	4	0.073

```
**** Wald tests for independent variables (N=337)
Ho: All coefficients associated with given variable(s) are 0.
```

occ	chi2	df	P>chi2
white	8.149	4	0.086
ed	84.968	4	0.000
exper	7.995	4	0.092

```
**** Hausman tests of IIA assumption (N=337)
Ho: Odds(Outcome-J vs Outcome-K) are independent of other alternatives.
```

Omitted	chi2	df	P>chi2	evidence
Menial	7.324	12	0.835	for Ho
BlueCol	0.320	12	1.000	for Ho
Craft	-14.436	12	1.000	for Ho
WhiteCol	-5.541	11	1.000	for Ho

```
**** Small-Hsiao tests of IIA assumption (N=337)
```

Ho: Odds(Outcome-J vs Outcome-K) are independent of other alternatives.

Omitted	lnL(full)	lnL(omit)	chi2	df	P>chi2	evidence
Menial	-227.792	-168.813	117.957	12	0.000	against Ho
BlueCol	-164.194	-150.870	26.647	12	0.009	against Ho
Craft	-202.582	-139.441	126.281	12	0.000	against Ho
WhiteCol	-207.622	-150.645	113.954	12	0.000	against Ho

```
**** Wald tests for combining alternatives (N=337)
```

Ho: All coefficients except intercepts associated with a given pair of alternatives are 0 (i.e., alternatives can be combined).

Alternatives tested	chi2	df	P>chi2
Menial- BlueCol	3.994	3	0.262
Menial- Craft	3.203	3	0.361
Menial-WhiteCol	11.951	3	0.008
Menial- Prof	48.190	3	0.000
BlueCol- Craft	8.441	3	0.038
BlueCol-WhiteCol	20.055	3	0.000
BlueCol- Prof	76.393	3	0.000
Craft-WhiteCol	8.892	3	0.031
Craft- Prof	60.583	3	0.000
WhiteCol- Prof	22.203	3	0.000

```
**** LR tests for combining alternatives (N=337)
```

Ho: All coefficients except intercepts associated with a given pair of alternatives are 0 (i.e., alternatives can be collapsed).

Alternatives tested	chi2	df	P>chi2
Menial- BlueCol	4.095	3	0.251
Menial- Craft	3.376	3	0.337
Menial-WhiteCol	13.223	3	0.004
Menial- Prof	64.607	3	0.000
BlueCol- Craft	9.176	3	0.027
BlueCol-WhiteCol	22.803	3	0.000
BlueCol- Prof	125.699	3	0.000
Craft-WhiteCol	9.992	3	0.019
Craft- Prof	95.889	3	0.000
WhiteCol- Prof	26.736	3	0.000

The tests of IIA do not include a test based on the base outcome. These are obtained with the base option:

```
. mlogtest, iia base
**** Hausman tests of IIA assumption (N=337)
```

Ho: Odds(Outcome-J vs Outcome-K) are independent of other alternatives.

Omitted	chi2	df	P>chi2	evidence
Menial	7.324	12	0.835	for Ho
BlueCol	0.320	12	1.000	for Ho
Craft	-14.436	12	1.000	for Ho
WhiteCol	-5.541	11	1.000	for Ho
Prof	-0.119	12	1.000	for Ho

```
**** Small-Hsiao tests of IIA assumption (N=337)
Ho: Odds(Outcome-J vs Outcome-K) are independent of other alternatives.
Omitted | lnL(full)  lnL(omit)   chi2   df   P>chi2   evidence
--------+---------------------------------------------------------
 Menial |  -166.326   -161.385   9.882   12   0.626   for Ho
BlueCol |  -137.095   -133.528   7.135   12   0.849   for Ho
  Craft |  -128.141   -121.944  12.395   12   0.414   for Ho
WhiteCol|  -159.113   -155.967   6.290   12   0.901   for Ho
   Prof |  -135.499   -132.190   6.617   12   0.882   for Ho
```

The set option allows you to test coefficients for two or more variables simultaneously:

```
. mlogtest, lr set(ed exper white ed white ed exper)
**** Likelihood-ratio tests for independent variables (N=337)
Ho: All coefficients associated with given variable(s) are 0.
        occ |     chi2    df   P>chi2
------------+--------------------------
      white |    8.095    4    0.088
         ed |  156.937    4    0.000
      exper |    8.561    4    0.073
------------+--------------------------
     set_1: |  166.087   12    0.000
         ed |
      exper |
      white |
         ed |
      white |
         ed |
      exper |
------------+
```

Saved results

mlogtest saves whichever of the following results were computed. The row and column labels of the matrix identify the specific elements of the matrices:

Matrices
    r(wald)   results of Wald test that all coefficients of an independent variable
                     equal zero
    r(lrtest)   results of likelihood-ratio test that all coefficients of an independent
                     variable equal zero
    r(combine)   results of Wald tests for combining categories
    r(lrcomb)   results of likelihood-ratio tests for combining categories
    r(hausman)   results of Hausman tests of IIA assumption
    r(smhsiao)   results of Small–Hsiao tests of IIA assumption

Acknowledgment

The code for the Small–Hsiao test is based on a program by Nick Winter.

A.11   mlogview

mlogview   Dialog box for using `mlogplot` interactively. For a detailed discussion, see sections 6.6.7–6.6.10.

Syntax

    mlogview

Description

`mlogview` creates a dialog box for creating an odds-ratio plot or discrete-change plot of the coefficients from a model fitted with `mlogit`. To make a discrete-change plot, you must run `prchange` before opening the dialog box. The dialog box settings are translated into the appropriate `mlogplot` command, which is executed when the user clicks one of the plot buttons in `mlogview`. The resulting `mlogplot` command is listed in the Results window and can be used later in a do-file.

Dialog box controls

The dialog box has the following controls for selecting variables and the amount of change to plot for each variable:

Select variables: select independent variables to include in plot.

Select amount of change: for a given independent variable, select the amount of change to be plotted:

+1: unit change.

+SD: standard deviation change.

0->1: change from 0 to 1 (for use with discrete change).

Don't plot: exclude variable from plot.

The buttons determine the type of plot to create:

DC Plot: draw discrete-change plot.

OR Plot: draw odds-ratio plot.

OR+DC Plot: draw odds-ratio plot in which the size of the letters indicates the discrete change.

For odds-ratio plots, only six variables can be plotted at one time. To automatically create a plot for the next six variables in the model, click

Next 6: list next six independent variables for an odds-ratio plot.

Characteristics of the graph are controlled by:

Note: title for plot to be printed at the top.

Number of ticks: number of tick marks on x-axis of plot.

Plot from: minimum and maximum values of x-axis for plot.

Connect if: for odds-ratio plots, connect categories if the coefficient for the odds of those two categories is not significant at the level specified in this box.

Base outcome: value of outcome to use as base outcome for odds-ratio plot.

Pack odds-ratio plot: eliminate extra vertical space in odds-ratio plot.

Use variable labels: use variable labels instead of names to identify variables in plot.

Finally, to print the graph,

Print: send plot to printer.

A.12   Overview of prchange, prgen, prtab, and prvalue

prchange, prgen, prtab, and **prvalue**   Setting values of the independent variables using x() and rest(). For a detailed discussion, see sections 3.6.3–3.6.4.

Syntax

prchange, prgen, prtab, and prvalue compute predicted probabilities for specified values of the independent variables. These values are set using x()and rest(). This entry provides a detailed description of how to use these options to set the values for the independent variables for any of these commands.

x(*variable1=value1* [*variable2=value2*]...) assigns *variable1* to *value1*, *variable2* to *value2*, and so on. You can assign values to as many or as few variables as you want. The assigned value is either a specific number (e.g., female=1) or a mnemonic specifying a descriptive statistic (e.g., phd=mean to set variable phd to the sample mean, and pub3=max to assign pub3 to the maximum value in the sample). The mnemonics for descriptive statistics are discussed below. Only numeric values can be used if a group statistic (i.e., statistics that begin with *gr*) has been specified with rest(). Although the equal signs are optional, they make the command easier to read.

rest(*stat*) sets the values of all variables not specified in x() to the sample statistic indicated by *stat*. If x() is not specified, all variables are set to *stat*. For example, rest(mean) sets all variables to their mean. The choices for *stat* can be unconditional or conditional on a group specified by x(). For example, x(female=0) rest(grmean) sets female to 0 and all other variables to their mean for those where female is 0. If rest() is not specified, it is assumed to be rest(mean). The available types of *stat* are

mean, median, min, and max refer to the mean, median, minimum, and maximum, conditional on any if or in conditions that are specified. Descriptive statistics are calculated using the estimation sample unless the all option has been specified, in which case all observations in memory are used.

previous sets the variables to their values from a previous command that set x() or rest(). For example, if the previous command was prvalue, x(a=1) rest(mean), then
prvalue, rest(previous) is equivalent to prvalue, x(a=1) rest(mean).

upper and lower can be used only with binary models, and refer to the values of the independent variables that yield either the highest (upper) or the lowest (lower) probability of a positive outcome.

grmean, grmedian, grmin, grmax refer to the group mean, median, minimum, and maximum. These statistics are calculated for whatever group of observations is specified in x(). For example, prvalue, x(male=0 white=0) rest(grmean) calculates predicted probabilities, where male and white are held to 0 and the other independent variables are held to the means of the observations in which male==0 and white==0. This is the same as typing prvalue if male==0 & white==0, x(male=0 white=0) rest(mean).

all specifies that all observations in memory are to be used to calculate descriptive statistics, excepting cases excluded by if or in conditions or the use of rest(gr*). The default is to use only the cases in the estimation sample.

Examples

```
. use http://www.stata-press.com/data/lf2/ordwarm2, clear
(77 & 89 General Social Survey)
. ologit warm male white age ed prst if yr89==1, nolog
(output omitted)
. * set independent variables to mean in estimation sample
. prvalue
(output omitted)
. * set independent variables to mean for all cases in memory
. prvalue, all
(output omitted)
. * set male to 0, ed to 12, age to 30, everything else to mean
. prvalue, x(male=0 ed=12 age=30) rest(mean)
(output omitted)
. * set male to 0, all others to mean for males
. prvalue, x(male=0) rest(grmean)
(output omitted)
. * set male to 0, age to median, all others to mean
. prvalue, x(male=0 age=median) rest(mean)
(output omitted)
```

```
. * set to statistics in full sample instead of estimation sample
. prvalue, x(male=0 age=median) rest(mean) all
  (output omitted)
```

A.13   praccum

praccum   Accumulates results from a series of calls to `prvalue`; these results can then be used to plot predictions.

Syntax

praccum , { <u>us</u>ing(*matrixname*) | <u>sa</u>ving(*matrixname*) } [xis(*value*)

    <u>gen</u>erate(*prefix*)]

Description

`praccum` accumulates the predictions generated by a series of calls to `prvalue` and optionally saves the accumulated values to new variables for plotting predicted values. The command allows you to plot predicted values in situations (e.g., nonlinearities) that cannot be handled by `prgen`. `praccum` works with `cloglog`, `cnreg`, `intreg`, `logistic`, `logit`, `mlogit`, `nbreg`, `ologit`, `oprobit`, `poisson`, `probit`, `regress`, `tobit`, `zinb`, and `zip`.

Predicted values of interest are produced by a series of paired calls to `prvalue` and `praccum`. The sequence begins with using `prvalue` to compute the first set of predicted values, followed by the initial call of `praccum`, where predicted values are saved in a new matrix defined by `saving()` or `using()`. After each subsequent call to `prvalue` and `praccum`, the probabilities are appended to the matrix specified by `using()`. When all the predicted probabilities have been computed, the `generate()` option is used to create new variables that contain the information from the matrix of saved results.

Options

<u>us</u>ing(*matrixname*) specifies the name of the matrix to which the predictions from the previous call to `prvalue` should be added. An error is generated if the matrix does not have the correct number of columns. This can happen if you try to append values to a matrix generated from calls to `praccum` based on a different model. Matrix *matrixname* will be created if it does not already exist.

<u>sa</u>ving(*matrixname*) specifies that a new matrix should be generated and should contain the predicted values from the previous call to `prvalue`. You use this option only when you initially create the matrix. After the matrix is created, you add to it with `using()`. The difference between `saving()` and `using()` is that `saving()` will overwrite *matrixname* if it exists, whereas `using()` will append results to it.

xis(*value*) indicates the value of the *x*-variable associated with the predicted values that are accumulated. If this is not specified, no new values will be added to the matrix.

generate(*prefix*) indicates that new variables should be generated from the matrix specified by using(). The names of the new variables begin with *prefix*. Details are given below.

Examples

First, we fit a model that includes age and age^2, and then we call prvalue:

```
. use http://www.stata-press.com/data/lf2/binlfp2,clear
(PSID 1976 / T Mroz)
. gen age2 = age*age
. logit lfp k5 k618 age age2 wc hc lwg inc, nolog
  (output omitted )
. prvalue, x(age 20 age2 400) rest(mean) brief
  Pr(y=inLF|x):        0.7324    95% ci: (0.3972,0.9191)
  Pr(y=NotInLF|x):     0.2676    95% ci: (0.0809,0.6028)
```

The first call of praccum saves the predictions generated by prvalue to the new matrix m_age:

```
. praccum, saving(m_age) xis(20)
```

Additional calls to praccum append to this matrix the new predictions that have been generated by new calls to prvalue:

```
. prvalue, x(age 25 age2 625) rest(mean) brief
  (output omitted )
. praccum, using(m_age) xis(25)
. prvalue, x(age 30 age2 900) rest(mean) brief
  (output omitted )
. praccum, using(m_age) xis(30)
. prvalue, x(age 35 age2 1225) rest(mean) brief
  (output omitted )
. praccum, using(m_age) xis(35)
. prvalue, x(age 40 age2 1600) rest(mean) brief
  (output omitted )
. praccum, using(m_age) xis(40)
. prvalue, x(age 45 age2 2025) rest(mean) brief
  (output omitted )
. praccum, using(m_age) xis(45)
. prvalue, x(age 50 age2 2500) rest(mean) brief
  (output omitted )
. praccum, using(m_age) xis(50)
. prvalue, x(age 55 age2 3025) rest(mean) brief
  (output omitted )
```

```
. praccum, using(m_age) xis(55)
. prvalue, x(age 60 age2 3600) rest(mean) brief
  (output omitted )
```

The last call to `praccum` generates new variables, which are listed:

```
. praccum, using(m_age) xis(60) gen(agsq)
New variables created by praccum:
```

Variable	Obs	Mean	Std. Dev.	Min	Max
agsqx	9	40	13.69306	20	60
agsqp0	9	.4282142	.1752595	.2676314	.7479599
agsqp1	9	.5717858	.1752595	.2520402	.7323686

```
. * these variables were created
. list agsq* in 1/9
```

	agsqx	agsqp0	agsqp1
1.	20	.2676314	.7323686
2.	25	.2682353	.7317647
3.	30	.2836163	.7163837
4.	35	.3152536	.6847464
5.	40	.3656723	.6343277
6.	45	.4373158	.5626842
7.	50	.5301194	.4698806
8.	55	.6381241	.3618759
9.	60	.7479599	.2520402

`praccum` is less cumbersome when it is used in conjunction with `forvalues`. Including the following in a do-file produces the same results as the example above:

```
capture matrix drop m_age
forvalues count = 20(5)60 {
    local countsq = "`"count"'"^2
    prvalue, x(age "`"count"'" age2 "`"countsq"'") rest(mean) brief
    praccum, using(m_age) xis("`"count"'")
}
praccum, using(m_age) gen(agsq)
```

Variables generated

The new variables created by `praccum` are the same as those created by `prgen` (see the table under the entry for `prgen`). The only difference is that whereas *name*x for `prgen` represents the values of x as specified by the `from()` and `to()` options, for `praccum` each value of *name*x must be specified by the user by specifying the `xis()` option.

A.14   prchange

prchange   Discrete and marginal change for regression models for categorical and count variables. For a detailed discussion, see section 3.6.5.

Syntax

prchange [*varlist*] [*if*] [*in*] [, x(*variable1*=*value1* [...]) rest(*stat*)

   outcome(#) fromto brief nobase nolabel help all uncentered

   delta(#) conditional]

See section A.12 for details on x() and rest().

Description

prchange computes marginal and discrete change for regression models for categorical and count variables: cloglog, cnreg, intreg, logistic, logit, mlogit, nbreg, ologit, oprobit, poisson, probit, regress, tobit, zinb, and zip. Marginal change is the partial derivative of the predicted probability or rate with respect to the independent variables. Discrete change is the difference in the predicted value as one independent variable changes values while all others are held at specified values. By default, changes are calculated holding all other variables at their mean. Values for the independent variables can also be set with the x() and rest() options (see section A.12 for details).

By default, discrete change is computed for a change in each variable from its minimum to its maximum (Min->Max), from 0 to 1 (0->1), from its specified value minus .5 units to its specified value plus .5 (-+1/2), and from its specified value minus .5 standard deviations to its value plus .5 standard deviations (-+sd/2). If the uncentered option is chosen, the last two quantities are computed from the specified value plus one unit (+1) and from the specified value plus 1 standard deviation (+sd). With the delta(#) option, changes of a standard deviation are replaced with changes of # units.

Options

outcome(#) specifies that changes should be printed only for the outcome indicated; for example, outcome(2). For mlogit, ologit, oprobit, and slogit, the default is to provide results for all outcomes. For the count models, the default is to present results with respect to the predicted rate; specifying an outcome number will present changes in the probability of that outcome. .

fromto specifies that the starting and ending probabilities from which the discrete change is calculated should be displayed.

brief prints only limited output.

nobase suppresses inclusion of the base values of independent variables.

nolabel uses values rather than value labels in the output.

help provides information explaining the headings in the output.

all specifies that calculations of means, medians, etc., should use the entire sample
   instead of the sample used to fit the model.

uncentered requests that changes of one unit and one standard deviation (or the amount
   specified by delta()) should begin at the value specified by rest() or x() rather
   than be centered around the specified value.

delta(#) indicates that the changes are to be # units rather than a standard deviation
   change.

conditional indicates that you want to generate conditional predictions rather than
   unconditional predictions for the ztp and ztnb models.

Examples

```
. * fit the model
. use http://www.stata-press.com/data/lf2/ordwarm2, clear
(77 & 89 General Social Survey)
. ologit warm male white age ed prst if yr89==1, nolog
  (output omitted )
. * holding all variables to the estimation sample means
. prchange age white

ologit: Changes in Predicted Probabilities for warm

age
              Avg|Chg|          SD           D           A          SA
Min->Max      .18353666    .11080101    .25627232   -.13218805   -.23488527
   -+1/2      .00257031    .00124054    .00390008   -.00145978   -.00368083
  -+sd/2      .04392809    .02133921     .066517    -.02479631   -.06305985
MargEfct      .00257032    .00124052    .00390013   -.00145983   -.00368082

white
              Avg|Chg|          SD           D           A          SA
   0->1       .03860572    .01742518    .05978625   -.01481366   -.06239779

                    SD           D           A          SA
Pr(y|x)      .05785507    .28683352    .45240733    .20290408

               male      white      age        ed       prst
      x=     .443107    .866521   45.1269   12.8523    41.5449
sd(x)=       .497025    .340278    17.129    3.0222    14.8981

. * specific set of base values
. prchange, x(male=1 white=1 age=30) rest(median)

ologit: Changes in Predicted Probabilities for warm

male
              Avg|Chg|          SD           D           A          SA
   0->1       .07925584    .03510679    .12340491   -.02130097   -.13721071

white
              Avg|Chg|          SD           D           A          SA
   0->1       .04032505    .01954348    .06110662    -.0226658   -.05798429
```

```
age
              Avg|Chg|          SD           D           A          SA
Min->Max      .1914724   .16630837   .21663642  -.21676943  -.16617538
   -+1/2      .00269591   .00144775   .00394407  -.00209403  -.00329781
  -+sd/2      .04605574   .02489513   .06721637  -.03557727   -.0565342
MargEfct      .00269594   .00144773   .00394415  -.00209407  -.00329781

ed
              Avg|Chg|          SD           D           A          SA
Min->Max      .10027184  -.05862336  -.14192033   .08446175   .11608193
   -+1/2      .00499106  -.00268043  -.00730169   .00387657   .00610557
  -+sd/2      .01508015   -.0081038  -.02205652   .01170659   .01845369
MargEfct      .00499121  -.00268031  -.00730212   .00387692   .00610551

prst
              Avg|Chg|          SD           D           A          SA
Min->Max      .07671434  -.04028213  -.11314654   .05304095   .10038772
   -+1/2      .00112195   -.0006025  -.00164139   .00087148   .00137244
  -+sd/2      .01670909  -.00898061  -.02443758   .01296943   .02044876
MargEfct      .00112195  -.00060249  -.00164141   .00087148   .00137243

                    SD          D           A          SA
Pr(y|x)   .06827433   .31734699    .4385626   .17581607

              male     white       age        ed      prst
    x=           1         1        30        12        40
sd(x)=   .497025   .340278    17.129    3.0222   14.8981
```

A.15   prcounts

prcounts   Predicted probabilities and rates for count models. For a detailed discussion, see sections 8.1.2, 8.1.3, and 8.7.1.

Syntax

prcounts *prefix* [*if*] [*in*] [, <u>m</u>ax(*max*) <u>p</u>lot]

Description

prcounts computes the predicted rate and probabilities of counts from 0 through the specified maximum count based on the last estimates from the count model fitted by poisson, nbreg, zip, zinb, ztnb, or ztp. The predictions for each observation are stored in new variables. Optionally, you can generate variables for the graphical comparison of observed and expected counts.

Options

max(*max*) is the maximum count for which predicted probabilities should be provided. The default is 9.

plot specifies that variables for plotting expected counts should be generated.

Variables generated

The following variables are generated, where *name* is the prefix specified with `prcounts`.
y is the count variable, and each prediction is conditional on the independent variables
in the regression.

name`rate` The predicted rate $E(y)$.

name`pr`k The predicted probability $\Pr(y = k)$ for $k = 0$ to *max*.

name`prgt` The predicted probability $\Pr(y > max)$.

name`cu`k The predicted cumulative probability $\Pr(y \leq k)$ for $k = 0$ to *max*.

For the zero-truncated models `ztp` and `ztnb`, predictions that are conditional on $y > 0$
are computed:

name`Crate` The conditional predicted rate $E(y \mid y > 0)$.

name`Cpr`k The predicted probability $\Pr(y = k \mid y > 0)$ for $k = 1$ to *max*.

name`Cprgt` The predicted probability $\Pr(y > max \mid y > 0)$.

name`Ccu`k The predicted cumulative probability $\Pr(y \leq k \mid y > 0)$ for $k = 1$ to *max*.

Examples

`prcounts` can be used to compute predictions for each observation in the sample,

```
. zinb art fem mar kid5 phd ment, inf(fem mar kid5 phd ment)
 (output omitted )
. prcounts znb
. summarize znb*
```

Variable	Obs	Mean	Std. Dev.	Min	Max
znbrate	915	1.697666	.7165823	.468232	9.678316
znball0	915	.0585168	.1215135	1.01e-30	.6024709
znbpr0	915	.3119487	.1249968	.0169493	.7454953
znbpr1	915	.2556639	.0418208	.0353106	.3098438
(output omitted)					
znbpr9	915	.0032089	.0061532	.0000589	.0610584
znbcu0	915	.3119487	.1249968	.0169493	.7454953
znbcu1	915	.5676126	.1256956	.0522599	.8773996
(output omitted)					
znbcu9	915	.9948109	.0214203	.5720913	.9999722
znbprgt	915	.0051891	.0214203	.0000278	.4279087

or it can be used to compute average predictions that can be graphed:

```
. poisson art fem mar kid5 phd ment
(output omitted)
. prcounts prm, max(8) plot
. summarize prm*
    Variable │      Obs        Mean    Std. Dev.        Min         Max
─────────────┼─────────────────────────────────────────────────────────
     prmrate │      915    1.692896    .6685824    .8883344    9.627207
      prmpr0 │      915    .2092071    .0794247    .0000659    .4113403
      prmpr1 │      915    .3098447    .0634931    .0006345    .3678775
(output omitted)
      prmpr8 │      915     .001877    .0094055    3.96e-06    .1206255
      prmcu0 │      915    .2092071    .0794247    .0000659    .4113403
      prmcu1 │      915    .5190518    .1395755    .0007004    .7767481
(output omitted)
      prmcu8 │      915    .9978884     .023188    .3763166    .9999995
     prmprgt │      915    .0021116     .023188    4.77e-07    .6236834
      prmval │        9           4    2.738613           0           8
     prmobeq │        9    .1101396    .1153559    .0010929    .3005464
     prmpreq │        9    .1108765    .1174511     .001877    .3098447
     prmoble │        9    .8150577    .2373893    .3005464    .9912568
     prmprle │        9    .8122127    .2760109    .2092071    .9978884
. line prmpreq prmobeq prmval
(graph omitted)
```

A.16   prgen

prgen   Generate variables with predicted values for regression models in a way that is useful for making plots. For a detailed discussion, see section 3.6.7 and section 4.6.4.

Syntax

prgen *varname* [*if*] [*in*], generate(*prefix*) [from(#) to(#) ncases(#)
    gap(#) x(*variable1=value1* [...]) rest(*stat*) maxcnt(#) brief all noisily
    marginal conditional ci *prvalueci_options*]

Description

prgen computes predicted values for regression models in a way that is useful for making plots. Predicted values are computed for the case in which *varname* varies over a specified range while other independent variables are held constant at values set by x() and rest(). New variables are added to the existing dataset that contain these predicted values. These new variables begin with the name *prefix*. The new variables contain data only for the first k observations in the dataset, where k is 11 by default or can be specified with the ncases() option.

Options

See section A.12 for details on x() and rest().

generate(*prefix*) sets the prefix for the new variables that are created. Choosing a prefix that is different from the beginning letters of any of the variables in your dataset makes it easier to examine the results. For example, if you choose the prefix abcd, you can use the command summarize abcd* to examine all newly created variables.

from(#) and to(#) are the start and end values for *varname*. The default is for *varname* to range from the observed minimum to the observed maximum of *varname*.

ncases(#) specifies the number of predicted values prgen computes as *varname* varies from the start value to the end value. The default is ncases(11).

gap(#) is an alternative to ncases(). You specify the gap or size between tick marks and prgen determines if the specified value divides evenly into the range specified with from() and to(). If it does, prgen determines the appropriate value for ncases().

maxcnt(#) is the maximum count value for which a predicted probability is computed for count models. The default is maxcnt(9).

brief suppresses output; variables are still generated.

all specifies that any calculations of means, medians, etc., should use the entire sample instead of the sample used to fit the model.

Options for confidence intervals and marginals

marginal requests that a variable or variables be created containing the marginal change in the outcome relative to *varname*, holding all other variables constant.

conditional indicates that you want to generate conditional predictions rather than unconditional predictions for the ztp and ztnb models.

ci generates confidence intervals for the predictions. All of the other options for constructing confidence intervals listed for prvalue can be used with prgen. Indeed, prgen simply passes the options along to prvalue—the command that is doing all the work.

prvalueci_options are any options available with prvalue for computing confidence intervals; see section A.18.

Examples

To compute predicted probabilities as age varies from 20 to 80, where list reproduces the values of some of the new variables created by prgen,

```
. use http://www.stata-press.com/data/lf2/ordwarm2, clear
(77 & 89 General Social Survey)
. oprobit warm yr89 male white age ed prst, nolog
  (output omitted)
. * predicted probabilities as age changes from 20 to 80
. prgen age, f(20) t(80) gen(wrm)
oprobit: Predicted values as age varies from 20 to 80.
             yr89        male       white         age          ed        prst
x=      .39860445   .46489315    .8765809   44.935456   12.218055   39.585259
. list wrmx wrmp1 wrmp2 wrmp3 wrmp4 in 1/11
```

	wrmx	wrmp1	wrmp2	wrmp3	wrmp4
1.	20	.0640175	.2609412	.4251544	.2498869
2.	26	.0737291	.2780399	.4210533	.2271777
3.	32	.0845284	.2948091	.4149821	.2056804
4.	38	.0964731	.3110608	.4070269	.1854392
5.	44	.1096137	.326604	.3972993	.166483
6.	50	.1239925	.3412487	.3859335	.1488253
7.	56	.1396419	.3548089	.3730837	.1324654
8.	62	.1565829	.3671075	.3589206	.1173891
9.	68	.1748236	.3779791	.3436272	.1035701
10.	74	.1943586	.3872746	.3273952	.0909716
11.	80	.2151676	.3948639	.3104213	.0795472

or, we can compute predictions at values that are average for male respondents, which can then be plotted:

```
. prgen age, x(male=1) rest(grmean) f(20) t(80) gen(mal)
oprobit: Predicted values as age varies from 20 to 80.
             yr89        male       white         age          ed        prst
x=      .37992495           1   .89212008   44.113508   12.337711   40.366792
. line malp1 malp2 malp3 malp4 malx, c(s ..)
  (graph omitted)
```

Variables generated

If ci is specified as an option for prgen, variables are created containing the upper and lower bounds of the confidence interval for the outcome. These variables have the same names as those in the table above, except for adding ub at the end for the variable with the upper bound and lb for the lower bound. If marginal is specified, variables are created that contain the marginal change in the outcome with respect to *varname*, holding all other variables constant. The variables containing marginal changes have the same names as those in the table above, except for adding a D prior to the outcome abbreviation and D*varname* after. For example, the marginal for *prefix*p0 is named *prefix*Dp0D*varname*. These are computed only for those models for which prchange computes the marginal change.

For which model	Name	Content
All models	*prefix*x	Values of *varname* from `from(#)` to `to(#)`
logit, probit	*prefix*p0	Predicted probability $\Pr(y = 0)$
	*prefix*p1	Predicted probability $\Pr(y = 1)$
mlogit, mprobit, ologit, oprobit, slogit	*prefix*p*k*	Predicted probability $\Pr(y = k)$, for all outcomes
	*prefix*s*k*	Cumulative probability $\Pr(y \leq k)$, for all outcomes
nbreg, poisson, zinb, zip, ztnb, ztp	*prefix*mu	Predicted rate μ
	*prefix*p*k*	Predicted probability $\Pr(y = k)$, for $0 \leq k \leq$ `maxcnt()`
	*prefix*s*k*	Cumulative probability $\Pr(y \leq k)$, for $0 \leq k \leq$ `maxcnt()`
zinb, zip	*prefix*inf	Predicted probability $\Pr(\text{Always } 0 = 1) = \Pr(\text{inflate})$
cnreg, intreg, regress, tobit	*prefix*xb	Predicted value of y

A.17   prtab

prtab   Construct tables of predicted values. For a detailed discussion, see sections 3.6.6 and 4.6.3.

Syntax

prtab *rowvar* [*colvar* [*supercolvar*]] [*if*] [*in*] [, by(*superrowvar*)
    x(*variable1*=*value1* [...]) <u>rest</u>(*stat*) <u>o</u>utcome(*string*) <u>nobase</u> <u>nol</u>abel
    <u>novarlbl</u> <u>brief</u> <u>cond</u>itional all]

Description

After fitting a regression model, `prtab` presents up to a four-way table of the predicted values, either probabilities or rates, for different combinations of values of the independent variables. The command works with `cloglog`, `cnreg`, `intreg`, `logistic`, `logit`, `mlogit`, `mprobit`, `nbreg`, `ologit`, `oprobit`, `poisson`, `probit`, `regress`, `slogit`, `tobit`, `zinb`, `zip`, `ztnb`, and `ztp`.

Options

See section A.12 for details on x() and rest().

by(*superrowvar*) specifies a numeric variable to be treated as a superrow. Only one superrow variable is allowed.

outcome(*string*) presents results for the specified outcome (e.g., outcome(2) requests values only for outcome 2). For ordered models or mlogit, the default is to provide results for all outcomes, each in a separate table; for count models, the default is to present changes in the predicted rate.

nobase suppresses the list of the values of the independent variables.

nolabel causes the numeric codes to be displayed rather than value labels.

novarlbl causes the variable name to be displayed rather than the variable label.

brief prints only limited output.

conditional indicates that you want to generate conditional predictions rather than unconditional predictions for the ztp and ztnb models.

all specifies that any calculations of means, medians, etc., should use the entire sample instead of the sample used to fit the model.

Examples

To compute predicted probabilities of labor force participation by whether the husband or wife attended college,

```
. use http://www.stata-press.com/data/lf2/binlfp2, clear
(Data from 1976 PSID-T Mroz)
. probit lfp k5 k618 age wc hc lwg inc, nolog
  (output omitted)
. prtab wc hc

probit: Predicted probabilities of positive outcome for lfp
```

Wife College: 1=yes 0=no	Husband College: 1=yes 0=no NoCol College	
NoCol	0.5149	0.5376
College	0.7004	0.7200

```
            k5       k618        age         wc         hc        lwg
x=     .2377158  1.3532537  42.537849   .2815405  .39176627  1.0971148
            inc
x=    20.128965
```

In models for ordinal or nominal outcomes, a separate table is produced for each outcome:

```
. use http://www.stata-press.com/data/lf2/ordwarm2, clear
(77 & 89 General Social Survey)
. ologit warm yr89 male white age ed prst, nolog
  (output omitted)
. prtab male white, x(prst=min)
ologit: Predicted probabilities for warm
Predicted probability of outcome 1 (SD)
```

Gender: 1=male 0=female	Race: 1=white 0=not white White Not Whit
Women	0.0695 0.0995
Men	0.1346 0.1870

```
Predicted probability of outcome 2 (D)
```

Gender: 1=male 0=female	Race: 1=white 0=not white White Not Whit
Women	0.2492 0.3094
Men	0.3588 0.4032

  (*and so on*)

For count models, by default the table contains predicted rates,

```
. use http://www.stata-press.com/data/lf2/couart2, clear
(Academic Biochemists / S Long)
. poisson art fem mar kid5 phd ment, nolog
```

Poisson regression   Number of obs   =   915
                                                      LR chi2(5)   =   183.03
                                                      Prob > chi2   =   0.0000
Log likelihood = -1651.0563   Pseudo R2   =   0.0525

art	Coef.	Std. Err.	z	P>|z|	[95% Conf. Interval]	
fem	-.2245942	.0546138	-4.11	0.000	-.3316352	-.1175532
mar	.1552434	.0613747	2.53	0.011	.0349512	.2755356
kid5	-.1848827	.0401272	-4.61	0.000	-.2635305	-.1062349
phd	.0128226	.0263972	0.49	0.627	-.038915	.0645601
ment	.0255427	.0020061	12.73	0.000	.0216109	.0294746
_cons	.3046168	.1029822	2.96	0.003	.1027755	.5064581

```
. prtab fem mar

poisson: Predicted rates for art

Gender:      | Married: 1=yes
1=female     |         0=no
0=male       | Single  Married

        Men  | 1.6109   1.8815
      Women  | 1.2869   1.5030

            fem         mar        kid5         phd        ment
x=   .46010929   .66229508   .49508197   3.1031093   8.7672131
```

or you can request predicted probabilities for a specified count:

```
. prtab fem mar, outcome(0)

poisson: Predicted probabilities of count = 0 for art

Gender:      | Married: 1=yes
1=female     |         0=no
0=male       | Single  Married

        Men  | 0.1997   0.1524
      Women  | 0.2761   0.2225

            fem         mar        kid5         phd        ment
x=   .46010929   .66229508   .49508197   3.1031093   8.7672131
```

A.18   prvalue

prvalue   Compute predicted values for specified values of the independent variables. For a detailed discussion, see section 3.6.4 and section 4.6.2.

Syntax

prvalue $\left[\textit{if} \right]$ $\left[\textit{in} \right]$ $\left[$, x(*variable1=value1* $\left[... \right]$) $\underline{r}$est(*stat*) $\underline{maxc}$nt($\#$) $\underline{s}$ave

diff ystar $\underline{noba}$se $\underline{nolabel}$ brief all $\underline{l}$evel($\#$) $\underline{del}$ta ept $\underline{boot}$strap

$\underline{r}$eps($\#$) $\underline{d}$ots match $\underline{s}$ize($\#$) $\underline{s}$aving(*filename*, ...)

$\left[\underline{bias}corrected \,|\, \underline{percentile} \,|\, \underline{normal} \right] \left. \right]$

Description

After fitting a regression model, prvalue computes the predicted values at specific values of the independent variables. When appropriate, predicted probabilities are provided. The command produces predictions and confidence intervals after cloglog, cnreg, intreg, logistic, logit, mlogit, nbreg, ologit, oprobit, poisson, probit, regress, tobit, zinb, and zip. The prvalue command produces predictions but not confidence intervals after mprobit, slogit, ztnb, and ztp.

Options

See section A.12 for details on x() and rest().

maxcnt(#) is the maximum count value for which the probability is computed in count models. The default is 9.

save preserves the current values of independent variables and predictions for later comparisons.

diff computes the differences between current predictions and those that were previously saved.

ystar prints the predicted value of y^* for binary and ordinal models.

nobase suppresses printing values of the independent variables.

nolabel uses values rather than value labels in output.

brief prints only limited output.

all specifies that any calculations of means, medians, etc., should use the entire sample instead of the sample used to fit the model.

Options for confidence intervals

prvalue computes confidence intervals for many predictions. When used with the save and diff options, it computes confidence intervals for discrete changes in predictions. Here we list the options related to confidence intervals. Details on each method for computing intervals are given in section 3.7.

level(#) specifies the confidence level, as a percentage, for confidence intervals. For example, level(95) requests a 95% confidence interval.

ept computes confidence intervals for predicted probabilities for cloglog, logit, and probit by endpoint transformation. This method cannot be used for changes in predictions.

delta calculates confidence intervals by the delta method using analytic derivatives.

bootstrap computes confidence intervals using the bootstrap method. This method takes roughly 1,000 times longer to compute than other methods.

Options used for bootstrapped confidence intervals

reps(#) specifies the number of bootstrap replications to be performed. The default is 1,000. The accuracy of a bootstrap estimate depends critically on the number of replications. Although sources differ on the recommended number of replications, Efron and Tibshirani (1993, 188) suggest 1,000 replications for confidence intervals. You can use bssize (Poi 2004) to calculate the number of bootstrap replications to be used. In our experience, this method often suggests more than 1,000 replications.

saving(*filename, save_options*) creates a data file with the estimates from each of the bootstrapped samples (i.e., one case for each replication). This option is useful when you need to examine the distribution of bootstrapped estimates. For example, this option is required if you plan to use `bssize` to calculate the number of replications to be used (Poi 2004).

For information about the *save_options*, type `help prefix_saving_option` in Stata.

`dots` is used with `bootstrap` to write a dot (.) at the beginning of each replication and periodically prints the percentage of total replications that have been completed. If computations appears to be stuck (i.e., new dots do not appear), it is likely that the estimation is not converging for the current bootstrap sample. We have found this to be most common with `zip` and `zinb` (see chapter 8 for a detailed discussion of what to do when nonconvergence is a problem). When this happens, you can click on the break symbol to stop computations for the current sample or wait until the maximum number of iterations have been computed (by default, the maximum number of iterations is 16,000). When a model does not converge for a given bootstrap sample, that sample is dropped.

`match` specifies that the bootstrap will resample within each category of the dependent variable in proportion to the distribution of the outcome categories in the estimation sample. If `match` is not specified, the proportions in each category of the bootstrap sample are determined entirely by the random draw, and it is possible to have samples with no cases in some categories. This option does not apply to `cnreg`, `intreg`, `nbreg`, `poisson`, `regress`, `tobit`, `zinb`, and `zip`.

`size(#)` specifies the number of cases to be sampled when bootstrapping. The default is the size of the estimation sample. If `size(#)` is specified, # must be less than or equal to the size of the estimation sample. In general, it is best to not specify `size()` (see *http://www.stata.com/support/faqs/stat/reps.html* for more information).

`percentile` computes the bootstrapped confidence interval using the percentile method. This is the default method.

`biascorrected` computes the bootstrapped confidence interval using the bias-corrected method.

`normal` computes the bootstrapped confidence interval using the normal approximation method.

Examples

```
. use http://www.stata-press.com/data/lf2/ordwarm2, clear
(77 & 89 General Social Survey)
. oprobit warm yr89 male white age ed prst, nolog
  (output omitted)
```

```
. * by default, hold all independent variables to their means
. prvalue

oprobit: Predictions for warm

Confidence intervals by delta method
                                    95% Conf. Interval
        Pr(y=SD|x):        0.1118    [ 0.0988,     0.1248]
        Pr(y=D|x):         0.3290    [ 0.3021,     0.3558]
        Pr(y=A|x):         0.3956    [ 0.3796,     0.4117]
        Pr(y=SA|x):        0.1636    [ 0.1483,     0.1790]

            yr89       male      white        age         ed        prst
x=    .39860445  .46489315   .8765809  44.935456  12.218055  39.585259

. *to compute all variables at their minimum
. prvalue, rest(min)

oprobit: Predictions for warm

Confidence intervals by delta method
                                    95% Conf. Interval
        Pr(y=SD|x):        0.1060    [ 0.0606,     0.1514]
        Pr(y=D|x):         0.3226    [ 0.2958,     0.3493]
        Pr(y=A|x):         0.4000    [ 0.3835,     0.4165]
        Pr(y=SA|x):        0.1714    [ 0.1088,     0.2339]

        yr89   male  white    age     ed   prst
x=         0      0      0     18      0     12

. * predictions for white females, with other variables at the median
. prvalue, x(white=1 male=0) rest(median)

oprobit: Predictions for warm

Confidence intervals by delta method
                                    95% Conf. Interval
        Pr(y=SD|x):        0.1012    [ 0.0851,     0.1172]
        Pr(y=D|x):         0.3169    [ 0.2903,     0.3435]
        Pr(y=A|x):         0.4036    [ 0.3867,     0.4206]
        Pr(y=SA|x):        0.1783    [ 0.1561,     0.2005]

        yr89   male  white    age     ed   prst
x=         0      0      1     42     12     37

. * or with other variables at median for white females
. prvalue, x(white=1 male=0) rest(grmedian)

oprobit: Predictions for warm

Confidence intervals by delta method
                                    95% Conf. Interval
        Pr(y=SD|x):        0.1056    [ 0.0891,     0.1220]
        Pr(y=D|x):         0.3220    [ 0.2953,     0.3488]
        Pr(y=A|x):         0.4004    [ 0.3838,     0.4169]
        Pr(y=SA|x):        0.1720    [ 0.1504,     0.1937]

        yr89   male  white    age     ed   prst
x=         0      0      1     44     12     37
```

```
. * to compare predictions for males and females
. prvalue, x(male=0) save

oprobit: Predictions for warm

Confidence intervals by delta method
                              95% Conf. Interval
         Pr(y=SD|x):    0.0791    [ 0.0669,     0.0913]
         Pr(y=D|x):     0.2867    [ 0.2611,     0.3123]
         Pr(y=A|x):     0.4182    [ 0.3992,     0.4372]
         Pr(y=SA|x):    0.2160    [ 0.1949,     0.2371]
              yr89       male      white       age         ed        prst
    x=  .39860445          0    .8765809  44.935456  12.218055  39.585259

. prvalue, x(male=1) diff

oprobit: Change in Predictions for warm

Confidence intervals by delta method
                       Current      Saved     Change    95% CI for Change
         Pr(y=SD|x):    0.1601     0.0791     0.0810    [ 0.0626,     0.0994]
         Pr(y=D|x):     0.3694     0.2867     0.0827    [ 0.0812,     0.0842]
         Pr(y=A|x):     0.3560     0.4182    -0.0622    [-0.0687,    -0.0558]
         Pr(y=SA|x):    0.1145     0.2160    -0.1015    [-0.1233,    -0.0797]
                   yr89       male      white       age         ed        prst
   Current=  .39860445          1    .8765809  44.935456  12.218055  39.585259
     Saved=  .39860445          0    .8765809  44.935456  12.218055  39.585259
      Diff=          0          1          0          0          0          0
```

Saved results

Scalars

r(xb)	the linear combination of the bs and the specified base value
r(xb_hi), r(xb_lo)	upper and lower limits of confidence interval for xb
r(p0), r(p1)	predicted probability $y = 0$ and $y = 1$ for binary models
r(p0_hi), r(p1_hi)	upper limits of predicted probabilities for binary models
r(p0_lo), r(p1_lo)	lower limits of predicted probabilities for binary models
r(mu)	predicted rate for count models
r(mu_hi), r(mu_lo)	upper and lower limits of confidence interval for predicted rate for poisson model
r(always0)	predicted probability of being in the always zero (inflate==1) category in zero-inflated count model

Macros

r(level)	level of confidence intervals (when appropriate)

Matrices

r(x)	base values for independent variables
r(x2)	base values for independent variables in the binary equation of zero-inflated count models
r(probs)	predicted probabilities
r(values)	category values corresponding to predicted probabilities in r(probs)

A.19   spex

spex   Fit example models, run commands, and load example datasets.

Syntax

spex [*commandname*] [*filename*] [, web user]

Description

spex allows users to easily use example files and models from this book. Typing spex *filename* will load the specified example dataset, whereas typing spex *modelname* will fit that model using data and an example specification from the book. Here is a partial list (since we are regularly adding new options) of commands and datasets that can be used with spex:

- *commandname* can be asmprobit, clogit, cloglog, cnreg, intreg, logit, misschk, mlogit, mprobit, nbreg, ologit, oprobit, poisson, probit, regress, rologit, slogit, tobit, zinb, zip, ztnb, and ztp.
- *filename* can be binlfp2, couart2, gsskidvalue2, lfpgraph2, nels_censored2, nomintro2, nomocc2, ordwarm2, recodedata2, regjob2, science2, sciwork, tobjob2, travel2, travel2case, and wlsrnk.

Options

web indicates that the data are to be accessed from their location on the SPost web site (*http://www.indiana.edu/~jslsoc/stata/spex_data/*).

user indicates that the data are to be accessed from the working directory or somewhere else along the user's adopath (i.e., somewhere accessed by sysuse).

When specifying an estimation command with spex, you can add other options (e.g., nolog), and these options will be passed along as additional options to the estimation command.

Examples

To have Stata to open the example dataset binlfp2.dta:

```
. spex binlfp2
  (output omitted)
```

To have Stata estimate the main logit example used in chapter 4:

```
. spex logit
  (output omitted)
```

B Description of datasets

The following datasets are used as examples in the book:

1. `binlfp2`: Data on labor force participation.

2. `couart2`: Data on scientific productivity.

3. `gsskidvalue2`: Data on parental values from the General Social Survey.

4. `nomocc2`: Data on occupations from the General Social Survey.

5. `ordwarm2`: Data on attitudes toward working mothers from the General Social Survey.

6. `science2`: Data on the careers of biochemists.

7. `travel2`: Data on travel mode choice.

8. `wlsrnk`: Data on job values.

Variable labels and descriptive statistics are provided for each dataset.

B.1 binlfp2: Data on labor force participation

```
. use http://www.stata-press.com/data/lf2/binlfp2, clear
(Data from 1976 PSID-T Mroz)

. describe
Contains data from http://www.stata-press.com/data/lfr/binlfp2.dta
  obs:           753                          Data from 1976 PSID-T Mroz
  vars:            8                          30 Apr 2001 16:17
  size:       13,554 (99.5% of memory free)   (_dta has notes)
```

| | storage | display | value | |
variable name	type	format	label	variable label
lfp	byte	%9.0g	lfplbl	Paid Labor Force: 1=yes 0=no
k5	byte	%9.0g		# kids < 6
k618	byte	%9.0g		# kids 6-18
age	byte	%9.0g		Wife´s age in years
wc	byte	%9.0g	collbl	Wife College: 1=yes 0=no
hc	byte	%9.0g	collbl	Husband College: 1=yes 0=no
lwg	float	%9.0g		Log of wife´s estimated wages
inc	float	%9.0g		Family income excluding wife´s

```
Sorted by:  lfp
```

```
. summarize
```

Variable	Obs	Mean	Std. Dev.	Min	Max
lfp	753	.5683931	.4956295	0	1
k5	753	.2377158	.523959	0	3
k618	753	1.353254	1.319874	0	8
age	753	42.53785	8.072574	30	60
wc	753	.2815405	.4500494	0	1
hc	753	.3917663	.4884694	0	1
lwg	753	1.097115	.5875564	-2.054124	3.218876
inc	753	20.12897	11.6348	-.0290001	96

B.2   couart2: Data on scientific productivity

```
. use http://www.stata-press.com/data/lf2/couart2, clear
(Academic Biochemists / S Long)
. describe
Contains data from http://www.stata-press.com/data/lfr/couart2.dta
  obs:          915                          Academic Biochemists / S Long
  vars:           6                          30 Jan 2001 10:49
  size:      11,895 (99.5% of memory free)   (_dta has notes)
```

variable name	storage type	display format	value label	variable label
art	byte	%9.0g		Articles in last 3 yrs of PhD
fem	byte	%9.0g	sexlbl	Gender: 1=female 0=male
mar	byte	%9.0g	marlbl	Married: 1=yes 0=no
kid5	byte	%9.0g		Number of children < 6
phd	float	%9.0g		PhD prestige
ment	byte	%9.0g		Article by mentor in last 3 yrs

```
Sorted by:  art
. summarize, sep(0)
```

Variable	Obs	Mean	Std. Dev.	Min	Max
art	915	1.692896	1.926069	0	19
fem	915	.4601093	.4986788	0	1
mar	915	.6622951	.473186	0	1
kid5	915	.495082	.76488	0	3
phd	915	3.103109	.9842491	.755	4.62
ment	915	8.767213	9.483916	0	77

B.3   gsskidvalue2: Data on parental values from the General Social Survey

```
. use http://www.stata-press.com/data/lf2/gsskidvalue2, clear
(1993 and 1994 General Social Survey)

. describe

Contains data from http://www.stata-press.com/data/lfr/gsskidvalue2.dta
  obs:         4,598                          1993 and 1994 General Social
                                              Survey
  vars:           11                          9 Jan 2001 10:28
  size:       91,960 (99.3% of memory free)
```

| | storage | display | value | |
variable name	type	format	label	variable label
year	int	%8.0g		gss year for this respondent
id	int	%8.0g		respondent id number
female	byte	%9.0g	dummy	Female
black	byte	%9.0g	dummy	Black
othrrace	byte	%9.0g	dummy	Nonblack/Nonwhite
degree	byte	%14.0g	EDdeg	rs highest degree
anykids	byte	%9.0g	dummy	R have any children?
kidvalue	byte	%9.0g	kidvalue	Which is most important for a child to learn?
income	long	%8.0g	income91	total family income
age	byte	%8.0g	age	age of respondent
income91	byte	%8.0g	income91	total family income

```
Sorted by:

. summarize
```

Variable	Obs	Mean	Std. Dev.	Min	Max
year	4598	1993.651	.4767952	1993	1994
id	4598	1254.447	818.4198	1	2992
female	4598	.5704654	.4950636	0	1
black	4598	.1233145	.3288336	0	1
othrrace	4598	.0437147	.2044817	0	1
degree	4584	1.430628	1.165915	0	4
anykids	4584	.7236038	.4472639	0	1
kidvalue	2989	2.22449	.8944211	1	4
income	4103	34790.7	22387.45	1000	75000
age	4598	46.12375	17.33162	18	99
income91	4598	19.44411	19.50733	1	99

(Continued on next page)

B.4   nomocc2: Data on occupations from the General Social Survey

```
. use http://www.stata-press.com/data/lf2/nomocc2, clear
(1982 General Social Survey)
. describe

Contains data from http://www.stata-press.com/data/lfr/nomocc2.dta
  obs:           337                          1982 General Social Survey
  vars:            4                          15 Jan 2001 15:24
  size:        2,696 (99.5% of memory free)   (_dta has notes)
```

variable name	storage type	display format	value label	variable label
occ	byte	%10.0g	occlbl	Occupation
white	byte	%10.0g		Race: 1=white 0=nonwhite
ed	byte	%10.0g		Years of education
exper	byte	%10.0g		Years of work experience

```
Sorted by:   occ
```

```
. summarize
```

Variable	Obs	Mean	Std. Dev.	Min	Max
occ	337	3.397626	1.367913	1	5
white	337	.9169139	.2764227	0	1
ed	337	13.09496	2.946427	3	20
exper	337	20.50148	13.95936	2	66

B.5   ordwarm2: Data on attitudes toward working mothers from the General Social Survey

```
. use http://www.stata-press.com/data/lf2/ordwarm2, clear
(77 & 89 General Social Survey)
. describe
Contains data from http://www.stata-press.com/data/lfr/ordwarm2.dta
  obs:         2,293                          77 & 89 General Social Survey
  vars:           10                          3 May 2001 09:54
  size:       32,102 (99.4% of memory free)   (_dta has notes)
```

variable name	storage type	display format	value label	variable label
warm	byte	%10.0g	SD2SA	Mom can have warm relations with child
yr89	byte	%10.0g	yrlbl	Survey year: 1=1989 0=1977
male	byte	%10.0g	sexlbl	Gender: 1=male 0=female
white	byte	%10.0g	racelbl	Race: 1=white 0=not white
age	byte	%10.0g		Age in years
ed	byte	%10.0g		Years of education
prst	byte	%10.0g		Occupational prestige
warmlt2	byte	%10.0g	SD	1=SD; 0=D,A,SA
warmlt3	byte	%10.0g	SDD	1=SD,D; 0=A,SA
warmlt4	byte	%10.0g	SDDA	1=SD,D,A; 0=SA

```
Sorted by:  warm
. summarize
```

Variable	Obs	Mean	Std. Dev.	Min	Max
warm	2293	2.607501	.9282156	1	4
yr89	2293	.3986044	.4897178	0	1
male	2293	.4648932	.4988748	0	1
white	2293	.8765809	.3289894	0	1
age	2293	44.93546	16.77903	18	89
ed	2293	12.21805	3.160827	0	20
prst	2293	39.58526	14.49226	12	82
warmlt2	2293	.1295246	.3358529	0	1
warmlt3	2293	.4448321	.4970556	0	1
warmlt4	2293	.8181422	.3858114	0	1

B.6   science2: Data on the careers of biochemists

```
. use http://www.stata-press.com/data/lf2/science2, clear
(Note that some of the variables have been artificially constructed.)
. describe
Contains data from http://www.stata-press.com/data/lfr/science2.dta
  obs:          308                       Note that some of the variables
                                            have been artificially
                                            constructed.
  vars:          35                       10 Mar 2001 05:51
  size:      17,556 (99.5% of memory free)  (_dta has notes)
```

variable name	storage type	display format	value label	variable label
id	float	%9.0g		ID Number.
cit1	int	%9.0g		Citations: PhD yr -1 to 1.
cit3	int	%9.0g		Citations: PhD yr 1 to 3.
cit6	int	%9.0g		Citations: PhD yr 4 to 6.
cit9	int	%9.0g		Citations: PhD yr 7 to 9.
enrol	byte	%9.0g		Years from BA to PhD.
fel	float	%9.0g		Fellow or PhD prestige.
felclass	byte	%9.0g	prstlb	* Fellow or PhD prestige class.
fellow	byte	%9.0g	fellbl	Postdoctoral fellow: 1=y,0=n.
female	byte	%9.0g	femlbl	Female: 1=female,0=male.
job	float	%9.0g		Prestige of 1st univ job.
jobclass	byte	%9.0g	prstlb	* Prestige class of 1st job.
mcit3	int	%9.0g		Mentor´s 3 yr citation.
mcitt	int	%9.0g		Mentor´s total citations.
mmale	byte	%9.0g	malelb	Mentor male: 1=male,0=female.
mnas	byte	%9.0g	naslb	Mentor NAS: 1=yes,0=no.
mpub3	byte	%9.0g		Mentor´s 3 year publications.
nopub1	byte	%9.0g	nopublb	1=No pubs PhD yr -1 to 1.
nopub3	byte	%9.0g	nopublb	1=No pubs PhD yr 1 to 3.
nopub6	byte	%9.0g	nopublb	1=No pubs PhD yr 4 to 6.
nopub9	byte	%9.0g	nopublb	1=No pubs PhD yr 7 to 9.
phd	float	%9.0g		Prestige of Ph.D. department.
phdclass	byte	%9.0g	prstlb	* Prestige class of Ph.D. dept.
pub1	byte	%9.0g		Publications: PhD yr -1 to 1.
pub3	byte	%9.0g		Publications: PhD yr 1 to 3.
pub6	byte	%9.0g		Publications: PhD yr 4 to 6.
pub9	byte	%9.0g		Publications: PhD yr 7 to 9.
work	byte	%9.0g	worklbl	Type of first job.
workadmn	byte	%9.0g	wadmnlb	Admin: 1=yes; 0=no.
worktch	byte	%9.0g	wtchlb	* Teaching: 1=yes; 0=no.
workuniv	byte	%9.0g	wunivlb	* Univ Work: 1=yes; 0=no.
wt	byte	%9.0g		
faculty	byte	%9.0g	faclbl	1=Faculty in University
jobrank	byte	%9.0g	joblbl	Rankings of University Job.
totpub	byte	%9.0g		Total Pubs in 9 Yrs post-Ph.D.

```
                                        * indicated variables have notes
```

```
Sorted by:
```

```
. summarize
```

Variable	Obs	Mean	Std. Dev.	Min	Max
id	308	58654.49	2283.465	57001	62420
cit1	308	11.60714	18.37658	0	137
cit3	308	14.97078	21.37068	0	196
cit6	308	18.37013	23.34766	0	143
cit9	308	21.07143	25.8195	0	214
enrol	278	5.564748	1.467253	3	14
fel	308	3.190877	.9872379	1	4.77
felclass	308	2.694805	1.019855	1	4
fellow	308	.3928571	.4891803	0	1
female	308	.3474026	.4769198	0	1
job	163	2.967117	.880396	1.01	4.69
jobclass	163	2.417178	.8945373	1	4
mcit3	306	20.95098	26.26862	0	133
mcitt	303	44.58086	56.05372	0	223
mmale	303	.9867987	.1143249	0	1
mnas	304	.0789474	.2701012	0	1
mpub3	306	11.02614	9.08571	0	48
nopub1	308	.25	.4337174	0	1
nopub3	308	.2045455	.4040255	0	1
nopub6	308	.1915584	.3941678	0	1
nopub9	308	.1980519	.3991801	0	1
phd	308	3.177987	1.012738	1	4.77
phdclass	308	2.675325	1.057827	1	4
pub1	308	2.545455	3.092685	0	24
pub3	308	3.185065	3.908752	0	31
pub6	308	4.165584	4.780714	0	29
pub9	308	4.512987	5.315134	0	33
work	302	2.062914	1.37829	1	5
workadmn	302	.089404	.2857995	0	1
worktch	302	.615894	.4871904	0	1
workuniv	302	.705298	.4566654	0	1
wt	308	3.402597	1.288989	1	6
faculty	302	.5298013	.4999395	0	1
jobrank	163	2.417178	.8945373	1	4
totpub	308	11.86364	12.77623	0	84

B.7 travel2: Data on travel mode choice

```
. use http://www.stata-press.com/data/lf2/travel2, clear
(Greene & Hensher 1997 data on travel mode choice)
. describe

Contains data from http://www.stata-press.com/data/lfr/travel2.dta
  obs:           456                       Greene & Hensher 1997 data on
                                             travel mode choice
  vars:           13                       10 Mar 2001 06:07
  size:       9,576 (99.5% of memory free)
```

variable name	storage type	display format	value label	variable label
id	int	%9.0g		Identification number
mode	byte	%9.0g	mode	Mode of transportation
train	byte	%9.0g		1=train 0=other mode
bus	byte	%9.0g		1=bus 0=other mode
car	byte	%9.0g		1=car 0=other mode
time	int	%9.0g		Total traveling time
invc	byte	%8.0g		In-vehicle costs
choice	byte	%8.0g		1=option selected 0=not selected
ttme	byte	%8.0g		Terminal time (0 for car)
invt	int	%8.0g		in-vehicle time
gc	int	%8.0g		Generalized costs includeing lost wages
hinc	byte	%8.0g		Household income
psize	byte	%8.0g		Size of traveling party

```
Sorted by:  id
. summarize
```

Variable	Obs	Mean	Std. Dev.	Min	Max
id	456	103.8882	61.03044	1	210
mode	456	2	.8173933	1	3
train	456	.3333333	.4719223	0	1
bus	456	.3333333	.4719223	0	1
car	456	.3333333	.4719223	0	1
time	456	632.1096	270.2547	180	1440
invc	456	33.95175	21.795	2	109
choice	456	.3333333	.4719223	0	1
ttme	456	25.24781	21.15744	0	99
invt	456	606.8618	265.8235	180	1440
gc	456	113.4912	53.7725	30	269
hinc	456	31.80921	19.25813	2	72
psize	456	1.809211	1.069457	1	6

B.8   wlsrnk: Data on job values

```
. use http://www.stata-press.com/data/lf2/wlsrnk, clear
(1992 Wisconsin Longitudinal Study data on job values)

. describe

Contains data from wlsrnk.dta
  obs:         3,226                        1992 Wisconsin Longitudinal
                                            Study data on job values
  vars:           16                        22 Jul 2005 12:39
  size:       77,424 (92.6% of memory free)
```

variable name	storage type	display format	value label	variable label
id	int	%9.0g		caseid
fem	byte	%9.0g		female
hn	float	%9.0g		henmon-nelson test score
value1	byte	%9.0g		esteem 1992
value2	byte	%9.0g		variety 1992
value3	byte	%9.0g		autonomy 1992
value4	byte	%9.0g		security 1992
hashi1	byte	%9.0g		job has high esteem
hashi2	byte	%9.0g		job has high variety
hashi3	byte	%9.0g		job has high autonomy
hashi4	byte	%9.0g		job has high security
haslo1	byte	%9.0g		job has low esteem
haslo2	byte	%9.0g		job has low variety
haslo3	byte	%9.0g		job has low autonomy
haslo4	byte	%9.0g		job has low security
noties	byte	%9.0g		no tied ranks in outcome

```
Sorted by:

. summarize
```

Variable	Obs	Mean	Std. Dev.	Min	Max
id	3226	2334.708	1344.129	1	4682
fem	3226	.5034098	.5000659	0	1
hn	3226	.0984121	.9872199	-2.84291	2.941332
value1	3226	2.898326	1.069181	1	4
value2	3226	1.620893	.8915367	1	4
value3	3226	2.058896	1.05107	1	4
value4	3226	1.902666	1.08684	1	4
hashi1	3226	.2600744	.4387429	0	1
hashi2	3226	.1469312	.3540922	0	1
hashi3	3226	.3508989	.4773254	0	1
hashi4	3226	.4649721	.4988489	0	1
haslo1	3226	.216367	.4118312	0	1
haslo2	3226	.2597644	.4385732	0	1
haslo3	3226	.1704898	.3761208	0	1
haslo4	3226	.1717297	.3772038	0	1
noties	3226	.1286423	.3348555	0	1

References

Akaike, H. 1973. Information theory and an extension of the maximum likelihood principle. In *Second International Symposium on Information Theory*, ed. B. Petrov and F. Csaki, 267–281. Budapest: Akademiai Kiado.

Allison, P. D. 2001. *Missing Data*. Thousand Oaks, CA: Sage.

Allison, P. D. and N. A. Christakis. 1994. Logit models for sets of ranked items. *Sociological Methodology* 24: 199–228.

Alvarez, R. M. and J. Nagler. 1998. When politics and models collide: Estimating models of multiparty elections. *American Journal of Political Science* 42: 55–96.

Amemiya, T. 1981. Qualitative response models: a survey. *Journal of Economic Literature* 19: 1483–1536.

Anderson, J. A. 1984. Regression and ordered categorical variables (with discussion). *Journal of the Royal Statistical Society Series B* 46: 1–30.

Arminger, G. 1995. Specification and estimation of mean structures: regression models. In *Handbook of Statistical Modeling for the Social and Behavioral Sciences*, ed. G. Arminger, C. C. Clogg, and M. E. Sobel, 77–183. New York: Plenum Press.

Beggs, S., S. Cardell, and J. Hausman. 1981. Assessing the potential demand for electric cars. *Journal of Econometrics* 16: 1–19.

Brant, R. 1990. Assessing proportionality in the proportional odds model for ordinal logistic regression. *Biometrics* 46: 1171–1178.

Bunch, D. S. 1991. Estimatibility in the multinomial probit model. *Transportation Research* 25B(1): 1–12.

Bunch, D. S. and R. Kitamura. 1989. Multinomial probit estimation revisited: Testing of new algorithms and evaluation of alternative model specifications for trinomial models of household car ownership. Transportation Research Group Research Report UCD-TRG-RR-4, University of California, Davis, CA.

Cameron, A. C. and P. K. Trivedi. 1986. Econometric models based on count data: Comparisons and applications of some estimators and tests. *Journal of Applied Econometrics* 1: 29–53.

—. 1998. *Regression Analysis of Count Data*, vol. 30 of *Econometric Society Monograph*. New York: Cambridge University Press.

—. 2005. *Microeconometrics: Methods and Applications*. New York: Cambridge University Press.

Cheng, S. and J. S. Long. 2005. Testing for IIA in the multinomial logit model. University of Connecticut, Working Paper.

Cleves, M., W. Gould, and R. Gutierrez. 2004. *An Introduction to Survival Analysis Using Stata*. rev. ed. College Station, TX: Stata Press.

Clogg, C. C. and E. S. Shihadeh. 1994. *Statistical Models for Ordinal Variables*. Thousand Oaks, CA: Sage.

Cook, R. D. and S. Weisberg. 1999. *Applied Regression Including Computing and Graphics*. New York: Wiley.

Cox, D. R. 1972. Regression models and life-tables (with discussion). *Journal of the Royal Statistical Society, Series B* 34: 187–220.

Cramer, J. S. 1986. *Econometric Applications of Maximum Likelihood Methods*. Cambridge, UK: Cambridge University Press.

Cytel Software Corporation. 2005. *LogXact Version 6*. Cambridge, MA.

Drukker, D. M. and V. Wiggins. 2004. Verifying the solution from a nonlinear solver: a case study: comment. *American Economic Review* 94: 397–399.

Efron, B. and R. J. Tibshirani. 1993. *An Introduction to the Bootstrap*. New York: Chapman & Hall.

Eliason, S. 1993. *Maximum Likelihood Estimation*. Newbury Park, CA: Sage.

Fahrmeir, L. and G. Tutz. 1994. *Multivariate Statistical Modeling Based on Generalized Linear Models*. New York: Springer.

Fienberg, S. E. 1980. *The Analysis of Cross-Classified Categorical Data*. 2nd ed. Cambridge, MA: MIT Press.

Fox, J. 1991. *Regression Diagnostics: An Introduction*. Newbury Park, CA: Sage.

Freese, J. 2002. Least likely observations in regression models for categorical outcomes. *Stata Journal* 2: 296–300.

Freese, J. and J. S. Long. 2000. sg155: Tests for the multinomial logit model. *Stata Technical Bulletin* 58: 19–25. In *Stata Technical Bulletin Reprints*, vol. 10, 247–255. College Station, TX: Stata Press.

Fry, T. R. L. and M. N. Harris. 1996. A Monte Carlo study of tests for the independence of irrelevant alternatives property. *Transportion Research Part B: Methodological* 30: 19–30.

—. 1998. Testing for independence of irrelevant alternatives: some empirical results. *Sociological Methods and Research* 26: 401–423.

Fu, V. K. 1998. sg88: Estimating generalized ordered logit models. *Stata Technical Bulletin* 44: 27–30. In *Stata Technical Bulletin Reprints*, vol. 8, 160–164. College Station, TX: Stata Press.

Gould, W., J. Pitblado, and W. Sribney. 2006. *Maximum Likelihood Estimation with Stata*. 3rd ed. College Station, TX: Stata Press.

Greene, W. H. 1994. Accounting for excess zeros and sample selection in Poisson and negative binomial regression models. Stern School of Business, New York University, Department of Economics, Working Paper no. 94-10.

—. 2003. *Econometric Analysis*. 5th ed. Upper Saddle River, NJ: Prentice Hall.

Greene, W. H. and D. Hensher. 1995. Multimonial logit and discrete choice models. In *LIMDEP Version 7.0*, ed. W. Greene. Bellport, NY: Econometric Software, Inc. CDA.

Grogger, J. T. and R. Carson. 1991. Models for truncated counts. *Journal of Applied Econometrics* 6: 225–238.

Guan, W. 2003. From the help desk: bootstrapped standard errors. *Stata Journal* 3: 71–80.

Gutierrez, R. G., S. L. Carter, and D. M. Drukker. 2001. sg160: On boundary-value likelihood-ratio tests. *Stata Technical Bulletin* 60: 15–18. In *Stata Technical Bulletin Reprints*, vol. 10, 269–273. College Station, TX: Stata Press.

Hagle, T. M. and G. E. Mitchell II. 1992. Goodness-of-fit measures for probit and logit. *American Journal of Political Science* 36: 762–784.

Halaby, C. 2003. Where job values come from: Family and schooling background, cognitive ability, and gender. *American Sociological Review* 68: 251–278.

Hardin, J. and J. Hilbe. 2001. *Generalized Linear Models and Extensions*. College Station, TX: Stata Press.

Hauser, R. M., W. H. Sewell, and T. S. Hauser. 1992/93. Wisconsin Longitudinal Study (WLS), machine-readable data file. Madison, WI: University of Wisconsin-Madison, WLS.
http://www.ssc.wisc.edu/~wls/documentation/

Hausman, J. and D. McFadden. 1984. Specification tests for the multinomial logit model. *Econometrica* 52: 1219–1240.

Hendrickx, J. 2000. sbe37: Special restrictions in multinomial logistic regression. *Stata Technical Bulletin* 56: 18–26. In *Stata Technical Bulletin Reprints*, vol. 10, 93–103. College Station, TX: Stata Press.

Hensher, D. 1986. Simultaneous Estimation of Hierarchical Logit Mode Choice Models. Working Paper 34, Macquarie University, School of Economic and Financial Studies.

Hosmer, D. W., Jr. and S. Lemeshow. 1980. Goodness of fit tests for the multiple logistic regression model. *Communications in Statistics* A10: 1043–1069.

——. 2000. *Applied Logistic Regression.* 2nd ed. New York: Wiley.

Jann, B. 2005. Making regression tables from stored estimates. *Stata Journal* 5: 288–308.

Keener, R. W. and D. M. Waldman. 1985. Maximum likelihood regression of rank-censored data. *Journal of the American Statistical Association* 80: 385–392.

Lall, R., S. J. Walters, K. Morgan, and MRC CFAS Cooperative Institute of Public Health. 2002. A review of ordinal regression models applied on health-related quality of life assessment. *Statistical Methods in Medical Research* 11: 49–67.

Lambert, D. 1992. Zero-inflated Poisson regression with an application to defects in manufacturing. *Technometrics* 34: 1–14.

Landwehr, J. M., D. Pregibon, and A. C. Shoemaker. 1984. Graphical methods for assessing logistic regression models. *Journal of the American Statistical Association* 79: 61–71.

Lemeshow, S. and D. W. Hosmer, Jr. 1982. The use of goodness-of-fit statistics in the development of logistic regression models. *American Journal of Epidemiology* 115: 147–151.

Little, R. J. A. and D. B. Rubin. 1987. *Statistical Analysis with Missing Data.* New York: Wiley.

Long, J. S. 1990. The origins of sex differences in science. *Social Forces* 68: 1297–1315.

——. 1997. *Regression Models for Categorical and Limited Dependent Variables*, vol. 7 of *Advanced Quantitative Techniques in the Social Sciences.* Thousand Oaks, CA: Sage.

Long, J. S. and L. H. Ervin. 2000. Using heteroscedasticity consistent standard errors in the linear regression model. *American Statistician* 54: 217–224.

McCullagh, P. 1980. Regression models for ordinal data (with discussion). *Journal of the Royal Statistical Society* 42: 109–142.

McCullagh, P. and J. A. Nelder. 1989. *Generalized Linear Models.* 2nd ed. New York: Chapman & Hall.

McFadden, D. 1973. Conditional logit analysis of qualitative choice behavior. In *Frontiers of Econometrics*, ed. P. Zarembka, 105–142. New York: Academic Press.

——. 1989. A method of simulated moments for estimation of discrete response models without numerical integration. *Econometrica* 57(5): 995–1026.

McKelvey, R. D. and W. Zavoina. 1975. A statistical model for the analysis of ordinal level dependent variables. *Journal of Mathematical Sociology* 4: 103–120.

Miller, P. W. and P. A. Volker. 1985. On the determination of occupational attainment and mobility. *Journal of Human Resources* 20: 197–213.

Mitchell, M. 2004. *A Visual Guide to Stata Graphics.* College Station, TX: Stata Press.

Mroz, T. A. 1987. The sensitivity of an empirical model of married women's hours of work to economic and statistical assumptions. *Econometrica* 55: 765–799.

Mullahy, J. 1986. Specification and testing of some modified count data models. *Journal of Econometrics* 33: 341–365.

Peterson, B. and F. E. Harrell, Jr. 1990. Partial proportional odds models for ordinal response variables. *Applied Statistics* 39: 205–217.

Poi, B. 2004. From the help desk: some bootstrapping techniques. *Stata Journal* 4: 312–328.

Powers, D. A. and Y. Xie. 2000. *Statistical Methods for Categorical Data Analysis.* San Diego: Academic Press.

Pregibon, D. 1981. Logistic regression diagnostics. *Annals of Statistics* 9: 705–724.

Pudney, S. 1989. *Modelling Individual Choice: The Econometrics of Corners, Kinks and Holes.* Cambridge, MA: Blackwell.

Punj, G. N. and R. Staelin. 1978. The choice process for business graduate schools. *Journal of Marketing Research* 15: 588–598.

Raftery, A. E. 1996. Bayesian model selection in social research. In *Sociological Methodology*, ed. P. V. Marsden, vol. 26, 111–163. Oxford: Blackwell.

Rothenberg, T. J. 1984. Hypothesis testing in linear models when the error covariance matrix is nonscalar. *Econometrica* 52: 827–842.

Sandor, Z. and P. Andra. 2004. Alternative sampling methods for estimating multivariate normal probabilities. *Journal of Econometrics* 120: 207–234.

Schafer, J. L. 1997. *Analysis of Incomplete Multivariate Data.* London: Chapman & Hall.

Small, K. A. and C. Hsiao. 1985. Multinomial logit specification tests. *International Economic Review* 26: 619–627.

Theil, H. 1970. On the estimation of relationships involving qualitative variables. *American Journal of Sociology* 76: 103–154.

Tobias, A. and M. J. Campbell. 1998. sg90: Akaike's information criterion and Schwarz's criterion. *Stata Technical Bulletin* 45: 23–25. In *Stata Technical Bulletin Reprints*, vol. 8, 174–177. College Station, TX: Stata Press.

Train, K. 2003. *Discrete Choice Methods with Simulation.* Cambridge: Cambridge University Press.

Vuong, Q. H. 1989. Likelihood ratio tests for model selection and non-nested hypotheses. *Econometrica* 57: 307–333.

Weisberg, S. 2005. *Applied Linear Regression.* 3rd ed. New York: Wiley.

White, H. 1982. Maximum likelihood estimation of misspecified models. *Econometrica* 50: 1–24.

Williams, R. 2005. gologit2: A program for generalized logistic regression/partial proportional odds models for ordinal variables. *http://www.nd.edu/~rwilliam/stata/gologit2.pdf.*

Windmeijer, F. A. G. 1995. Goodness-of-fit measures in binary choice models. *Econometric Reviews* 14: 101–116.

Winship, C. and R. D. Mare. 1984. Regression models with ordinal variables. *American Sociological Review* 49: 512–525.

Winship, C. and L. Radbill. 1994. Sampling weights and regression analysis. *Sociological Methods and Research* 23: 230–257.

Wolfe, R. 1998. sg86: Continuation-ratio models for ordinal response data. *Stata Technical Bulletin* 44: 18–21. In *Stata Technical Bulletin Reprints*, vol. 8, 149–153. College Station, TX: Stata Press.

Wolfe, R. and W. Gould. 1998. sg76: An approximate likelihood-ratio test for ordinal response models. *Stata Technical Bulletin* 42: 24–27. In *Stata Technical Bulletin Reprints*, vol. 7, 199–204. College Station, TX: Stata Press.

Wooldridge, J. M. 2002. *Econometric Analysis of Cross Section and Panel Data.* Cambridge, MA: MIT Press.

Xu, J. and J. S. Long. 2005. Confidence intervals for predicted outcomes in regression models for categorical outcomes. Indiana University, Working Paper.

Author index

Subject index